The Mating Mind

THE MATING MIND

How Sexual Choice
Shaped the Evolution
of Human Nature

GEOFFREY MILLER

DOUBLEDAY
New York London Toronto Sydney Auckland

PUBLISHED BY DOUBLEDAY
a division of Random House, Inc.
1540 Broadway, New York, New York, 10036

DOUBLEDAY and a portrayal of an anchor with a dolphin are trademarks of Doubleday, a division of Random House, Inc.

Library of Congress Cataloging-in-Publication Data applied for

ISBN 0-385-49516-1

Printed in the United States of America
First Edition: May 2000

1 3 5 7 9 10 8 6 4 2

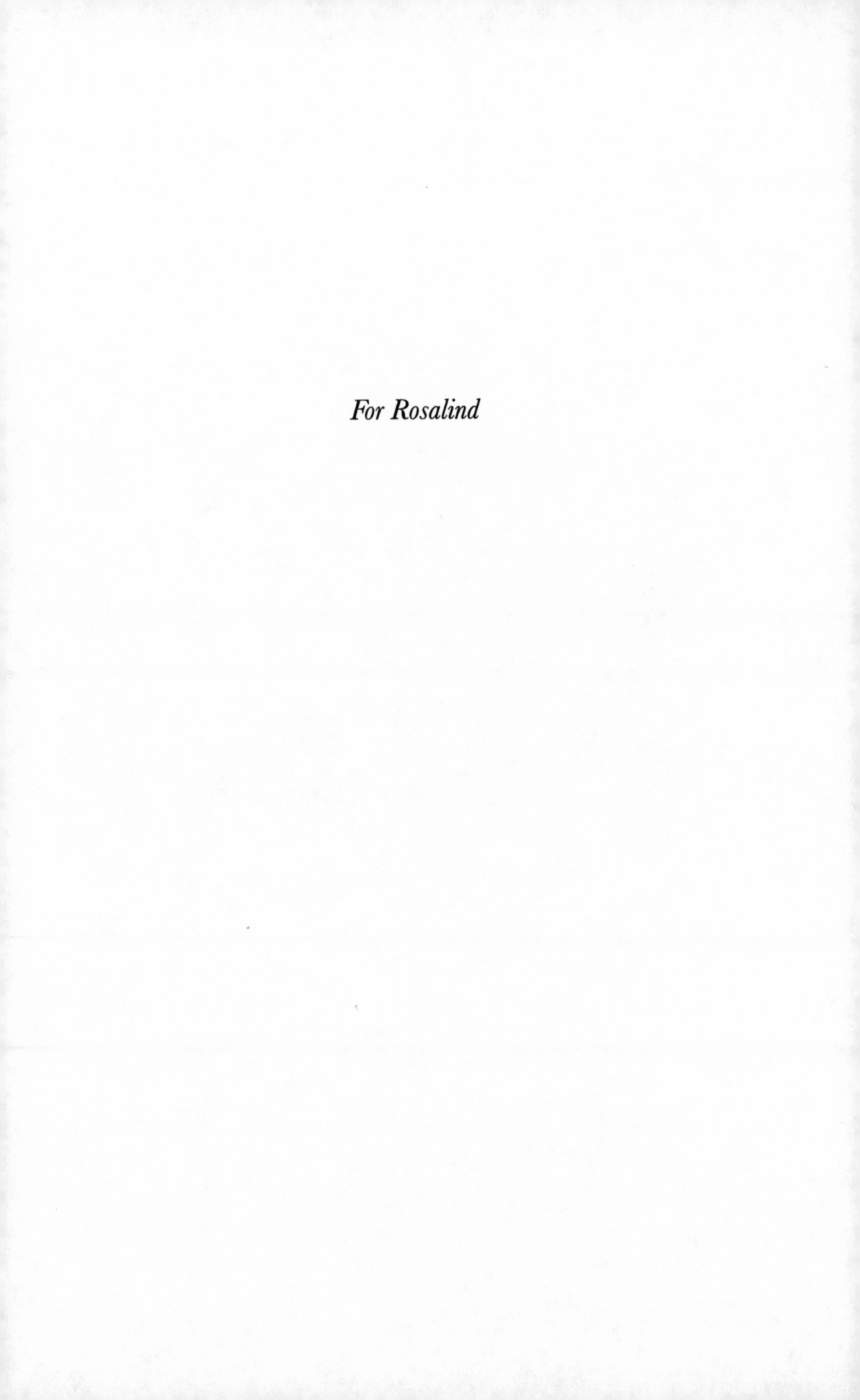

For Rosalind

Contents

1

Central Park

Central Park divides two of Manhattan's greatest treasure collections. On the West Side stands the American Museum of Natural History, with its dinosaur fossils, stuffed African elephants, dioramas of apes, and displays of ancient human remains. On the East Side stands the Metropolitan Museum of Art, with its Rembrandt self-portraits, peacock-shaped sitar, gold rapiers, Roman temple, Etruscan mirrors, and Jacques Louis David's *Death of Socrates.*

These works symbolize our unique human capacities for art, music, sports, religion, self-consciousness, and moral virtue, and they have troubled me ever since my student days studying biology at Columbia University. It was easy enough for me to take a taxi along the West Seventy-ninth Street transverse (the natural history museum) to East Eighty-first Street (the Met). It was not so easy for our ancestors to cross over from the pre-human world of natural history to the world of human culture. How did they transform themselves from apes to New Yorkers? Their evolutionary path seems obscure.

Yet we know there must have been a path. The human mind evolved somehow. The question scientists have asked for over a century is: How? Most people equate evolution with "survival of the fittest," and indeed most theories about the mind's evolution have tried to find survival advantages for everything that makes humans unique. To extend the metaphor, one kind of theory suggests our problem was not following the transverse to a collection of decorative arts, but traveling a different route to some useful inventions. Perhaps the human mind evolved for military prowess, symbolized by the Sea-Air-Space Museum on the aircraft carrier *USS Intrepid*, docked at Pier 86. Or perhaps our minds evolved for reciprocal economic advantage, symbolized by the World Trade Center and Wall Street, or through a thirst for pure knowledge, as housed in the New York Public Library. The survival advantages of better technology, trade, and knowledge seem obvious, so many believe the mind's evolution must have been technophilic and survivalist.

Ever since the Darwinian revolution, this survivalist view has seemed the only scientifically respectable possibility. Yet it remains unsatisfying. It leaves too many riddles unexplained. Human language evolved to be much more elaborate than necessary for basic survival functions. From a pragmatic biological viewpoint, art and music seem like pointless wastes of energy. Human morality and humor seem irrelevant to the business of finding food and avoiding predators. Moreover, if human intelligence and creativity were so useful, it is puzzling that other apes did not evolve them.

Even if the survivalist theory could take us from the world of natural history to our capacities for invention, commerce, and knowledge, it cannot account for the more ornamental and enjoyable aspects of human culture: art, music, sports, drama, comedy, and political ideals. At this point the survivalist theories usually point out that along the transverse lies the Central Park Learning Center. Perhaps the ornamental frosting on culture's cake arose through a general human ability to learn new things. Perhaps our big brains, evolved for technophilic survivalism, can be co-opted for the arts. However, this side-effect view is equally unsatisfying. Temperamentally, it reflects nothing more than a Wall Street trader's contempt for leisure. Biologically, it predicts that other big-brained species like elephants and dolphins should have invented their own versions of the human arts. Psychologically, it fails to explain why it is so much harder for us to learn mathematics than music, surgery than sports, and rational science than religious myth.

I think we can do better. We do not have to pretend that everything interesting and enjoyable about human behavior is a side-effect of some utilitarian survival ability or general learning capacity. I take my inspiration not from the Central Park Learning Center on the north side of the transverse but from the Ramble on the south side. The Ramble is a 37-acre woodland hosting 250 species of birds. Every spring, they sing to attract sexual partners. Their intricate songs evolved for courtship. Could some of our puzzling human abilities have evolved for the same function?

A Mind for Courtship

This book proposes that our minds evolved not just as survival machines, but as courtship machines. Every one of our ancestors managed not just to live for a while, but to convince at least one sexual partner to have enough sex to produce offspring. Those proto-humans that did not attract sexual interest did not become our ancestors, no matter how good they were at surviving. Darwin realized this, and argued that evolution is driven not just by natural selection for survival, but by an equally important process that he called *sexual selection through mate choice*. Following his insight, I shall argue that the most distinctive aspects of our minds evolved largely through the sexual choices our ancestors made.

The human mind and the peacock's tail may serve similar biological functions. The peacock's tail is the classic example of sexual selection through mate choice. It evolved because peahens preferred larger, more colorful tails. Peacocks would survive better with shorter, lighter, drabber tails. But the sexual choices of peahens have made peacocks evolve big, bright plumage that takes energy to grow and time to preen, and makes it harder to escape from predators such as tigers. The peacock's tail evolved through mate choice. Its biological function is to attract peahens. The radial arrangement of its yard-long feathers, with their iridescent blue and bronze eye-spots and their rattling movement, can be explained scientifically only if one understands that function. The tail makes no sense as an adaptation for survival, but it makes perfect sense as an adaptation for courtship.

The human mind's most impressive abilities are like the peacock's tail: they are courtship tools, evolved to attract and entertain sexual partners. By shifting our attention from a survival-centered view of evolution to a courtship-centered view, I shall try to show how, for the first time, we can understand more of the richness of human art, morality, language, and creativity.

A 1993 Gallup Poll showed that almost half of all Americans accept that humans evolved gradually over millions of years. Yet only about 10 percent believe that natural selection, alone and unguided, can account for the human mind's astounding abilities. Most think that the mind's evolution must have been guided by some intelligent force, some active designer. Even in more secular nations such as Britain, many accept that humans evolved from apes, but doubt that natural selection suffices to explain our minds.

Despite being a committed Darwinian, I share these doubts. I do not think that natural selection for survival can explain the human mind. Our minds are entertaining, intelligent, creative, and articulate far beyond the demands of surviving on the plains of Pleistocene Africa. To me, this points to the work of some intelligent force and some active designer. However, I think the active designers were our ancestors, using their powers of sexual choice to influence—unconsciously—what kind of offspring they produced. By intelligently choosing their sexual partners for their mental abilities, our ancestors became the intelligent force behind the human mind's evolution.

Evolutionary Psychology Turns Dionysian

The time is ripe for more ambitious theories of human nature. Our species has never been richer, better educated, more numerous, or more aware of our common historical origin and common planetary fate. As our self-confidence has grown, our need for comforting myths has waned. Since the Darwinian revolution, we recognize that the cosmos was not made for our convenience.

But the Darwinian revolution has not yet captured nature's last citadel—human nature. In the 1990s the new science of evolutionary psychology made valiant attempts. It views human nature as a set of biological adaptations, and tries to discover which problems of living and reproducing those adaptations evolved to solve. It grounds human behavior in evolutionary biology.

Some critics believe that evolutionary psychology goes too far and attempts to explain too much. I think it does not go far enough. It has not taken some of our most impressive and distinctive abilities as seriously as it should. For example, in his book *How the Mind Works*, Steven Pinker argued that human art, music, humor, fiction, religion, and philosophy are not real adaptations, but biological side-effects of other evolved abilities. As a cognitive scientist, Pinker was inclined to describe the human mind as a pragmatic problem-solver, not a magnificent sexual ornament: "The mind is a neural computer, fitted by natural selection with combinational algorithms for causal and probabilistic reasoning about plants, animals, objects and people."

Although he knows that reproductive success is evolution's bottom line, he overlooked the possible role of sexual selection in shaping conspicuous display behaviors such as art and music. He asked, for example, "If music confers no survival advantage, where does it come from and why does it work?" Lacking any manifest survival function, he concluded that art and music must be like cheesecake and pornography—cultural inventions that stimulate our tastes in evolutionarily novel ways, without improving our evolutionary success. His views that the arts are "biologically frivolous" has upset many performing artists sympathetic to evolutionary psychology. In a televised BBC debate following the publication of *How the Mind Works*, the theatrical director and intellectual polymath Jonathan Miller took Pinker to task for dismissing the arts as non-adaptations without considering all their possible functions. One of my goals in writing this book has been to see whether evolutionary psychology could prove as satisfying to a performing artist as to a cognitive scientist. It may be economically important to consider how the mind works, but it is also important to consider how the mind mates.

The view of the mind as a pragmatic, problem-solving survivalist has also inhibited research on the evolution of human creativity, morality, and language. Some primate researchers have suggested that human creative intelligence evolved as nothing more than a way to invent Machiavellian tricks to deceive and manipulate others. Human morality has been reduced to a tit-for-tat accountant that keeps track of who owes what to whom. Theories of language evolution have neglected human story-telling, poetry, wit, and song. You have probably read accounts of evolutionary psychology in the popular press, and felt the same unease that it is missing something important. Theories based on the survival of the fittest can nibble away at the edges of human nature, but they do not take us to the heart of the mind.

Moreover, the ritual celibacy of these survivalist doctrines seems artificial. Why omit sexual desire and sexual choice from the pantheon of evolutionary forces that could have shaped the human mind, when biologists routinely use sexual choice to explain behavioral abilities in other animals? Certainly, evolutionary psychology is concerned with sex. Researchers such as David Buss and Randy Thornhill have gathered impressive evidence that we have evolved sexual preferences that favor pretty faces, fertile bodies, and high social status. But evolutionary psychology in general still views sexual preferences more often as outcomes of evolution than as causes of evolution. Even where the sexual preferences of our ancestors have been credited with the power to shape mental evolution, their effects have been largely viewed as restricted to sexual and social emotions—to explain, for example, higher male motivations to take risks, attain social status, and demonstrate athletic prowess. Sexual choice has not been seen as reaching very deep into human cognition and communication, and sexuality is typically viewed as irrelevant to the serious business of evolving human intelligence and language.

In reaction to these limitations, I came to believe that the Darwinian revolution could capture the citadel of human nature only by becoming more of a sexual revolution—by giving more credit to sexual choice as a driving force in the mind's evolution. Evolutionary psychology must become less Puritan and more Dionysian. Where others thought about the survival problems our ancestors faced during the day, I wanted to think about the courtship problems they faced at night. In poetic terms, I wondered whether the mind evolved by moonlight. In scientific terms, sexual selection through mate choice seemed a neglected factor in human mental evolution. Through ten years of researching sexual selection and human evolution, since the beginning of my Ph.D., it became clear to me that sexual selection theory offered valuable intelligence about aspects of human nature that are important to us, and that cry out for evolutionary explanation, but that have been ignored, dismissed, or belittled in the past.

Trying a Different Tool

The human brain and its diverse capacities are so complex, and so costly to grow and maintain, that they must have arisen through direct selection for some important biological function. To date, it has proven very difficult to propose a biological function for human creative intelligence that fits the scientific evidence. We know that the human mind is a collection of astoundingly complex adaptations, but we do not know what biological functions many of them evolved to serve.

Evolutionary biology works by one cardinal rule: to understand an adaptation, one has to understand its evolved function. The analysis of adaptations is more than a collection of just-so stories, because according to evolutionary theory there are only two fundamental kinds of functions that explain adaptations. Adaptations can arise through natural selection for survival advantage, or sexual selection for reproductive advantage. Basically, that's it.

If you have two tools and one doesn't work, why not try the other? Science has spent over a century trying to explain the mind's evolution through natural selection for survival benefits. It has explained many human abilities, such as food preferences and fear of snakes, but it consistently fails to explain other abilities for decorative art, moral virtue, and witty conversation. It seems reasonable to ask whether sexual selection for reproductive benefits might account for these leftovers. This suggestion makes sexual selection sound like an explanation of last resort. It should not be viewed that way, because sexual selection has some special features as an evolutionary process. As we shall see, sexual selection is unusually fast, powerful, intelligent, and unpredictable. This makes it a good candidate for explaining any adaptation that is highly developed in one species but not in other closely related species that share a similar environment.

What Makes Sexual Selection So Special?

In the 1930s, biologists redefined natural selection to include sexual selection, because they did not think sexual selection was very important. Following their precedent, modern biology textbooks define natural selection to include every process that leads some genes to out-compete other genes by virtue of their survival or reproductive benefits. When one biologist says "evolution through natural selection," other biologists hear "evolution for survival or reproductive advantage." But non-biologists, including many other scientists, still hear "survival of the fittest." Many evolutionary psychologists, who should know better, even ask what possible "survival value" could explain some trait under discussion. This causes enormous confusion, and ensures that sexual selection continues to be neglected in discussions of human evolution.

In this book I shall use the terms "natural selection" and "sexual selection" as Darwin did: natural selection arising through competition for survival, and sexual selection arising through competition for reproduction. I am perfectly aware that this is not the way professional biologists currently use these terms. But I think it is more important, especially for non-biologist readers, to appreciate that selection for survival and selection for attracting sexual partners are distinct processes that tend to produce quite different kinds of biological traits. Terms should be the servants of theories, not the masters. By reviving Darwin's distinction between natural selection for survival and sexual selection for reproduction, we can talk more easily about their differences.

One difference is that sexual selection through mate choice can be much more intelligent than natural selection. I mean this quite literally. Natural selection takes place as a result of challenges set by an animal's physical habitat and biological niche. The habitat includes the factors that matter to farmers: sunlight, wind, heat, rain, and land quality. The niche includes predators and prey, parasites and germs, and competitors from one's own species. Natural selection is just something that happens as a side-effect of these factors influencing an organism's survival chances. The habitat is inanimate and doesn't care about those it affects. Biological competitors just care about making their own livings. None of these selectors cares whether it imposes evolutionary selection pressures that are consistent, directional, efficient, or creative. The natural selection resulting from such selectors just happens, willy-nilly.

Sexual selection is quite different, because animals often have very strong interests in acting as efficient agents of sexual selection. The genetic quality of an animal's sexual partner determines, on average, half the genetic quality of their offspring. (Most animals inherit half their genes from mother and half from father.) As we shall see, one of the main reasons why mate choice evolves is to help animals choose sexual partners who carry good genes. Sexual selection is the professional at sifting between genes. By comparison, natural selection is a rank amateur. The evolutionary pressures that result from mate choice can therefore be much more consistent, accurate, efficient, and creative than natural selection pressures.

As a result of these incentives for sexual choice, many animals are sexually discriminating. They accept some suitors and reject others. They apply their faculties of perception, cognition, memory, and judgment to pick the best sexual partners they can. In particular, they go for any features of potential mates that signal their fitness and fertility.

In fact, sexual selection in our species is as bright as we are. Every time we choose one suitor over another, we act as an agent of sexual selection. Almost anything that we can notice about a person is something our ancestors might have noticed too, and might have favored in their sexual choices. For example, some of us fall in love with people for their quick wits and generous spirits, and we wonder how these traits could have evolved. Sexual choice theory suggests that the answer is right in front of us. These traits are sexually attractive, and perhaps simpler forms of them have been attractive for hundreds of thousands of years. Over many generations, those with quicker wits and more generous spirits may have attracted more sexual partners, or higher-quality partners. The result was that wits became quicker and spirits more generous.

Of course, sexual selection through mate choice cannot favor what its agents cannot perceive. If animals cannot see the shapes of one another's heart ventricles, then heart ventricles cannot be directly shaped by sexual selection—vivisection is not a practical method for choosing a sexual partner. A major theme of this book is that before language evolved, our ancestors could not easily perceive one another's thoughts, but once language had arrived, thought itself became subject to sexual selection. Through language, and other new forms of expression such as art and music, our ancestors could act more like psychologists—in addition to acting like beauty contest judges—when choosing mates. During human evolution, sexual selection seems to have shifted its primary target from body to mind.

This book argues that we were neither created by an omniscient deity, nor did we evolve by blind, dumb natural selection. Rather, our evolution was shaped by beings intermediate in intelligence: our own ancestors, choosing their sexual partners as sensibly as they could. We have inherited both their sexual tastes for warm, witty, creative, intelligent, generous companions, and some of these traits that they preferred. We are the outcome of their million-year-long genetic engineering experiment in which their sexual choices did the genetic screening.

Giving so much credit to sexual choice can make sexual selection sound almost too powerful. If sexual selection can act on any trait that we can notice in other individuals, it can potentially explain any aspect of human nature that scientists can notice too. Sexual selection's reach seems to extend as far as psychology's subject matter. So be it. Scientists don't have to play fair against nature. Physics is full of indecently powerful theories, such as Newton's laws of motion and Einstein's theory of general relativity. Darwin gave biology two equally potent theories: natural selection and sexual selection. In principle, his two theories explain the origins of all organic complexity, functionality, diversity, and beauty in the universe. Psychologists generally believe that so far they have no theories of comparable power. But sexual selection can also be viewed as a psychological theory, because sexual choice and courtship are psychological activities. Psychologists are free to use sexual selection theory just where it is most needed: to explain mental abilities that look too excessive and expensive to have evolved for survival.

This sexual choice view also sounds rather circular as an explanation of human mental evolution. It puts the mind in an unusual position, as both selector and selectee in its own evolution. If the human mind catalyzed its own evolution through mate choice, it sounds as though our brains pulled themselves up by their own bootstraps. However, most positive-feedback processes look rather circular, and a positive-feedback process such as sexual selection may be just what we need to explain unique, highly elaborated adaptations like the human mind. Many theorists have accepted that some sort of positive-feedback process is probably required to explain why the human brain evolved to be so large so quickly. Sexual selection, especially a process called runaway sexual selection, is the best-established example of a positive-feedback process in evolution.

Positive-feedback systems are very sensitive to initial conditions. Often, they are so sensitive that their outcome is unpredictable. For example, take two apparently identical populations, let them undergo sexual selection for many generations, and they will probably end up looking very different. Take two initially indistinguishable populations of toucans, let them choose their sexual partners over a thousand generations, and they will evolve beaks with very different colors, patterns, and shapes. Take two populations of primates, and they will evolve different hairstyles. Take two populations of hominids (bipedal apes), and one may evolve into us, and the other into Neanderthals. Sexual selection's positive-feedback dynamics make it hard to predict what will happen next in evolution, but they do make it easy to explain why one population happened to evolve a bizarre ornament that another similar population did not.

Sexual Selection and Other Forms of Social Selection

In the 1990s evolutionary psychologists reached a consensus that human intelligence evolved largely in response to social rather than ecological or technological challenges. Some primate researchers have suggested that the transition from monkey brains to ape brains was driven by selection for "Machiavellian intelligence" to outsmart, deceive, and manipulate one's social competitors. Anthropologist Robin Dunbar has suggested that large primate brains evolved to cope with large numbers of primate social relationships. He views human language, especially gossip, as an extension of primate grooming behavior. Many researchers have suggested that acquiring our ability to attribute beliefs and desires to others, which they call our "Theory of Mind," was a key stage in human evolution.

Scientists became excited about social competition because they realized that it could have become an endless arms race, requiring ever more sophisticated minds to understand and influence the minds of others. An arms race for social intelligence looks a promising way to explain the human brain's rapid expansion and the human mind's rapid evolution.

The human mind is clearly socially oriented, and it seems likely that it evolved through some sort of social selection. But what kind of social selection, exactly? Sexual selection is the best-understood, most powerful, most creative, most direct, and most fundamental form of social selection. From an evolutionary perspective, social competition centers around reproduction. Animals compete socially to acquire the food, territory, alliances, and status that lead to reproduction. Sexual selection is the most direct form of social selection because mate choice directly favors some traits over others, and immediately produces offspring that are likely to inherit the desired traits.

In other forms of social selection, the link between behavior and reproduction is much less direct. For example, the ability to form and maintain social alliances leads to easier foraging, better protection against predators, and better sexual access to desired mates. This in turn may lead to higher reproductive success, if the desired mates are willing. Other forms of social selection are important, but mostly because they change the social scenery behind sexual selection. Social selection is like the political tension between the Montagues and Capulets. It matters largely because it influences the sexual prospects of Romeo and Juliet.

Sexual selection is the premier example of social selection, and courtship is the premier example of social behavior. Theories of human evolution through social selection without explicit attention to sexual selection are like dramas without romance. Prehistoric social competition was not like a power struggle between crafty Chinese eunuchs or horticulturally competitive nuns: it was a complex social game in which real males and real females played for real sexual stakes. They played sometimes with homicidal or rapacious violence, and sometimes with Machiavellian strategizing, but more often with forms of psychological warfare never before seen in the natural world: conversation, charm, and wit.

What Makes Sexually Selected Traits So Special?

Apart from sexual selection being a special sort of evolutionary process, the adaptations that it creates also tend to show some special features. Adaptations for courtship are usually highly developed in sexually mature adults but not in youth. They are usually displayed more conspicuously and noisily by males than by females. They produce sights and sounds that prove attractive to the opposite sex. They often reveal an animal's fitness by being difficult to produce if the animal is sick, starving, injured, or full of harmful mutations. They show conspicuous differences between individuals, and those differences are often genetically heritable. ("Heritable" implies that some proportion of the differences between individuals in a particular trait are due to genetic differences between individuals.) As we shall see, the human mind's most distinctive features, such as our capacities for language, art, music, ideology, humor, and creative intelligence, fit these criteria quite well.

However, traits with these features are sometimes not considered legitimate biological adaptations. Evolutionary psychologists Steven Pinker and John Tooby have argued that our science should focus on human universals that have been optimized by evolution, no longer showing any significant differences between individuals, or any genetic heritability in those differences. That is a good rule of thumb for identifying survival adaptations. But, as we shall see, it rules out all sexually selected adaptations that evolved specifically to advertise individual differences in health, intelligence, and fitness during courtship. Sexual selection tends to amplify individual differences in traits so that they can be easily judged during mate choice. It also makes some courtship behaviors so costly and difficult that less capable individuals may not bother to produce them at all. For art to qualify as an evolved human adaptation, not everyone has to produce art, and not everyone has to show the same artistic ability. On the contrary, if artistic ability were uniform and universal, our ancestors could not have used it as a criterion for picking sexual partners. As we shall see, the same reasoning may explain why people show such wide variation in their intelligence, language abilities, and moral behavior.

While sexually selected adaptations can be distinguished from survival adaptations from the outside, they may not feel any different from the inside. In particular, they may not feel very sexual when we're using them. Sexual selection is a theory of evolutionary function, not a theory of subconscious motivation. When I argue that a particular human ability evolved to attract sexual partners, I am not claiming that there is some sort of Freudian sex drive at work behind the scenes. Peacock tails do not need a sexual subconscious in order to be sexually attractive, and neither do our instincts for art, generosity, or creativity.

Why Now?

If sexual selection is so great, why hasn't it been used before now to explain the most distinctive aspects of human nature? In the next chapter, I trace the reasons why sexual selection theory was neglected for a century after Darwin and why it was revived only in the 1980s. The century of neglect is important to appreciate, because virtually all of 20th-century science has tried to explain human mental evolution using natural selection alone. Even now, sexual selection is usually invoked only to explain the differences between women and men, not those between humans and other primates. Although evolutionary biologists and evolutionary psychologists all know about sexual selection, its power, subtlety, and promise for explaining human mental traits have been overlooked.

The idea that sexual choice was an important factor in the human mind's evolution may sound radical, but it is firmly grounded in current biology. Twenty years ago, this book could not have been written. Only since then have scientists come to realize how profoundly mate choice influences evolution. There has been a renaissance of interest in sexual selection, with an outpouring of new facts and ideas. Today, the world's leading biology journals are dominated by technical papers on sexual selection theory and experiments on how animals choose their mates. But this has been a secret renaissance, hidden from most areas of psychology and the humanities, and largely unrecognized by the general public.

Prudery has also marginalized sexual selection—which is, after all, about sex. Many people, especially scientists, are ambivalent about sex: fascinated but embarrassed, obsessed yet guilty, alternately ribald and puritanical. Scientists still feel awkward teaching sexual selection to students, talking about it with journalists, and writing about it for the public. Science is not so different from popular culture in this respect. Just as there are very few good films that explicitly show sexual penetration, there have been very few good theories of human mental evolution that depict our ancestors as fully sexual beings capable of intelligent mate choice.

The sexual choice idea is also timely because it counters the charge that evolutionary psychology is some sort of "biological reductionism" or "genetic determinism." Many critics allege that evolutionary psychology tries to reduce psychology to biology, by explaining the mind's intricacies in terms of the brute replication of genes. In general, there is nothing wrong with reductionism—it is a powerful and successful strategy for understanding the world, and a cornerstone of the scientific method. However, there are serious problems with biological reductionism in the sense of trying to account for all of human nature in terms of the survival of the fittest. Often this strategy has led scientists to dismiss far too glibly many important human phenomena, such as creativity, charity, and the arts. This book tries very hard to avoid that particular type of reductionism. My theory suggests that our most cherished abilities were favored by the most sophisticated minds ever to have emerged on our planet before modern humans: the minds of our ancestors. It doesn't reduce psychology to biology, but sees psychology as a driving force in biological evolution. It portrays our ancestors' minds as both products and consumers evolving in the free market of sexual choice. My metaphors for explaining this theory will come more from marketing, advertising, and the entertainment industry than from physics or genetics. This is probably the least reductionistic theory of the mind's evolution one could hope for that is consistent with modern biology.

The Gang of Three

This sexual choice theory did not start out as a way of Darwinizing the humanities or trying to explain human creativity.

It began as an attempt to solve three basic problems concerning human mental evolution. These problems crop up as soon as we ask why we evolved certain abilities that other species did not evolve.

The first problem is that really large brains and complex minds arose very late in evolution and in very few species. Life evolved relatively quickly after the Earth cooled from a molten blob to a planet with a stable surface and some pools of water. Then it was another three billion years before any animal evolved a brain heavier than one pound. Even then, brains heavier than a pound evolved only in the great apes, in several varieties of elephants and mammoths, and in a few dozen species of dolphins and whales. Chimpanzee brains weigh one pound, our brains weigh three pounds, bottlenose dolphin brains weigh four pounds, elephant brains weigh eleven pounds, and sperm whale brains weigh eighteen pounds. But over 99 percent of animal species thrive with brains much smaller than a chimpanzee's. Far from showing any general trend towards big-brained hyper-intelligence, evolution seems to abhor our sort of intelligence, and avoids it whenever possible. So, why would evolution endow our species with such large brains that cost so much energy to run, given that the vast majority of successful animal species survive perfectly well with tiny brains?

Second, there was a very long lag between the brain's expansion and its apparent survival payoffs during human evolution. Brain size tripled in our ancestors between two and a half million years ago and a hundred thousand years ago. Yet for most of this period our ancestors continued to make the same kind of stone handaxes. Technological innovation was at a standstill during most of our brain evolution. Only long after our brains stopped expanding did any tradition of cumulative technological progress develop, or any global colonization beyond the middle latitudes, or any population growth beyond a few million individuals. Arguably, one could not ask for a worse correlation between growth in a biological organ and evidence of its supposed survival benefits. Our ancestors of a hundred thousand years ago were already anatomically modern humans with bodies and brains just like ours. Yet they did not invent agriculture for another ninety thousand years, or urban civilization for another ninety-five thousand years. How could evolution favor the expansion of a costly organ like the brain, without any major survival benefits becoming apparent until long after the organ stopped expanding?

The third problem is that nobody has been able to suggest any plausible survival payoffs for most of the things that human minds are uniquely good at, such as humor, story-telling, gossip, art, music, self-consciousness, ornate language, imaginative ideologies, religion, and morality. How could evolution favor such apparently useless embellishments? The fact that there are no good theories of these adaptations is one of science's secrets. Linguistics textbooks do not include a good evolutionary theory of language origins, because there are none. Cultural anthropology textbooks present no good evolutionary theories of art, music, or religion, because there are none. Psychology textbooks do not offer any good evolutionary theories of human intelligence, creativity, or consciousness, because there are none. The things that we most want to explain in any evolutionary framework seem the most resistant to any such explanation. This has been one of the greatest obstacles to achieving any real coherence in human knowledge, to building any load-bearing bridges between the natural sciences, the social sciences, and the humanities.

These three problems compound one another. They roam around like a gang, knocking the sense out of any innocent young theories that happen to stroll along. If a new theory overcomes problem three by claiming a previously unrecognized survival benefit for art or language, then problem one raises the objection, "Why do we not see hundreds of species taking advantage of that survival benefit by growing larger brains with these abilities?" Or, suppose a new theory tackles problem two by emphasizing the success if our early *Homo erectus* ancestors in spreading from equatorial Africa across similar latitudes in Asia. Then problem three can point out that many smaller-brained mammals such as cats and monkeys expanded in similar ways, without evolving such mental embellishments.

Most theories of human evolution attempt to solve only one of these three problems. A few might solve two. None has ever solved all three. This is because the three problems create a paradox that cannot be solved by thinking in terms of survival of the fittest. Many human mental abilities are unique to our species, but evolution is opportunistic and even-handed. It doesn't discriminate between species. If our unique abilities must be explained through some survival benefit, we can always ask why evolution did not confer that same benefit on many other species. Adaptations that have large survival benefits typically evolve many times in many different lineages, in a process called convergent evolution. Eyes, ears, claws, and wings have evolved over and over again in many different lineages at many different points in evolutionary history. If the human mind evolved mostly for survival benefits, we might expect convergent evolution to have driven many lineages toward human-type minds. Yet there is no sign of convergent evolution toward human-style language, moral idealism, humor, or representational art.

In *The Language Instinct*, Steven Pinker claimed that the elephant's trunk raises some of the same problems as human language: it is a large, complex adaptation that arose relatively recently in evolution, in only one group of mammals. Yet the elephant's trunk does not really raise any of our three problems. There was convergent evolution towards grasping tentacle-like structures among octopi and squid. The evolution of the trunk allowed the ancestors of the elephant to split apart very quickly into dozens of species of mammoths, mastodons, and elephants, in an evolutionary pattern called an "adaptive radiation." These species all had trunks, and they thrived all over the globe until our ancestors hunted them to extinction. An elephant uses its trunk every day to convey leaves from trees to mouth, showing clear survival benefits during foraging. The trouble with our unique human abilities is that they do not show the standard features of survival adaptations—convergent evolution, adaptive radiation, and obvious survival utility—and so are hard to explain through natural selection.

Sexual selection cuts through this Gordian knot. Biologists recognize that sexual selection through mate choice is a fickle, unpredictable, diversifying process. It takes species that make their livings in nearly identical ways and gives them radically different sexual ornaments. It never happens the same way twice. It drives divergent rather than convergent evolution. There are probably half a million species of beetle, but no two have the same kind of sexual ornamentation. There are more than three hundred species of primates, but no two have the same shape and color of facial hair. If the human mind's most unusual capacities evolved originally as courtship ornaments, their uniqueness comes as no surprise. Nor should we be surprised at the lack of survival benefits while brain size was tripling. The brain's benefits were mainly reproductive.

We get confused about the human mind's biological functions because of a historical accident called human history. The courtship ornaments that our species happened to evolve, such as language and creativity, happened to yield some completely unanticipated survival benefits in the last few thousand years: agriculture, architecture, writing, metalworking, firearms, medicine, and microchips. The usefulness of these recent inventions tempts us to credit the mind with some general survival advantage. From the specific benefits of specific inventions, we infer a generic biological benefit from the mind's "capacity for culture." We imagine evolution toiling away for millions of years, aiming at human culture, confident that the energetic costs of large brains will someday pay off with the development of civilization. This is a terrible mistake. Evolution does not have a Protestant work ethic. It does not get tax credits for research and development. It cannot understand how a costly investment in big brains today may be justified by cultural riches tomorrow.

To understand the mind's evolution, it is probably best to forget everything one knows about human history and human civilization. Pretend that the last ten thousand years did not happen. Imagine the way our species was a hundred thousand years ago. From the outside, they would look like just another group of large primates foraging around Africa, living in small bands, using a few simple tools. Even their courtship looks uneventful: a male and a female just sit together, their eyes meet, and they breathe at each other in odd staccato rhythms for several hours, until they start kissing or one gives up and goes away. But if one could understand their quiet, intricately patterned exhalations, one could appreciate what is going on. Between their balloon-shaped skulls pass back and forth a new kind of courtship signal, a communication system unlike anything else on the planet. A language. Instead of dancing around in physical space like normal animals, these primates use language to dance around in mindscapes of their own invention, playing with ideas.

Talking about themselves gave our ancestors a unique window into one another's thoughts and feelings, their past experiences and future plans. Any particular courtship conversation may look trivial, but consider the cumulative effects of millions of such conversations over thousands of generations. Genes for better conversational ability, more interesting thoughts, and more attractive feelings would spread because they were favored by sexual choice. Evolution found a way to act directly on the mental sophistication of this primate species, not through some unique combination of survival challenges, but through the species setting itself a strange new game of reproduction. They started selecting one another for their brains. Those brains won't invent literature or television for another hundred thousand years. They don't need to. They have one another.

The intellectual and technical achievements of our species in the last few thousand years depend on mental capacities and motivations originally shaped by sexual selection. Trained by years of explicit instruction, motivated by sophisticated status games, and with cultural records that allow knowledge to accumulate across generations, our sexually selected minds can produce incredible things such as Greek mathematics, Buddhist wisdom, British evolutionary biology, and Californian computer games. These achievements are not side-effects of having big brains that can learn everything, but of having minds full of courtship adaptations that can be retrained and redirected to invent new ideas even when we are not in love.

Fossils, Stories, and Theories

Anyone presenting a theory about human mental evolution is usually expected to present a speculative chronology of what evolved when, and to show how the current fossil and archeological data support that chronology. I will attempt neither, because I think these expectations have too often led theorists to miss the forest for the trees. The human mind is a collection of biological adaptations, and an evolutionary theory of the mind must, above all, explain what selection pressures constructed those adaptations. Chronology is of limited use, because knowing when an adaptation arose is often not very informative about why it arose. Fossil and archeological evidence has been enormously important in showing how many pre-human species evolved, when they evolved, where they lived, and what tools they made. This sort of evidence is crucial in putting human evolution in its biological and geological context, but it has not proven terribly useful in explaining why we have the mental adaptations that we do—and in some cases it can be misleading.

For example, an overreliance on archeological data may lead scientists to underestimate the antiquity of some of our most distinctive abilities. Many have assumed that if there is no archeological evidence for music, art, or language in a certain period, then there cannot have been any. Historically, European archeologists tended to focus on European sites, but we now know that our human ancestors colonized Europe tens of thousands of years after they first evolved in Africa a hundred thousand years ago. This Eurocentric bias led to the view that music, art, and language must be only about 35,000 years old. Some archeologists such as John Pfeiffer claimed there was an "Upper Paleolithic symbolic revolution" at this date, when humans supposedly learned how to think abstractly and symbolically, leading to a rapid emergence of art, music, language, ritual, religion and technological innovation. If these human abilities emerged so recently in Europe, we would not expect to find them among African or Australian peoples—yet there is plenty of anthropological evidence that all humans everywhere in the world share the same basic capacities for visual, musical, linguistic, religious, and intellectual display. The same over-conservative reasoning would lead us to say that human language must be only 4,000 years old, because the archeological evidence for writing goes back only that far.

Also, the fossil and archeological evidence is still very patchy and is accumulating very quickly, with new discoveries often undermining our interpretations of old findings. Physical evidence about human origins seems the most secure place to begin in theorizing about human evolution, but this security is largely illusory. Since 1994 at least four new species of hominid have been discovered. Every year brings new bones or stones that necessitate a major rethinking about the times, places, and products associated with human origins. The result is often theories as transient as the evidence they cite. Most human evolution theories of twenty or fifty years ago are barely worth reading now because, by tying themselves too closely to the physical evidence then available, they aimed too much for empirical respectability at the expense of theoretical coherence. The theories that remain relevant are those derived from fundamental principles of evolutionary biology and commonsense observations about the human mind. Darwin's thoughts on the human mind's evolution in *The Descent of Man* are still useful because he did not overreact to the new discoveries of Neanderthal skulls and living gorillas that fascinated Victorian London. Classic selection pressures are more important than classic fossils.

A final limitation is that fossil and archeological evidence has proven much more informative about how our ancestors could afford the energy costs of large brains, than about what they actually used their brains for. Evidence in the last decade has revealed how our ancestors evolved the ability to exploit energy-rich foods such as game animals that could be hunted for meat, and underground tubers that could be dug up and cooked. These energy-rich foods could also be digested using shorter intestines than other apes have. As anthropologist Leslie Aiello has argued, since guts use a lot of energy, our smaller guts also increased our energy budget above what is available to other apes. The ability to exploit these new food sources, at a lower gut-cost, could have allowed our ancestors to afford larger bodies, larger brains, more milk production, or whatever other costly traits evolution might have favored. But a higher energy budget does not in itself explain why our brains expanded, or why any of our distinctive human abilities evolved. Sexual selection principles, not fossil evidence, may explain why we wasted so much of our energy on biological luxuries like talking, dancing, painting, laughing, playing sports, and inventing rituals.

An evolutionary account of the human mind cannot be constructed directly from fossils and stone artifacts. As archeologist Steven Mithen argued in his thoughtful book *The Prehistory of the Mind*, the physical evidence of prehistory must be interpreted in a much more sophisticated evolutionary psychology framework. Yet many scientists still have a special reverence for archeological evidence which is out of all proportion to what it can tell us about mental evolution. Fossils were certainly critical in convincing people that we had actually evolved in continuous stages from primate ancestors—almost 50 percent of Americans now accept the fossil evidence for human evolution. But evidence supporting the fact of human evolution is not always the best evidence for the mechanism of human evolution. A more fruitful place to start theorizing about the past is the present: the current capacities of the human mind (the adaptations to be explained) and the principles of current evolutionary biology (the selection pressures that can explain them). Bones and stones can be valuable sources of evidence, but they become most useful when combined with studies of other primates, and with studies of humans in tribal societies, modern societies, and psychology laboratories.

This may sound like a radical change in scientific method, but it isn't. In broadening the focus from stones and bones to the comparative analysis of present adaptations, I am in fact proposing something rather conservative: that the evolutionary psychology of the human mind can play by the same scientific rules as the evolutionary biology that studies any other adaptation in any other species. It can present a bold theory about the function of the adaptation and the selection pressures that produced it, and see whether the adaptation has special features consistent with that function and those origins. Paleontology makes useful contributions to such studies, but it is not the most important source of data on the design and functions of biological adaptations. The details of an adaptation as it currently exists are often more informative than the fossilized remnants of its earlier forms. In this book I shall draw upon the fascinating discoveries of fossil-hunters and archeologists where appropriate, but I believe that the features of the modern human mind are often the best clues to its origin.

Show Me the Genes

From the 1980s, DNA evidence has become almost as important as fossil and archeological evidence in understanding human evolution. In the coming decades it is likely to become hugely more important, especially in tracing the human mind's origins. This is because evolved mental capacities depend on genes, even when they leave no fossil or archeological records. After the Human Genome Project identifies all 80,000 or so human genes in the next couple of years, we can look forward to three further developments that will allow much more powerful tests of my theory and other theories of mental evolution.

Neuroscientists will start to identify which genes underlie which mental capacities, by analyzing the proteins they produce, and the role those proteins play in brain development and brain functioning. (Of course there is no single gene for language or art—these are complex human abilities that probably depend on hundreds or thousands of genes.) Behavior geneticists will also identify different forms of particular genes that underlie individual differences in mental abilities such as artistic ability, sense of humor, and creativity. Psychologist Robert Plomin and his collaborators have already identified the first specific gene associated with extremely high intelligence (a form of the gene labelled "IGF2R" on chromosome 6). Very little such work has been done so far, but the genes that underlie our unique human capabilities will be identified sooner or later, and evolutionary psychology will benefit.

Also, geneticists will find out more about which genes we share with other apes. Research centers in Atlanta and Leipzig are already pushing for the development of a Chimpanzee Genome Project. Since 1975, geneticists have been using a method called DNA hybridization to show that our DNA is roughly 98 percent similar to that of chimpanzees (compared to only 93 percent with most monkeys). However, this method is fairly crude, and we will not know exactly which of our genes are unique until the results of the Chimpanzee Genome Project can be compared to those of the Human Genome Project. Geneticists already know there are some significant differences: humans have 23 pairs of chromosomes whereas other apes have 24 pairs, and the genes on human chromosomes 4, 9 and 12 appear to have been reshuffled significantly compared with their arrangement on the chimpanzee chromosomes. There are plenty of genetic differences to account for our distinctive mental capacities, and the more we know about the unique human genes, the more we can infer about their evolutionary origins and functions.

Finally, it may be possible to recover more DNA from our extinct fossil relatives. DNA decays fairly quickly, and it is very hard to recover DNA from fossils older than about 50,000 years ago (*Jurassic Park* notwithstanding). However, Neanderthals survived until about 30,000 years ago, and a German team led by Svante Pääbo has already succeeded in recovering a DNA fragment from a Neanderthal's arm bone. This fragment, just 379 DNA base pairs long, showed 27 differences compared with modern humans, and 55 differences compared with chimpanzees. This substantial difference between humans and Neanderthals suggests that our lineages split apart at least 600,000 years ago—

much earlier than previously thought. It also shows that humans did not evolve from Neanderthals. Potentially, the same techniques could be applied to *Homo erectus* specimens from Asia, which also persisted until about 30,000 years ago, but which split off from our ancestors even earlier. It might even be possible, at some future date, to show which other hominids shared the genes underlying our apparently unique mental abilities. For example, if Neanderthals are found to share some of the same genes for language, art, music, and intelligence that modern humans have, then we could infer that those capacities evolved at least 600,000 years ago. Although behavior does not fossilize, some of the DNA underlying behavior does, and it can sometimes last long enough for us to analyze.

The DNA revolution will unveil many more aspects of human evolution and human psychology. I cannot yet show you the many genes that must underlie each of the human mental adaptations analyzed in this book. However, the genetic evidence that will emerge in the coming years will probably render my ideas—even the apparently most speculative ones—fully testable in ways I cannot anticipate. My sexual choice theory sometimes sounds as if it could explain anything, and hence explains nothing. This overlooks the fact that biologists are developing ever more sophisticated ways of testing which adaptations have evolved through sexual selection, and many of these methods—including a range of new genetic analyses—can be applied to human mental traits. Indeed, one goal of this book is to inspire other scientists to join me in testing these ideas.

What We Can Expect From a Theory of Human Mental Evolution

Any theory of human mental evolution should, I think, strive to meet three criteria: evolutionary, psychological, and personal. The evolutionary criteria are paramount. Any theory of human mental evolution should play by the rules of evolutionary biology, using accepted principles of descent, variation, selection, genetics, and adaptation. It is best not to introduce speculative new processes of the sort that have been touted recently, such as "gene–culture co-evolution," "cognitive fluidity as a side-effect of having a large brain," or "quantum consciousness." Complex adaptations such as human mental capacities need to be explained by cumulative selection for a function that promotes survival or reproduction.

This evolutionary criterion makes it much more important to identify the selection pressures that shaped each adaptation than to identify how the adaptation went through some series of structural changes, having started from some primitive state. Complex adaptations are explained by identifying functional features and specifying their fitness costs and benefits in a biological context. The emphasis is on what and why, rather than how, when, or where. For every theory of every adaptation, there is one demand that modern biologists make: show me the fitness! That is, show how this trait promoted survival or reproduction.

Psychologically, the human mind as explained by the theory should bear some resemblance to the minds of ordinary women and men as we know them. The mental adaptations described in the theory should fit our understanding of normal human abilities and personalities. If you're married, imagine your in-laws. If you commute by public transport, visualize your traveling companions. They're the kinds of minds the theory should account for: ordinary people, in all their variation. We should not worry too much about the minds of exceptional geniuses such as theoretical physicists and management consultants. We are not really trying to explain "the human mind" as a single uniform trait, but human minds as collections of adaptations with details that vary according to age, sex, personality, culture, occupation, and so forth. Still, differences within our species are minor compared with differences across species, so it can be useful to analyze "the human mind" as distinct from "the chimpanzee mind" or "the mind of the blue-footed booby."

Finally, any theory of human origins should be satisfying at a personal level. It should give us insight into our own consciousness. It should seem as compelling in our rare moments of personal lucidity as it is when we are mired in that mixture of caffeine, television, habit, and self-delusion that in modern society we call "ordinary consciousness." It is so easy, when engaged in abstract theorizing about mental evolution, to forget that we are talking about the origins of our own genes, from our own parents, that built our own minds, over our own lifetime. Equally, we are talking about the origins of the genes that built the mind and body of the first person you ever fell in love with, and the last person, and everyone in between. A theory that can't give a satisfying account of your own mind, and the minds you've loved, will never be accepted as providing a scientific account of the other six billion human minds on this planet. Theories that don't fulfill this human hunger for self-explanation may win people's minds, but it will not win their hearts. The fact that 47 percent of Americans still think humans were created by God in the last ten thousand years suggests that evolutionary theories of human origins, however compelling at the rational level, have not proved satisfying to many people. We might as well admit that this is a third demand to impose on theories of human mental evolution, and see whether we can fulfill it. This criterion should not take precedence over evolutionary principles or psychological evidence, but I think it can be a useful guide in developing testable new ideas. If we cannot fulfill this criterion, perhaps we'll just have to live with the existential rootlessness that Jean-Paul Sartre viewed as an inevitable part of the human condition.

Working Together

In facing these three challenges, I have found my professional training as an experimental cognitive psychologist of limited value. What I learned about the psychology of judgment and decision-making was helpful in thinking about sexual choice. But most experimental psychology views the human mind exclusively as a computer that learns to solve problems, not as an entertainment system that evolved to attract sexual partners. Also, psychology experiments usually test people's efficiency and consistency when interacting with a computer, not their wit and warmth when interacting with a potential spouse. These attitudes have carried over into fashionable new areas such as cognitive neuroscience.

Because cognitive psychology and neuroscience usually ignore human courtship behavior, this book discusses very little of the research areas I was trained to pursue. Such research reveals how human minds process information. But evolution does not care about information processing as such: it cares about fitness—the prospects for survival and reproduction. Experiments that investigate how minds process arbitrary visual and verbal information shed very little light on the fitness costs and benefits of the human abilities that demand evolutionary explanation, such as art and humor. Conversely, some less well-funded research on individual differences, personality, intelligence, and behavior genetics has proven surprisingly useful to me. Such research bears directly on the key questions in sexual selection: how do traits differ between individuals, how can those differences be perceived during mate choice, how are those differences inherited, and how are they related to overall fitness? Its conclusions are not always what we refer to nowadays as "politically correct." I would have been more comfortable combining evolutionary biology with a politically correct neuroscience that ignores human sexuality, individual differences, and genes. But in evolutionary psychology we have to deal with evolution, and that means paying attention to genetically heritable individual differences that give survival or reproductive advantages over other individuals.

Many recent books about the human mind's evolution have offered radical new ideas about how evolution works, but have described the mind's capacities very conservatively. That approach suggests that modern evolutionary theory is a castle built on sand, whereas modern psychology is the Rock of Gibraltar. I take the opposite view. Mostly, my sexual choice theory relies on conservative, well-established evolutionary principles, but it takes a rather playful, irreverent view of human behavior.

This book also draws on a wide range of facts and ideas from many areas of science, including psychology, anthropology, evolutionary theory, primatology, archeology, cognitive science, game theory, and behavior genetics. I also borrow a number of ideas from contemporary feminism and cultural theory, and from some of my intellectual heroes such as Friedrich Nietzsche and Thorstein Veblen. I won't pretend to be expert in all these topics. Outside our own areas of expertise, scientists keep up to date by reading the same popular science books and magazine articles as other people do. This makes us vulnerable to the same intellectual fads that sweep through academic and popular culture; it also makes us dependent on the popularizers of other sciences, who sometimes have idiosyncratic views. I have tried to minimize such distortions by being fairly conservative about which ideas and data I rely on. I try to identify which of my arguments are well supported by the current evidence as I understand it, and which still need to be evaluated with further research.

There are also limits to my practical understanding of our mental adaptations. I know less about art than most artists, less about language than political speech writers, and less about comedy than Matt Groening, originator of *The Simpsons.* If you find that you know more about some aspect of the human mind than I do, my errors and omissions could be considered your opportunities. There is plenty of room in evolutionary psychology for contributions by people with all sorts of expertise.

This book presents one possible way to apply sexual selection theory in evolutionary psychology, but there are countless other ways. There is no pretense here of having a complete theory of the human mind, human evolution, or human sexual relationships. This is a snapshot of a provisional theory under construction. My aim is to stimulate discussion, debate, and further research, not to win people over to some doctrine set in stone.

An Ancestral Romance

This book's most unusual challenge is that readers will sometimes be asked to imagine what it was like for our ancestors to fall in love with beings considerably hairier, shorter, poorer, less creative, less articulate, and less self-conscious than ourselves. This is best done without visualizing such beings too concretely. I have never managed to feel genuine desire for any museum model of an Australopithecine female, however realistically their sloping foreheads, thick waists, and furry buttocks have been rendered. Nor have I found it easy to imagine feeling genuine love when gazing into the eyes of one of these ancestors from three million years ago. Our sexual preferences seem too hard-wired to permit these imaginative leaps. The limits of our contemporary sexual imaginations have always been an obstacle to appreciating the role of sexual choice in human evolution.

On the other hand, ancestral romance is not so hard to understand at a slightly more abstract level. Indeed, it may be intuitively easier to understand human evolution through sexual selection than through natural selection. While our ancestors faced very different survival problems than we do today, the problems of sexual rejection, heartbreak, jealousy, and sexual competition remain almost unchanged. Few of us have any experience digging tubers, butchering animals, escaping from lions, or raiding other tribes. But our past sexual relationships may prove a useful guide to understanding the sexual choices that shaped our species.

Each of our romantic histories goes back only a few years, but the romantic history of our genes goes back millions. We are here only because our genes enjoyed an unbroken series of successful sexual relationships in every single generation since animals with eyes and brains first evolved half a billion years ago. In each generation, our genes had to pass through a gateway called sexual choice. Human evolution is the story of how that gateway evolved new security systems, and how our minds evolved to charm our way past the ever more vigilant gatekeepers.

2

Darwin's Prodigy

The idea of sexual selection has a peculiar history that embodies the best and the worst of science. The best, because it follows the classic heroic model. A lone genius (Charles Darwin), working from his country home without any official academic position, proposes a bold theory that explains diverse, previously baffling facts. Despite presenting the theory in a lucid, engaging best-seller (*The Descent of Man, and Selection in Relation to Sex*), the theory is immediately attacked, mocked, reviled, and dismissed by his narrow-minded colleagues. The theory falls into obscurity, but, as decades pass, more and more supporting evidence accumulates in ways that could never have been anticipated by the original thinker. Finally, over a century after it is first proposed, the theory gradually becomes accepted as a major, original contribution. Sexual selection theory has returned like the prodigal son. Science shows once again how truth wins out against historical contingency and ideological hostility.

Yet this history also shows the worst of science. Over a century passed before biologists took seriously Darwin's most provocative ideas about mate choice. The delay resulted not just from rational skepticism, but from a set of reactionary prejudices deriving from sexism, anthropocentrism, and a misguided type of reductionism. These prejudices were so strong that, for more than fifty years after Darwin, virtually no biologists or psychologists bothered to put his mate choice ideas to a good experimental test (though such tests have subsequently proven fairly easy to do, usually with positive results).

This chapter introduces some basic sexual selection ideas

through a narrative history. The history is important because the century when sexual selection was in exile was the century when the origins of the human mind seemed the most inexplicable. Before Darwin, religious myths accounted for human origins; after Darwin, evolution satisfactorily accounted for the human body, but not the human mind. In the 20th century, a unique scientific fascination with human psychology coexisted with an unprecedented bafflement about its origins. By considering the 19th-century origins of sexual selection theory, we may better understand aspects of human nature that were overlooked for most of the 20th century.

Ornaments of Gold

As a child, Charles Darwin was fascinated by nature. He collected beetles avidly, and was once so determined to capture a specimen, despite having his hands full, that he placed it in his mouth to carry home. His reward was a mouthful of defensive beetle-acid, but his enthusiasm remained intact. His family estate, The Mount, near Shrewsbury, had an excellent library full of his father's natural history books, a greenhouse stocked with exotic plants, an aviary for the fancy pigeons his mother kept, and access to a bank of the River Severn. Young Charles preferred nature's sights and sounds to the rote learning of Latin at the local Shrewsbury School.

By age 23, Darwin had left Shrewsbury for South America. His round-the-world voyage on the *Beagle* introduced him to the astounding volume and diversity of nature's ornaments. England had passerine birds with intricate songs, and pheasants with stately colors, but nothing prepared the young naturalist for the richly ornamented flora and fauna of the tropics: iridescent humming birds visiting outlandish flowers; beetles with carapaces of gold, sapphire, and ruby; enigmatic orchids; screaming parrots; butterflies like two blue hands clapping; monkeys with red, white, black, and tan faces; exotic Brazilian fruits on market stalls. On a single day during a foray from Rio, Darwin caught no less than 68 species of beetle. His diaries record his "transports of pleasure" and the "chaos of delights" inspired by the jungle's baroque extravagance—"like a view in the Arabian Nights."

Darwin wanted an explanation for this rich array of diversity. Two decades before Darwin's trip, theologians such as William Paley had argued that God ornaments the world to inspire man's wonder and devotion. Darwin may have wondered why God would put tiny golden bugs in the heart of a sparsely populated jungle, a thousand miles from the nearest church. Were nature's ornaments really for our eyes only? Between the *Beagle*'s voyage and his notebooks of 1838, Darwin had worked out the principle of evolution by natural selection. He realized that bugs must be golden for their own purposes, not to delight our eyes or to symbolize divine providence.

Animal ornaments must have evolved for some reason, but Darwin could not see how his new theory of natural selection could account for these seemingly useless luxuries. He had seen that many animals, especially males, have colorful plumage and melodious songs. These are often complex and costly traits. They usually have no apparent use in the animals' daily routine of feeding, fleeing, and fighting. The animals do not strive to display these ornaments to humans when we appear to need some spiritual inspiration. Instead, they display their beauty to the opposite sex. Usually, males display more. Peacocks spread their tails in front of peahens. In every European city, male pigeons harass female pigeons with relentless cooing and strutting. If the females go away, the male displays stop. If the female comes back, the males start again. Why?

Once his travels had confronted Darwin with the enigma of animal ornamentation, he could never take it for granted again. After his return, it seemed to him that English gardens were awash with peacocks. Their tails kept the problem in the forefront of Darwin's mind, sometimes with nauseating effect. Darwin once confided to his son Francis that "The sight of a feather in a peacock's tail, whenever I gaze at it, makes me sick!" The peacocks seemed to mock Darwin's theory that natural selection shapes every trait to some purpose.

Science by Stealth

Darwin cured his peacock-nausea by developing the theory of sexual selection. We do not know exactly when or how he developed it, because historians of science have not tried very hard to find out. They have written at least a thousand times as much about the discovery of natural selection as they have about the discovery of sexual selection. Even today, there is only one good history of sexual selection theory—Helena Cronin's *The Ant and the Peacock.* But we do know this: at some point between the *Beagle*'s voyage in the 1830s and the publication of *The Origin of Species* in 1859, Darwin started to understand animal ornamentation. In that epoch-making book he felt comfortable enough about sexual selection to devote three pages to it, but not confident enough to give it a whole chapter.

From that acorn grew the oak: his 900-page, two-volume *The Descent of Man, and Selection in Relation to Sex* of 1871. The title is misleading. Less than a third of the book—only 250 pages—concerns our descent from ape-like ancestors. The rest concentrates on sexual selection, including 500 pages on sexual selection in other animals, and 70 pages on sexual selection in human evolution. Darwin was no longer troubled by tiny gold bugs or peacock feathers. He considered his sexual selection idea to be so important that he featured it in the one book he was sure humans would read: his summary of the evidence for human evolution.

However, Darwin was a subtle and strategic writer, often hiding his intentions. His introduction to *The Descent* claimed that "The sole object of this work is to consider, firstly, whether man, like every other species, is descended from some pre-existing form; secondly, the manner of his development; and thirdly, the value of the differences between the so-called races of man." Later in the introduction he pretended that his only reason for considering sexual selection was its utility in explaining human racial differences. He apologizes that "the second part of the present work, treating of sexual selection, has extended to an inordinate length, compared with the first part, but this could not be avoided." Immediately after claiming that he lacked the editorial self-control to leave sexual selection for another book, he complained that lack of space required him to leave for another book his essay *The Expression of the Emotions in Man and Animals.* What was Darwin thinking? *The Expression of the Emotions* provided direct evidence of psychological similarities between humans and other animals. One would think it belonged in *The Descent,* if the book's sole object was to consider man's biological similarities to other animals. Yet Darwin left his best evidence of similarity for another book, and inserted almost 600 pages on sexual selection. I suspect that this was science by stealth. Perhaps Darwin intended to smuggle into popular consciousness his outrageous claim that mate choice guides evolution, while his relatively predictable views on human evolution would draw the fire of his critics. As we shall see, this clever plan was not entirely successful.

The Grand Gateway of Sex

So how does sexual selection explain ornamentation? Darwin's problem was the ubiquity of large, costly, complex traits like peacock's tails that seem to contribute nothing to an animal's survival ability. Natural selection, as Darwin defined it, arises from individual differences in survival ability. It cannot favor traits opposed to survival. Since most ornaments decrease an individual's survival ability, they presumably could not have evolved by natural selection for survival.

This means that evolution must include some form of creative, trait-shaping selection other than natural selection. Darwin reasoned that in a sexually reproducing species, any traits that help in competing for sexual mates will tend to spread through the species. These traits may evolve even if they reduce survival ability. While natural selection adapts species to their environments, sexual selection shapes each sex in relation to the other sex. In *The Origin,* Darwin argued that sexual selection depends "not on a struggle for existence in relation to other organic beings or to external conditions, but on a struggle between the individuals of one sex, generally the males, for the possession of the other sex. The result is not death to the unsuccessful competitor, but few or no offspring."

Darwin didn't know about genes or DNA. But he understood that in a sexually reproducing species, the only way to pass a trait from one generation to the next was, by definition, through sexual reproduction. If an animal doesn't have sex, its heritable traits will die with it, and it will leave no hereditary trace in the next generation. As far as evolution is concerned, the animal may as well have died in infancy. Survival without reproduction means evolutionary oblivion. On the other hand, reproduction followed by death can still translate into evolutionary success. Sexual inheritance puts sexual reproduction at the heart of evolution. The concept of sexual selection is simply a way of describing how differences in reproductive success lead to evolutionary change.

Sexual, Natural, Artificial

To explain sexual selection, Darwin used the familiar metaphor of artificial selection. Victorian England was still mostly agricultural and pastoral. People knew about artificial selection, in which farmers domesticate plants and animals by allowing some individuals to breed and others not. Darwin had already used this barnyard type of artificial selection as a metaphor to explain how natural selection worked. Sexual selection he compared to a rather different sort of artificial selection more familiar to the leisured classes, and more relevant to gorgeous ornamentation: breeding pet birds to make them look unusual and attractive. In *The Origin* he argued that "if man can in a short time give beauty and an elegant carriage to his bantams, according to his standard of beauty, I can see no good reason to doubt that female birds, by selecting, during thousands of generations, the most melodious or beautiful males, according to their standard of beauty, might produce a marked effect."

The analogy between artificial selection by human breeders and sexual selection by female animals may seem strained. But for Darwin there was no essential difference between human minds and animal minds: both could work as selective forces in evolution. As a dog-lover and an experienced horseman, Darwin felt comfortable attributing intelligence to animals. He reasoned that if humans can breed dogs, cats, and birds according to our aesthetic tastes, why shouldn't these animals be able to breed themselves according to their own sexual tastes?

Biology students now are usually taught that sexual selection is a subset of natural selection, and that natural selection is only loosely analogous to artificial selection by human breeders. This was not Darwin's view: he saw sexual selection as an autonomous process that was midway between natural and artificial selection. Darwin was fairly careful about his terms. For him, artificial selection meant the selective breeding of domesticated species by humans for their economic, aesthetic, or alimentary value. Natural selection referred to competition within or between species that affects relative survival ability. Sexual selection referred to sexual competition within a species that affects relative rates of reproduction. Darwin knew that Herbert Spencer's term "survival of the fittest" could be misleading. Heritable differences in reproduction ability were as important in evolution as heritable differences in survival ability.

However, whereas natural and artificial selection can apply equally well to mushrooms, lemon trees, and oysters, Darwin believed that sexual selection acts most strongly in the higher animals. This is because courtship behavior and selective mate choice behavior are best carried out by mobile animals with eyes, ears, and nervous systems. The mate choice mechanisms that drive sexual selection are much more similar to artificial selection by humans than to blind forms of natural selection by physical or ecological environments. Darwin understood that sexual selection's dependence on active choice might create distinct evolutionary patterns such as fashion cycles and rapid divergence between closely related species.

Males Court, Females Choose

Darwin was more interested in explaining ornamentation than in explaining sex differences. Still, he could not help but notice that male animals are almost always more heavily ornamented than females. He also noticed that most of the differences between males and females are either specializations for making eggs or sperm, or specializations in the weaponry and ornamentation used during sexual competition. Sexual selection was not only useful in explaining ornamental traits that natural selection could not explain. It could also account for almost all differences between the sexes.

This made a rather neat story. Males usually compete to inseminate females. They do this by intimidating other males with weaponry, and by attracting females with ornaments. Females exercise sexual choice, picking the stronger and more attractive males over the weaker and plainer. Over generations, male weaponry evolves to be more intimidating and male ornamentation evolves to be more impressive. There are two results. First, within each sexual species, males diverge from the female norm. Mature males become more strongly differentiated, compared with females, compared with young animals, and compared with their own ancestors. The other result is very fast divergence between species. The weaponry and ornamentation of one species can go off in a very different direction from the weaponry and ornamentation of a closely related species. Thus, Darwin's sexual selection idea could explain three enigmas: the ubiquity across many species of ornaments that do not help survival, sex differences within species, and rapid evolutionary divergence between species.

Darwin had no real explanation of why males court and females choose. Why aren't males choosier? Why don't females evolve weapons and ornaments equally? The fact was that they don't. Darwin felt obligated to report his findings even though, as he admitted, his sexual selection theory was incomplete. *The Descent of Man* is mostly a report on sex differences in ornamentation in non-human animals. Darwin gathered hundreds of examples of males growing larger ornaments than females, and fighting for sexual access to females. He offered a staggering amount of evidence that this typical pattern of sex differences holds from insects through humans. As we shall see, however, critics tended to ignore Darwin's evidence and focus on the gaps in his theory.

What Females Want

Darwin envisioned two main processes of sexual selection: competition among males for the "possession" of female mates, and selection by choosy females among male suitors. Male weapons and pugnacity evolved for fighting other males, and male ornaments and courtship displays evolved for attracting females. The second process, sexual selection through female choice, interested him far more than male contests of strength. The hypothesis of female choice was, Darwin knew, among his most daring and unanticipated. The theory of sexual selection was an intellectual bolt from the blue, and sexual selection through female choice was especially shocking. Darwin understood that his hypothesis of female choice among animals would challenge Victorian social attitudes.

To bolster the case for female choice in *The Descent*, Darwin relied heavily on the analogy with artificial selection. His two-volume study of domestication in 1868 showed how human breeders of chickens, horses, or dogs can select over many generations for greater egg yield, running speed, or emotional stability. If human choice can have such dramatic evolutionary effects, then surely female animals choosing mates can unconsciously select for longer tails, louder songs, or brighter colors in their male suitors. In *The Descent,* Darwin argued that female choice could produce traits as extravagant as those shaped by artificial selection:

> All animals present individual differences, and as man can modify his domesticated birds by selecting the individuals which appear to him the most beautiful, so the habitual or even occasional preference by the female of the more attractive males would almost certainly lead to their modification; and such modifications might in the course of time be augmented to almost any extent, compatible with the existence of the species.

This was a strong claim: sexual selection through mate choice alone, according to the aesthetic preference of female animals, could drive traits to a very high degree of elaboration.

The only limit is extinction: if the courtship trait becomes so costly that it imperils the survival of too many individuals, the species may simply die out. Darwin presented this conclusion with admirable sang-froid: so be it. Sexual selection may drive species to extinction, but that is no argument against its existence. Species do go extinct, appallingly often. Perhaps the ancient Irish elk went extinct because their sexual ornaments—antlers over six feet wide—proved too burdensome. There is no balance of nature that keeps this from happening. The extinction process merely lets us make this prediction: the sexual ornaments of species that have not yet gone extinct are not yet so costly that they kill off almost every male in every generation. Only if the costs of ornamentation result in the deaths of an extremely high proportion of males does a species have trouble maintaining its numbers.

Darwin did not speculate about how female preferences evolve, but he did pay considerable attention to how they apparently work in selecting mates. His analysis of the plumage of the Argus pheasant, spanning almost ten pages of *The Descent*, is a tour de force. The male Argus grows feathers with eyespots like that of the peacock. But each Argus eyespot, though spread out in a fan shape, is shaded to give a spherical appearance, as if illuminated from above. The direction of shading on each eyespot, relative to the feather's axis of growth, must vary in accordance with the typical angle at which the feather is displayed. Darwin thought it extraordinary that evolution could render such an optical illusion so perfectly on a bird's plumage, but he was confident that generations of female choice could account for it:

> The case of the male Argus is eminently interesting, because it affords good evidence that the most refined beauty may serve as a charm for the female, and for no other purpose. . . . Many will declare that it is utterly incredible that a female bird should be able to appreciate fine shading and exquisite patterns. It is undoubtedly a marvellous fact that she should possess this almost human degree of taste, though perhaps she admires the

> general effect rather than each separate detail. He who thinks he can safely gauge the discrimination and taste of the lower animals, may deny that the female Argus pheasant can appreciate such refined beauty; but he will then be compelled to admit that the extraordinary attitudes assumed by the male during the act of courtship, by which the wonderful beauty of his plumage is fully displayed, are purposeless; and this is a conclusion which I for one will never admit.

Darwin remained true to his conviction. Despite heavy opposition to the idea of female choice from his scientific peers, Darwin maintained that the biological evidence was overwhelming, and documented hundreds of male traits that seemed inexplicable in any other way. He reasoned that the function of an evolved adaptation is often revealed in its manifest use by the organism. If an eye is used conspicuously by an animal to see things, and for no other purpose, then the eye probably evolved for vision. If a male animal uses its horns to fight other males, and for no other purpose, then the horns probably evolved for male competition. If a tail is wagged energetically and saliently during courtship, and under no other conditions, and if the tail shows special features that render it visually impressive (e.g. bright coloration, complex patterning, large size), and if the females of the species prefer males with more impressive tails, then the tail probably evolved to court potential mates. The adaptationist logic is the same in each case. But where Darwin was willing to apply the same pragmatic standards of evidence and argument to courtship traits that he applied to other evolved adaptations, his more skeptical colleagues would demand much stronger evidence for female choice than they ever asked for natural selection.

Darwin's evidence for female choice was indirect because Victorian biology lacked methods for experimentally testing animal preferences. Wilhelm Wundt's experimental psychology laboratory in Leipzig, the first in the world, was not established until shortly before Darwin's death in 1882. For indirect evidence of female choice, Darwin had to analyze the marks such choice left on males. In hundreds of species, he analyzed the bodily and behavioral ornaments of males that may have been shaped through female choice. *The Descent* presented such overwhelming evidence for the use of male ornaments in courtship to attract females, that it seems incredible that Darwin's peers doubted the power of female choice. The main biological questions after Darwin should have been, "Why does mate choice evolve, why are females choosier than males, and what kinds of adaptations can be produced by mate choice?" The main psychological question should have been "What role did mate choice play in the evolution of the human mind?" Instead, most biologists after Darwin have asked, "How can we possibly believe that female animals choose with whom they mate?" The history of sexual selection theory is largely a history of this skepticism.

The skepticism about female choice is doubly odd because Darwin took such pains to explain what he meant by female choice. Again and again in *The Descent* he said that mate choice by females need not be conscious and deliberative, but can still be quite accurate, perceptive, and finely tuned. Most biologists accepted that predators choose which prey to chase, that birds choose where to build nests, and that apes choose where to look for food. Are such decisions "conscious"? It doesn't much matter whether we call animal decision-making conscious; what matters is the evolutionary effects of the choice on the animal's own fitness and on the reproductive success of others. Since Darwin freed himself from human prejudices about conscious decision-making, he could see that female choice probably extends to every animal species with a reasonably complicated nervous system. He wrote about female choice in crustaceans, spiders, and insects. The whole point of having a nervous system is to make important adaptive decisions. What decision could be more important than with whom to combine one's inheritance to produce one's offspring?

Mate choice is limited by an animal's senses. Darwin knew that some species have senses quite different from ours. To appreciate their sexual ornaments, we sometimes have to overcome our assumptions about what is perceivable and what is attractive. Usually, we can appreciate the beauty of sexual ornaments in other species only because our senses happen to respond to some of the same stimuli as the senses of those other species. Our primate color vision overlaps in sensitivity with that of many birds, so we can appreciate the colors and forms of bird plumage. But, as Darwin pointed out, our noses may be insensitive to the appealing scents that have been sexually selected in other mammals. We mistakenly perceive most mammals as relatively unornamented.

Even where our senses coincide with those of other species, our aesthetic tastes may differ. Darwin explained that some bird songs sound unmelodic and harsh to our ears, but may still seem attractive to females of the species. Male bitterns (relatives of herons) produce mating calls that sound like guttural gulping, belching, braying, and booming, giving rise to their vernacular names "thunder pumper" and "stake driver." Humans do not enjoy listening to bitterns, but Darwin understood that our tastes are irrelevant in the evolution of bittern mating calls; what matters is the tastes of female bitterns. Their tastes have been forceful enough over time that the male bittern esophagus used to produce their gulpy belches has evolved to thicken every spring just in time for courtship.

Darwin the Radical Psychologist

Sexual selection was a revolutionary idea in several respects. First, it was a truly novel concept. Darwin's theory that species evolve had been anticipated by many 18th- and 19th-century thinkers such as Jean-Baptiste de Lamarck, Étienne Geoffroy Saint-Hilaire, Frédéric Cuvier, and Robert Chambers. Darwin's own grandfather, Erasmus Darwin, had written rather erotic poems about the evolution of flowers. Darwin's theory of natural selection was co-discovered by Alfred Russel Wallace. Sexual selection was quite different. Darwin's notion that mate choice could shape organic form was without scientific precedent.

Second, sexual selection embodied Darwin's conviction that

evolution was a matter of differences in reproduction rather than just differences in survival. Animals expend their very lives in the pursuit of mates, against all the expectations of natural theology. Far from a Creator benevolently fitting each animal to prosper in its allotted niche, Nature shaped animals for exhausting sexual competition that may be of little benefit to the species as a whole.

Finally, Darwin recognized that the agents of sexual selection are literally the brains and bodies of sexual rivals and potential mates, rather than the mindless pressures of a physical habitat or a biological niche. Psychology haunts biology with the specter of half-conscious mate choice shaping the otherwise blind course of evolution. This psychologizing of evolution was Darwin's greatest heresy. It was one thing for a generalized Nature to replace God as the creative force. It was much more radical to replace an omniscient Creator with the pebble-sized brains of lower animals lusting after one another. Sexual selection was not only atheism, but indecent atheism.

Perhaps the least appreciated irony of Darwin's life is that, despite being recognized as the major advocate of natural selection, he seems to have lost interest in the process after publishing *The Origin* in 1859. Perhaps the ease with which the young naturalist Alfred Russel Wallace independently discovered natural selection during a bout of Malaysian malaria, and the need to acknowledge Wallace as a co-discoverer, may have soured Darwin's attitude to his most famous brainchild. In any case, Darwin did not follow up *The Origin* with the sort of research his Victorian colleagues expected. He did not produce a series of detailed case studies of natural selection showing how the external conditions of organic life shape the adaptations of animals and plants.

Instead, he embarked on a seemingly peculiar quest. He wanted to understand how the senses, minds, and behaviors of organisms influence evolution. His 1862 book *On the Various Contrivances by Which British and Foreign Orchids Are Fertilized by Insects* showed how the perceptual and behavioral abilities of pollinators shape the evolution of flower color and form. In 1868 his massive two-volume work *The Variation of Animals and Plants under Domestication* was published, in which he detailed how human needs and tastes have shaped the evolution of useful and ornamental features in domesticated species. Most provocatively, he combined sex with mind and the enigma of human evolution in his two-volume masterpiece *The Descent of Man, and Selection in Relation to Sex*. The trend continues with further works on animal emotions in 1872 and on the behavior of climbing plants in 1875. Even Darwin's final, wry insult to the doctrine of bodily resurrection, his 1881 book on how worms eat the dead to produce fertile soil, was obsessed with the evolutionary and ecological effects of animal behavior.

From *The Origin* until his death, Darwin was as much an evolutionary psychologist as an evolutionary biologist. Except for seven revisions of *The Origin* that successively weakened the role of natural selection in evolution, Darwin wrote little on natural selection. He was confident that he had established the fact of evolution (descent from a common ancestor) and the mechanism of adaptation (cumulative selection on minor heritable variations). He was also confident that other biologists would continue his work on natural selection. So Darwin turned to the really hard problem: how the mysteries of mind and matter interact over the depths of evolutionary time to produce the astonishing pinnacles of beauty manifest in nature, such as flowers, animal ornamentation, and human music.

His theory of sexual selection through mate choice was the crowning achievement of these investigations—yet it was the one most vehemently rejected by his contemporaries. In the last passage that Darwin wrote on sexual selection in *The Descent*, he portrayed mate choice as a psychological process that guides organic evolution:

> He who admits the principle of sexual selection will be led to the remarkable conclusion that the cerebral system not only regulates most of the existing functions of the body, but has indirectly influenced the progressive development of various bodily structures and of certain mental qualities. Courage,

> pugnacity, perseverance, strength and size of body, weapons of all kinds, musical organs, both vocal and instrumental, bright colors, stripes, and marks, and ornamental appendages, have all been indirectly gained by the one sex or the other, through the influence of love and jealousy, through the appreciation of the beautiful in sound, color or form, and through the exertion of a choice; and these powers of the mind manifestly depend on the development of the cerebral system.

Modern critics who accuse Darwin of reducing all of nature's beauty to the blind, dumb action of natural selection could not have read this far. Darwin spent decades thinking about aesthetic ornamentation in nature, realizing that natural selection cannot explain most of it, and developing his sexual selection ideas precisely to describe how animal psychology leads to the evolution of animal ornamentation.

Wallace Versus Female Choice

Alfred Wallace was an unlikely critic of Darwin's sexual selection theory. He independently discovered the principle of natural selection while Darwin was still reluctant to publish. He was even more of a hard-core adaptationist than Darwin, constantly emphasizing the power of selection to explain biological structures that seem inexplicable. He was the world's expert on animal coloration, with widely respected theories of camouflage, warning coloration, and mimicry. He was more generous than Darwin in attributing high intelligence to "savages." Where Darwin was of the landed gentry and fell into an easy marriage to a rich cousin, the working-class Wallace struggled throughout his early adulthood to secure a position sufficiently reputable that he could attract a wife. One might think that Wallace would have been more sensitive to the importance of sexual competition and female choice in human affairs. One might have expected Wallace to use those insights into human sexuality to appreciate the importance of female choice in shaping animal ornamentation. Yet Wallace was utterly hostile to Darwin's theory of sexual selection through mate choice.

The fallacious criticisms developed by Wallace are worth outlining because they continue to be reinvented even now. Wallace distinguished between ornaments that grow in both sexes, and those that grow only in males. The first he explained as identification badges to help animals recognize which species others belonged to. This species-recognition function continues to be advocated by most biologists today, to explain ornaments that show minimal sex differences. On the other hand, Wallace did not consider male ornaments to be proper adaptations that evolved for some real purpose. Instead, he suggested that they were unselected side-effects of an exuberant animal physiology that has a naturally predilection for bright colors and loud songs unless inhibited by the sensible restraint of natural selection.

Take a random animal, cut it in half, and you may see some brightly colored internal organs. Wallace pointed out that internal coloration cannot usually result from mate choice because skin is usually opaque. He argued that organs have a natural tendency to assume bright colors just because of their chemistry and physiology. Ordinarily, natural selection favors camouflage on the outside, so animals often look dull and drab.

Wallace then made an additional claim: the more active an organ, the more colorful it tends to be. He observed that males are generally more vigorous, and, confusing correlation with causation, he proposed that this explains why males are brighter. Male ornamentation for Wallace was the natural physiological outcome of inherently greater male health and vigor. In his 1889 book *Darwinism*, he argued, "The enormously lengthened plumes of the birds of paradise and the peacock . . . have been developed to so great an extent [because] there is a surplus of strength, vitality, and growth-power which is able to expend itself in this way without injury." Males become even more worked up in the mating season, which he thought explains why their ornaments grow more colorful just at the time when females happen to be looking at them. The surplus of energy that males build up in the mating season also tends to get released in ardent songs and extravagant dances.

Females, Wallace thought, are under stronger natural selective pressures to remain discreetly camouflaged because they are so often found near their vulnerable offspring. For example, he showed that female birds that brood in open nests have usually evolved dull camouflage, whereas those that brood in enclosed nests tend to have colors as bright as the males of the species. In Wallace's view, this implied that sexual courtship by males—one of the riskiest, most exhausting, most complex activities in the animal world—must be the default state of the organism, and that the camouflaged laziness shown by young animals, female animals, and males outside the breeding season is something maintained by natural selection. He seems to have envisioned all organic tissue as bursting with color, form, song, dance, and self-expression, which the prim headmistress of natural selection must keep under control.

Wallace understood camouflage and warning coloration. He knew that the perceptual abilities of predators could influence the evolution of prey appearance. So why was he so hostile to female choice, in which the perceptual abilities of females influence the evolution of male appearance? He seems to have forgotten that half of all predators are female. If a female predator can choose to avoid prey that have bright warning colors, why should she be unable to choose a sexual partner based on his bright ornamentation?

Moreover, Wallace's alternative to mate choice begged important questions. Why would males automatically be stronger and more vital than females? Why would they waste surplus energy in such displays? Wallace's arguments along these lines were implausible, ad hoc, and untested. Yet many Victorian biologists considered them at least as plausible as Darwin's mate choice theory. Even more strangely, Wallace's energy-surplus idea foreshadowed Freud's speculation that human artistic display results from a sublimation of excess sexual energy. They also foreshadowed Stephen Jay Gould's claim, first sketched out in his 1977 book *Ontogeny and Phylogeny*, that human creative intelligence is a side-effect of surplus brain size. However, these energy-surplus arguments make little evolutionary sense. In most species surplus energy is converted into fat, not creativity. Surplus brain-mass that yielded no survival or reproductive advantages would quickly be eliminated by selection.

If Darwin had found that male animals choose female mates selectively and that many females are highly ornamented to attract male attention, would Wallace and his contemporaries have been so skeptical about sexual choice? I think not. For male Victorian scientists, it was taken for granted that young single ladies should wear brilliant dresses and jewels to attract the attention of eligible bachelors. Male scientists had direct personal insight into male mate choice. They might easily have sympathized with male animals had Darwin credited them with powers of sexual discernment. They did sympathize with male animals engaged in violent contests with other males for the "possession" of females, which is presumably why they were able to accept Darwin's theory that male weaponry evolved for sexual competition. They simply did not like to think of males as sexual objects accepted or rejected by female choice. (This point is often overlooked by Darwin's feminist critics, who unfairly portray him as embodying Victorian social attitudes.)

The rejection of Darwin's female choice theory was, I think, due to ideological biases in 19th-century natural history, especially the unthinking sexism of most biologists other than Darwin. The rejection was cloaked in scientific argumentation, but the motivations for rejection were not scientific. Many male scientists at the time wrote as if female humans were barely capable of cognition and choice in any domain of life. Female animals were held in even greater contempt, as mere egg repositories to be fought over by males. Male scientists were willing to believe that combat between males, analogous to careerist economic competition in capitalist society, could account for many bodily and behavioral features of male animals. But they could not accept that the sexual whims of female animals could influence the stately progress of evolution.

Wallace paid a high price for his rejection of female choice. He recognized that the human mind contains many biological adaptations, such as elaborate language, music, and art, that seem impossible to explain as outcomes of natural selection for survival value. With more field experience among the primitive tribes of Oceania than Darwin ever amassed on his *Beagle* voyage, Wallace appreciated more acutely than Darwin how striking these adaptations were. He held the musical talents of the Pacific Islanders and African tribal peoples in the highest regard, but could find no survival value in their songs and dances. By rejecting sexual selection for ornamentation, he rejected the one process that might have explained such adaptations. Wallace found himself allied with anti-Darwinians who claimed that evolution could never account for human consciousness, intelligence, or creativity. Though he remained an evolutionist about everything else, Wallace became a creationist about the "human spirit." He went to seances. He developed interests in mesmerism and spiritualist charlatans. He died convinced that science could never fathom the origins or nature of the human mind.

Mendelian Exile

The years 1871 to 1930 were one long dry spell for sexual selection theory. Wallace's criticisms were especially damaging, and gave female choice a bad name. Within a few years of Darwin's death in 1882, sexual selection had already come to be regarded by most biologists as a historical curiosity. Especially hard hit was Darwin's claim that sexual choice played a major role in human evolution. Edward Westermark's *History of Human Marriage* of 1894 spent hundreds of pages trying to undermine the idea that premodern humans were free to choose their sexual partners. He thought that traditional arranged marriages destroyed any possibility of sexual selection. Like most anthropologists of his era, he saw women as pawns in male power games, and young lovers as dominated by matchmaking parents. He founded the tradition of seeing marriage primarily as a way of cementing alliances between families, a view that dominated anthropology until the last years of the 20th century.

Not all biologists were hostile to sexual selection. August Weismann, a leading Darwinian at the University of Freiburg in Germany, included a positive chapter on sexual selection in his *The Evolution Theory* of 1904. After discounting Wallace's surplus-energy theory, and supporting and adding to Darwin's examples of sexual ornamentation, Weismann concluded that "sexual selection is a much more powerful factor in transformation than we should at first be inclined to believe." He added, "Darwin has shown convincingly that a surprising number of characters in animals, from worms upwards, have their roots in sexual selection, and has pointed out the probability that this process has also played an important part in the evolution of the human race." Nonetheless, Weismann's thoughtful assessment was swept away in the rising tide of genetics.

The rediscovery around 1900 of Mendel's work on genetics distracted biologists from Darwin's ideas. For young biologists at the turn of that century, genes were the way forward. Sexual selection was dead, and even natural selection was an unfashionable hobby of the older generation. Biology entered a reductionistic phase of empiricism. Laboratory experiments on mutations attracted more attention and respect than grand theories of natural history. One of the leaders of the new genetics was Thomas Hunt Morgan, a Nobel prize-winner for his work on fruit fly mutations. In his 1903 book *Evolution and Adaptation*, Morgan dismissed sexual selection, concluding that "the theory meets with fatal objections at every turn." He proposed that sex hormones account for all sex differences in ornamentation, failing to realize that the sex hormones and their sex-specific effects themselves require an evolutionary explanation. Morgan's brave new world of mutated flies bred in bottles won over Darwin's world of ornamented butterflies breeding in the wild.

The Fisher King

It was several decades later that the novelty of breeding mutated fruit flies wore off, and some biologists rediscovered Darwin's ideas. One of these young thinkers was Ronald Fisher, whose career spanned the first half of the 20th century. Fisher was a polymath whose insights shaped many fields. To biologists, he was an architect of the "modern synthesis" that used mathematical models to integrate Mendelian genetics with Darwin's selection theories. To psychologists, Fisher was the inventor of various statistical tests that are still supposed to be used whenever possible in psychology journals. To farmers, Fisher was the founder of experimental agricultural research, saving millions from starvation through rational crop breeding programs. In each case, Fisher brought his powerful mathematical brain to bear on questions that had previously been formulated only vaguely and verbally.

Fisher considered Darwin's theory of mate choice to be one vague idea worth trying to formalize. In his first paper on sexual choice in 1915, Fisher enthused that "Of all the branches of biological science to which Charles Darwin's life-work has given us the key, few if any, are as attractive as the subject of sexual selection." Fisher understood that to make sexual selection scientifically respectable, he had to explain the origins of sexual preferences. In particular, Darwin failed to offer any explanation for female choice. Why should females bother to select male mates for their ornaments? Fisher's breakthrough was to view sexual preferences themselves as legitimate biological traits that can vary, that can be inherited, and that can evolve. In his 1915 paper he faced the problem squarely: "The question must be answered 'Why have the females this taste? Of what use is it to the species that they should select this seemingly useless ornament?'" Later, in a 1930 book, Fisher emphasized that "the tastes of organisms, like their organs and faculties, must be regarded as the product of evolutionary change, governed by the relative advantages which such tastes confer." While Darwin had left sexual preferences as mysterious causes of sexual selection, Fisher asked how sexual preferences themselves evolved.

In thinking about the evolution of sexual preferences, Fisher developed the two major themes of modern sexual selection theory. The first idea is the more intuitive, and concerns the information conveyed by sexual ornaments. In the 1915 paper, Fisher speculated thus:

> Consider, then, what happens when a clearly marked pattern of bright feathers affords . . . a fairly good index of natural superiority. A tendency to select those suitors in which the feature is best developed is then a profitable instinct for the female bird, and the taste for this "point" becomes firmly established . . . Let us suppose that the feature in question is in itself valueless, and only derives its importance from being associated with the general vigor and fitness of which it affords a rough index.

Fisher proposed that many sexual ornaments evolved as indicators of fitness, health, and energy. Suppose that healthier males have brighter plumage. Females may produce more and healthier offspring if they mate with healthier males. If they happen to have a sexual preference for bright plumage, their offspring will automatically inherit better health from their highly fit fathers. Over time, the sexual preference for bright plumage would become more common because it brings reproductive benefits. Then, even if bright male plumage is useless in all other respects, it will become more common among males simply because females prefer it. Fisher understood that preferences for fitness indicators could hasten the effect of natural selection, and could potentially affect both sexes. Unfortunately, Fisher's fitness-indicator idea was forgotten until the 1960s.

Fisher's other idea, the concept of runaway sexual selection, attracted more interest because it sounded much stranger. In fact, it was so strange that Thomas Hunt Morgan had first aired the idea in 1903 as a counter argument against sexual selection. Morgan asked what would happen if female birds had a tendency to prefer plumage slightly brighter than the males of their species currently possess. He realized that the males would evolve brighter plumage under the pressure of female choice, but that the females would still not be satisfied. They would just move the goal posts, demanding still more extreme ornamentation. Morgan mocked, "Shall we assume that . . . the two continue heaping up the ornaments on one side and the appreciation of these ornaments on the other? No doubt an interesting fiction could be built up along these lines, but would anyone believe it, and, if he did, could he prove it?" To Morgan, the possibility of an endless arms race between female preferences and male ornaments was an evolutionary impossibility that exposed the whole idea of sexual selection as a fallacy. But Fisher was used to integrating equations for exponential growth, and understood the speed and power of positive-feedback processes. He realized that an arms race between female preferences and male ornaments, far from undermining the theory of sexual selection, could offer an exciting possibility for explaining sexual ornamentation.

The idea of runaway sexual selection appeared in Fisher's masterpiece of 1930, *The Genetical Theory of Natural Selection.* Whenever attractive males can mate with many females and leave many offspring, the sexual preferences of females can drive male ornaments to extremes. Fisher suggested that when this happens, female preferences will evolve to greater extremes as well. This is because a female who prefers a super-ornamented male will tend to produce super-ornamented sons, who will be super-attractive to other females, and who will therefore produce more grandchildren. Evolution will favor super-choosy females for this reason. Yet the choosier the females become, the more extreme the male ornamentation will become in response. Both sexes end up on an evolutionary treadmill. The female preferences and male ornaments become caught up in a self-reinforcing cycle, a positive-feedback loop.

Fisher speculated that whenever the most ornamented individuals gain a large reproductive advantage, there is "the potentiality of a runaway process, which, however small the beginnings from which it arose, must, unless checked, produce great effects, and in the later stages with great rapidity." This runaway process, Fisher claimed, could make ornaments evolve with exponentially increasing speed. They would evolve until the ornaments become so cumbersome that their massive survival costs finally outweigh their enormous sexual benefits: "both the feature preferred and the intensity of preference will be augmented together with ever-increasing velocity, causing a great and rapid evolution of certain conspicuous characteristics, until the process can be arrested by the direct or indirect effects of Natural Selection." I shall explore the runaway process more thoroughly in the next chapter.

Like many mathematical geniuses presenting startling ideas, Fisher thought that runaway sexual selection was so obviously plausible that he did not need to present a detailed proof that it could work. He left that as an exercise for the reader. However, most mathematically talented scientists of the 1930s probably took up the challenge of quantum physics rather than evolutionary biology, and of those who went into biology, nobody took up Fisher's challenge.

Modern Exile

Sexual selection theory has been haunted by unconstructive critics. Whenever a new sexual selection idea raised its head, there was always an eminent biologist ready to knock it down. Wallace attacked female choice in animals, and Westermark attacked female choice in humans. After Fisher proposed his ideas about fitness indicators and the runaway process, the eminent biologist Julian Huxley attacked those too, in two widely influential papers criticizing sexual selection in 1938.

In the space of a few pages, Huxley managed to confuse sexual selection with natural selection, and failed to distinguish natural selection due to competition between individuals and natural selection due to competition between species. He argued that sexual ornaments are immoral because they undermine the good of the species, and if they are immoral, they must not really be sexual ornaments after all, but threat displays, or signals to prevent breeding between species, or perhaps something else. More damage was done by Huxley's popular 1942 textbook *Evolution: The Modern Synthesis*, which cast sexual selection in a marginal, even criminal role in evolution. After mentioning that biologists used to presume that bright colors displayed in courtship were products of sexual selection, Huxley observed that "It was rather the opposite of the presumption of British law that a prisoner is to be regarded as innocent until definite proof of guilt is adduced." Huxley apparently despised sexual selection because he thought it was bad for species, and he thought evolution should be for the good of species. He defined evolutionary progress as "improvement in efficiency of living" and "increased control over and independence of the environment." Since sexual ornaments had high costs that undermined survival chances and did not help an animal cope with the hostile environment, Huxley viewed them as anti-progressive, degenerate indulgences. His contempt for sexual selection combined Puritan prudery and socialist idealism with anxieties about the supposed degeneration of North European races—an ideological cocktail popular among biologists at the time.

After Huxley, the cause of sexual selection foundered again. The years from 1930 to about 1980 saw it exiled to the hinterlands of biology. Unlike the turn-of-the-century exile, this later rejection was not due to a general neglect of evolutionary theory. On the contrary, the Modern Synthesis of the 1930s and 1940s revived Darwinian selection ideas by showing how they could be reconciled with Mendelian genetics. In many ways, this was a golden age for evolutionary theory. Biologists now had proofs and mathematical insights, just as physicists did. Theoretical population genetics was thriving. Darwin was every biologist's hero again—but he was now regarded as a fallible hero, prone to endearing blunders like the hypothesis that female animals select their sexual partners by aesthetic criteria.

Science Troubled by Mate Choice

Biologists could have revived sexual selection in the 1930s by building upon Fisher's work. If they had, the benefits to the behavioral sciences would have been enormous. Anthropologists could have studied real mate choice in primitive cultures instead of concentrating on incest taboos and inter-tribal marriages. Psychotherapists might have rejected Freud's Lamarckian theories about our ancestors inheriting acquired memories of sexually competitive patricide and incest. Psychologists might have overcome the Behaviorist obsession with maze learning by rats, and found a more fruitful way to study human nature. The pioneering sex researchers Alfred Kinsey, William Masters, and Virginia Johnson could have interpreted their questionnaire studies from a richer evolutionary perspective. Archeologists interested in human evolution might not have been so concerned with hunting and warfare, and so baffled by cave paintings and Venus statuettes. Yet none of this happened.

Sexual selection's modern neglect owed more to scientific problems than to ideological biases. One problem is that sexual selection is hard to model mathematically. When a species is adapting to a fixed environment through natural selection, it is possible to predict how a given gene with a given survival effect will spread through a population. With sexual selection, however, the pressures come from other members of the species, which are themselves evolving. It is hard to know where to begin an analysis of sexual selection, because the feedback loops between sexual preferences and sexual ornaments make evolution hard to model and hard to predict. Only in the 1980s did some brilliant mathematical biologists finally start to develop workable models of sexual selection.

Also, the biologists of the Modern Synthesis were consumed by the problem of speciation—how a lineage splits into two distinct species that no longer interbreed. Sexual selection was seen as a possible explanation for speciation, rather than as an explanation for ornamentation. Mate preferences were viewed as nothing more than a way of making sure that individuals mate only with members of their own species. The boundaries of the species were defined by mate preferences, but these preferences were not viewed as ranking individual attractiveness within the species. For many biologists, such as Ernst Mayr, this led to the assumption that most sexual ornaments were nothing more than marks showing what species an animal is. Following Wallace, they were considered to be "species recognition signals."

Sexual selection also suffered at the hands of the early 20th-century doctrines of behaviorism in psychology and reductionism in the sciences generally. These warned against attributing any mental capacities to animals, and this made biologists feel uncomfortable talking about the evolution of female choice mechanisms. Even animal behavior researchers such as Konrad Lorenz and Niko Tinbergen viewed copulation as a stereotyped behavior that is "released" by a few simple stimuli. They did not view mate choice as a complex strategic decision with high stakes. Behaviorist psychologists were not willing to credit even humans with free will or the capacity for choice, so it seemed unscientific to talk about "mate choice" in animals rather than "sexual stimuli." The mid-20th century was the era of B. F. Skinner's manifesto *Science and Human Behavior*, in which people were portrayed as robots driven by conditioned associations. Only with the rise of cognitive psychology in the 1970s did it once again become intellectually respectable in psychology to talk about judgment and decision-making in humans or animals. By then, most psychologists had forgotten all about Darwin. When they thought of sex, they thought of Freud and his theories of subconscious drives and neurotic complexes. Human sexuality, with its alleged existential intricacies, had been set apart from animal sexuality, with its supposedly stereotyped copulation reflexes. A science of mate choice applicable to both animals and humans seemed an absurd conceit.

Moreover, many evolutionary biologists before the 1970s had a very limited concept of adaptation. To them, evolution basically solved problems of survival posed by the external environment. Evolution was supposed to be about the survival of the fittest and the good of the species. Sexual selection was neither progressive nor respectable. Certainly, runaway sexual selection was a theoretical possibility, but bizarre ornaments were not considered to be real adaptations. They impaired individual survival and predisposed species to extinction. Mere ornamentation was not a proper role for a genuine adaptation.

This narrow definition of adaptation was perhaps reinforced by 20th-century aesthetics, which held conspicuous, costly ornamentation in low regard. The modernist reaction against Victorian ornamentation may have spilled over into a reaction against Darwin's sexual selection theory. The Modern Synthesis coincided with the peak of an austere, modernist machine aesthetic. In the 1920s Walter Gropius and other theorists of the Bauhaus movement in Germany had argued that, in a socialist utopia, working people would not waste time and energy hand-decorating objects for purchase by the rich, merely so the rich could show how much wasteful ornamentation they could afford. Form should follow function. Ornament was viewed as morally decadent and politically reactionary, while simplicity and efficiency were considered progressive. This anti-ornament aesthetic seems to have spilled over from culture into nature, leading 1930s biologists to express their contempt for sexual selection's baroque excesses. For example, the socialist biologist J. B. S. Haldane suggested that with sexual selection, "the results may be biologically advantageous for the individual, but ultimately disastrous for the species." In one of his 1938 papers, Julian Huxley declared sexual selection a selfish process because it may "favour the evolution of characters which are useless or even deleterious to the species as a whole." Similar views were held by leading biologists such as Konrad Lorenz, George Simpson, and Ernst Mayr right through to the 1960s. They believed that evolved adaptations, like modernist design, should serve their economic purposes simply, efficiently, and plainly. Sexual ornamentation served no legitimate species-benefiting purpose, so must be ignored or derogated.

Darwin's sexual selection theory was kept in exile by these five factors: mathematical difficulties, an overemphasis on ornaments as species-recognition markers, a mechanistic view of animal psychology, a narrow definition of biological adaptation, and a modernist machine aesthetic. In other words, Darwin's favorite idea was not ignored because there was evidence against it. On the contrary, the mountain of evidence presented in *The Descent of*

Man was never seriously challenged. Sexual selection was ignored because biology was not ready—ideologically, conceptually, or methodologically—to deal with it.

A Second Chance

Sometimes an idea needs to be published twice so that a second generation can judge whether it makes sense. In 1958, almost three decades after the first edition, Fisher produced a second edition of *The Genetical Theory of Natural Selection*. This time it took root in the minds of a new, more mathematically skilled generation of young biologists such as John Maynard Smith and Peter O'Donald. They saw what Fisher was getting at: one could think seriously about the evolutionary origins of sexual preferences, and their evolutionary effects. Maynard Smith set about studying the courtship dances of fruit flies. He found that highly inbred, unfit males could not keep up with healthy females, so would be rejected as mates. The females seemed to be choosing for male fitness as evidenced by dancing ability. Maynard Smith also spent the next several decades wondering why sex evolved in the first place. O'Donald explored the mathematics of sexual selection throughout the 1960s and 1970s, trying to develop proofs of Fisher's intuitions.

A rivulet of interest in sexual selection started to flow through the minds of leading biologists. In his widely read *Adaptation and Natural Selection* of 1966, the young theorist George Williams used Fisher's sexual selection ideas to interrogate the concept of an evolved adaptation. Sexual selection was found not guilty of debauching evolution and making species degenerate. Williams put ornaments on an equal footing with other adaptations, giving sexual selection a status equal to that of selection for survival. In expanding and clarifying the definition of biological adaptation, Williams helped to overcome the machine aesthetic of the Modern Synthesis, and its emphasis on ornaments as species-recognition markers.

Finally, the reductionistic behaviorism of previous decades gave way to cognitive psychology in the 1970s. Once again it became respectable to talk about the mind. Cognition, choice, judgment, decision-making, and planning became part of psychology once again. This laid the foundation for the modern understanding of mate choice in general.

An increased acceptance of the role of female choice may have also been due to social trends. The sexual revolution of the 1960s and the rise of feminism led to more women studying and contributing to biology, and to a new appreciation of female choice in human social, sexual, and political life. Married male biologists could no longer take for granted the obedient support of their wives. They faced a new world in which women made choices more consciously and took more control of their lives. Although evolutionary theory was still extremely male-dominated, individual males were feeling more pressure from female choice. Female biologists doing field-work also drew more attention to female choice among the animals they studied. This was especially important in primatology, as women such as Jane Goodall, Dian Fossey, Sarah Hrdy, Jeanne Altmann, Alison Jolly, and Barbara Smuts explored female social and sexual strategies. Dismissing the idea that female choice could influence the direction of evolution began to look both sexist and unscientific. By drawing attention to the evolution of social and sexual behavior in animals, the sociobiology of the 1970s did for the study of animal sexuality what feminism did for the study of human sexuality. It empowered thinkers to ask "Why does sex work like this, instead of some other way?"

The Handicap Principle Raises the Stakes

The mathematical difficulties with sexual selection were the last barrier to crumble. In 1975, Israeli biologist Amotz Zahavi turned to sexual selection theory and proposed a strange new idea that he called the "handicap principle." It revived Fisher's fitness-indicator idea in a counter-intuitive way. Zahavi suggested that the high costs of many sexual ornaments are what keep the ornaments reliable as indicators of fitness. Peacock tails require a lot of energy to grow, to preen, and to carry around. Unhealthy, unfit peacocks can't afford big, bright tails. The ornament's cost guarantees the ornamented individual's fitness, and this is why costly ornaments evolve.

Zahavi promoted his idea actively and ambitiously, suggesting that the handicap principle applies not only to sexual ornaments, but to warning coloration, threat displays, and many aspects of human culture. Within a year of Zahavi's first paper, Richard Dawkins realized the handicap principle was potentially important, and gave it a remarkably balanced appraisal in his influential 1976 bestseller *The Selfish Gene*. But to other biologists such as John Maynard Smith, Zahavi's principle seemed so confused that it could not possibly explain sexual ornamentation. Mathematically inclined biologists thought the handicap principle was an easy target, and attacked it vigorously.

The controversy over Zahavi's idea marked the true revival of sexual selection theory. Within ten years of his 1975 paper, more research was published on sexual selection than in the previous hundred years. Fisher's fitness-indicator idea was finally in play, its share value boosted by Zahavi's takeover bid. Soon Fisher's runaway process attracted more intellectual capital as well. In 1980 Peter O'Donald published *Genetic Models of Sexual Selection*, summarizing twenty years of thinking about the mathematics of sexual selection. This inspired a spate of new mathematical modeling. In the early 1980s Russell Lande and Mark Kirkpatrick showed that Fisher's runaway process could indeed work. The genes underlying female choice really could get swept up in a positive-feedback loop with the genes underlying male sexual ornaments. Species could even split apart into new species entirely as a result of diverging sexual preferences. Critics attacked these runaway models, leading to the kind of rapid revision and rethinking that marks the most productive epochs of science.

Evolutionary controversies attract experimental biologists. For most of the 20th century, the experimental techniques existed for testing Darwin's basic idea that females choose their mates for their ornamentation. Experimental psychology had developed sophisticated methods and statistical tests for investigating how people make choices. These could have easily been applied to animals. But the work was not done, because biologists thought that sexual selection had been dismissed by the leading theorists. Once the theorists revived the ideas of fitness indicators and runaway processes, the experimenters took a fresh look at mate choice. In species after species, females were seen to show preferences for one male over another, for beautiful ornaments over bedraggled ones, for a higher level of fitness over a lower. Female choice was observed by Linda Partridge in fruit flies, by Malte Andersson in widowbirds, and by Michael Ryan in Tungara frogs. David Buss even showed evidence of mate choice in humans. Wherever males had sexual ornaments, females seemed to show sexual choice, just as Darwin predicted.

Sexual Selection Triumphant

Within a few years, sexual selection became the hottest area of evolutionary biology and animal behavior research. Before this revival, sexual selection was caught in a double bind. Nobody did experiments on mate choice because theorists doubted its existence. And nobody did theoretical work on sexual selection because there was no experimental evidence for mate choice. Once this vicious circle was broken by John Maynard Smith, George Williams, Amotz Zahavi, Robert Trivers, and other pioneers, Darwin's favorite idea was free to succeed.

Sexual selection's revival has been swift, dramatic, and unique. It may be the only major scientific theory to have become accepted after a century of condemnation, neglect, and misinterpretation. Throughout the 1990s, sexual selection research became one of the most successful and exciting areas of biology, dominating the leading evolution journals and animal behavior conferences. Helena Cronin's *The Ant and the Peacock* put sexual selection in its historical context, reminding biologists where it came from and where it might go. Malte Andersson's 1994 textbook *Sexual Selection* reviewed the state of the art for a new generation of scientists. Sexual selection became the most fruitful idea in the emerging science of evolutionary psychology. After a hundred years of neglect, *The Descent of Man* was once more being read—and not just for what it has to say on human evolution.

What Sexual Selection's Exile Costs the Human Sciences

Sexual selection's century of exile from biology had substantial costs for other sciences. Anthropologists paid little attention to human mate choice in the tribal peoples they studied for most of this century. By the time mate choice was accepted as an important evolutionary factor, most of those tribal peoples had been exterminated or assimilated. Psychologists had little evolutionary insight into human sexuality, and their discipline was dominated for decades by Freudianism. Almost all of 20th-century psychology developed without considering the possibility that sexual selection through mate choice might have played a role in the evolution of human behavior, the human mind, human culture, or human society. Following Marx, the social sciences saw a culture's mode of production as more important than its mode of reproduction. Economists had no explanation for the importance of "positional goods" that advertise one's wealth and rank in comparison to sexual rivals. In the other human sciences as well—archeology, political science, sociology, linguistics, cognitive science, neuroscience, education, and social policy—there was a blind spot where the theory of sexual selection should have been.

When these sciences did try to trace the evolutionary roots of human behavior, they have usually come up with theories based on "survival of the fittest" and "the goods of the species." Mate choice was simply not on the intellectual map as an evolutionary force. Darwin's broader vision, in which most of nature's ornamentation arises through sexual courtship, was never used to explain the ornamental aspects of human behavior and culture.

For example, without sexual selection theory, 20th-century science had great difficulty in explaining the aspects of human nature most concerned with display, status, and image. Economists could not explain our thirst for luxury goods and conspicuous consumption. Sociologists could not explain why men seek wealth and power more avidly than women. Educational psychologists could not explain why students became so rebellious and fashion-conscious after puberty. Cognitive scientists could not fathom why human creativity evolved. In each case, apparent lack of "survival value" made human behavior appear irrational and maladaptive.

More generally, the sciences concerned with human nature have often lamented their incompleteness, fragmentation, and isolation. People are certainly complicated entities to study, but other sciences such as organic chemistry, climate modeling, and computer science have coped with high degrees of complexity. The limited success of the human sciences may not have resulted from the complexity of human behavior, but from overlooking Darwin's crucial insight about the importance of sexual competition, courtship, and mate choice in human affairs.

Today, evolutionary biology is proclaiming that the old map of evolution was wrong. It put too much weight on the survival of the fittest and, until the 1980s, virtually ignored sexual selection through mate choice. Yet in the human sciences we are still using the old map, and we still do not know where we came from, or where we are going. The next few chapters offer a new map of evolution to help us find our way.

3

The Runaway Brain

The worlds of academia, high fashion, religion, and modern art produce sublime wonders, and sometimes monstrous absurdities. They can afford such creative freedom because their systems of self-regulation and self-perpetuation are insulated from the mundane pragmatics of the outside world. Their autonomy endows them with liberty and creative power. They are free to evolve under their own momentum, along lines of their own choosing, without having to justify themselves at every step to outside critics.

Sexual selection can work similarly. One of sexual selection's central processes allows species to evolve in arbitrary directions under their own momentum. We shall see how this process, Fisher's runaway process, can provide a pretty good first model for how the human mind evolved.

Evolution's Autarch

Under natural selection, species adapt to their environments. When the environment refers to a species' physical habitat, this seems simple enough. If a species lives in the Arctic, it had better evolve some warm fur. Under sexual selection, species adapt too, but they adapt to themselves. Females adapt to males, and males adapt to females. Sexual preferences adapt to the sexual ornaments available, and sexual ornaments adapt to sexual preferences.

This can make things quite confusing. In sexual selection, genes do not code just for the adaptations used in courtship, such as sexual ornaments. They also code for the adaptations used in mate choice, the sexual preferences themselves. What the physical environment is to natural selection, sexual preferences are to sexual selection. They are not only the tastes to which sexual ornaments must appeal, but the environment to which they must adapt.

With sexual selection, genes act as both the fashion models and the fashion critics, both the apostates and the inquisitors. This creates the potential for the same kind of feedback loops that drive progress in high fashion and modern theology. These feedback loops are the source of sexual selection's speed, creativity, and unpredictability. Yet they also raise the classic problem of runaway corruption in autarchies: who watches the watchmen? How can mere genes be trusted as both selectors and selectees in evolution under sexual selection? The world of mate choice plays by its own rules, and though survival is a prerequisite for mating (as it is for scholarship, fashion, and faith), the principles of sexual selection cannot be reduced to the principles of survival. The biologist seems to have no point of entry into this protean wonderland where genes build brains and bodies, which pick the genes that build the next generation's brains and bodies, which in turn pick the genes that pick the genes . . .

Imagine the headaches if natural selection worked that way. Organisms would select which environments exist, as well as environments selecting which organisms exist. Strange, unpredictable feedback loops would arise. Would the feedback loop between polar bears and Arctic tundra result in a tundra of Neptunian frigidity where bears have fur ten feet thick, or a tundra of Brazilian sultriness where bears run nude? Would migratory birds select for more convenient winds, lower gravity, and more intelligible constellations? Or just an ever-full moon that pleasingly resembles an egg? Evolutionary prediction seems impossible under these conditions. Yet this is just what happens with sexual selection: species capriciously transform themselves into their own sexual amusements.

Introducing sexual selection in this way is more than just an attempt to encourage you to share my belief that it is one of the weirdest and more wonderful of nature's phenomena. That I could achieve simply by presenting the standard catalog of sexual selection's "greatest hits": the peacock's tail, the nightingale's song, the bowerbird's nest, the butterfly's wing, the Irish elk's antlers, the baboon's rump, and the first three Led Zeppelin albums. By presenting sexual selection as a strange world of genes selecting other genes, I have tried to provoke a different question: How could one ever make a science out of sexual selection? Darwin showed that sexual selection exists and documented its effects, but it took another century before biologists had the scientific tools for explaining why sexual selection produces certain kinds of traits and not others. To understand how sexual selection shaped human mental evolution, we need to become familiar with this new toolbox of ideas and models. Let's first have a better look at Fisher's runaway process. It is the best example of how sexual selection exercises a power distinct from natural selection.

How Runaway Works

When Fisher's runaway process first appeared in print in 1930, other scientists greeted it with suspicion. Runaway did not fit the prevailing emphasis on the good of the species, the efficiency of survival adaptations, and the modernist machine aesthetic. Yet despite its frosty initial reception, runaway has finally been invited back to the center of the evolutionary stage. Theoretical biologists in the 1980s showed that Fisher was right: runaway can work. Indeed, it works so well that it is hard to avoid when sexual selection is in play. Because runaway may have had an important role in the evolution of the human mind, it is important to understand it as fully as possible. What follows is the simplest example of runaway I can offer, although the theory is subtle, and demands some concentrated attention.

Imagine a population of birds with short tails, in which the males contribute nothing to raising the offspring. Although this makes life hard for females after mating, it allows females to choose any male they want, even a male who has been chosen by many other females already. The most attractive male could mate with many females. He has no reason to turn down a sexual invitation from any female, because copulation costs so little time and energy.

Within this population, different males inevitably have different tail lengths, just as they have different wingspans, and different leg lengths. All biological traits show variation. Usually, much of that variation is heritable (that is, due to genetic differences between individuals), so longer-tailed males will tend to produce longer-tailed offspring. In other words, tail length varies and tail length is heritable, satisfying two out of Darwin's three requirements for evolution.

Now, suppose that some of the females become sexually attracted to tails that are longer than average. (It doesn't matter why they evolve this preference—perhaps there was a mutation affecting their sexual preferences, or their vision happened to respond more positively to large than to small objects.) Once this female preference for long tails arises, we have the third requirement for evolution: selection. In this case, it is sexual selection through mate choice. The choosy females who prefer long tails will tend to mate with long-tailed males, who are happy to copulate with all their admirers. The non-choosy females mate randomly, usually ending up with an average-tailed male.

After mating, the choosy females start producing offspring. Their sons have longer-than-average tails that they inherited from their fathers. (Their daughters may also inherit longer tails—a phenomenon we shall consider later.) The non-choosy females produce sons whose tails are about the same length as those of their fathers—but these mediocre tails are no longer average. They are now below average, because the average tail length has increased in this generation, due to sexual selection through mate choice. The genes for long tails have spread.

The question is, will they keep spreading? Fisher's key insight was that the offspring of choosy females will inherit not just longer tails, but also the genes for the sexual preference—the taste for long tails. Thus, the genes for the sexual preference tend to end up in the same offspring as the genes for the sexually selected trait. When genes for different traits consistently end up in the same bodies, biologists say the traits have become "genetically correlated." Fisher's runaway process is driven by this genetic correlation between sexual traits and sexual preferences in offspring, which arises through the sexual choices their parents made. This genetic correlation effect is subtle and counterintuitive, which is one reason why biologists took fifty years to prove that Fisher's idea could work.

Of course, when the sons of choosy females inherit the genes underlying their mother's sexual attraction to long tails, they may not express this preference in their own mating decisions. But they can pass their mother's sexual preferences on to their own daughters. Since their long tails make them sexually attractive, they tend to produce not only more sons than average, but more daughters as well. In this way, the sexual preference for long tails can genetically piggyback on the very trait that it prefers. This gives the runaway process its positive-feedback power, its evolutionary momentum.

The Runaway Brain

Did the runaway sexual selection process play a role in the evolution of the human brain? To see how this would work, take the previous example, and in the place of "bird," substitute "hominid"—meaning one of our ape-like ancestors that walked erect. For "long tail," substitute "creative intelligence." If hominid males varied in their creative intelligence, and if that creative intelligence was genetically heritable, two out of three prerequisites for sexual selection would be present.

The only other requirement would be for hominid females to develop a sexual preference for creative intelligence, for whatever reason. If they did, then males with higher creative intelligence would attract more sexual partners and produce more offspring, assuming our ancestors were not completely monogamous. Those offspring would inherit higher-than-average creative intelligence, and would also inherit the sexual preference for creative intelligence. Intelligence would become genetically correlated with the sexual taste for intelligence. The sexual taste would piggyback on the evolutionary success of the sexual trait that it favors. The sexual trait and the sexual preference would both spread through the population. The hominids would become more creatively intelligent, and demand more creative intelligence of their sexual partners. The key here is that creative intelligence need not have given the hominids any survival advantages whatsoever, but through runaway it could evolve as a pure sexual ornament.

In the early 1990s, the runaway process seemed to me ideally suited to explaining why the human brain evolved so quickly, and to such an extreme size, during a period when it seemed to make our ancestors no better at making tools or competing against other species of African hominids. It became the focus of my research and the subject of my 1993 Ph.D. thesis at Stanford, which was titled "Evolution of the Human Brain through Runaway Sexual Selection." The human brain's evolution clearly looked as if it was driven by some sort of positive-feedback process. Other theorists proposed other candidates for the positive feedback. In 1981, E. O. Wilson suggested that larger brains permitted more complex cultures, which in turn selected for larger brains. This could initiate an evolutionary feedback loop between brain size and cultural complexity. Richard Dawkins has supported this view, seeing the human brain as a repository of learned cultural units called "memes." Larger brains permit more memes, which in turn favor bigger brains.

Two other positive-feedback ideas have proven influential in evolutionary psychology. In 1976 Nicholas Humphrey proposed that pressures for social intelligence could have turned into a positive-feedback process that drove human brain evolution. In 1988 Andy Whiten and Richard Byrne extended this idea by focusing on the survival advantages of social deception and manipulation. Their "Machiavellian intelligence" hypothesis has been accepted by many primate researchers and psychologists interested in human social intelligence. Apart from social competition within groups, another positive-feedback possibility was competition between groups. In 1989 Richard Alexander proposed that perhaps tribal warfare turned into an arms race for ever greater technological and strategic intelligence. This military competition could drive brain size and intelligence upwards.

These theories all have some validity. Cultural, social, and military selection pressures were probably significant. But these positive-feedback loops seemed too speculative. They had not been admitted into the pantheon of evolutionary forces by biologists, and were not routinely used to explain interesting traits in other species. They were slightly ad hoc hypotheses restricted to primate and human evolution. The runaway process was different: it was part of mainstream evolutionary theory, one of the leading contenders for explaining complex, costly, ornamental traits in other species. Yet it had never been proposed as the driving force behind the evolution of the human brain.

This seemed a peculiar oversight in need of vigorous correction, and for several years I gave dozens of talks about the idea of human mental evolution through runaway sexual selection. Matt Ridley kindly gave the idea some attention in the final chapter of his book *The Red Queen*. However, I now think that the runaway brain idea is only partly successful. It has some strengths that can help account for some of the sex differences in human behavior and some of the differences between our species and other primates. However, it also has some serious problems, so it will constitute only a small part of my overall theory.

The Requirements of the Runaway Process

One possible problem is that runaway sexual selection demands polygyny—a mating pattern in which some males mate with two or more females. For runaway to work, some males must prove so attractive that they can copulate with several females to produce several sets of offspring. The least attractive males, as a rule, must be left single, heartbroken, and childless. Sexual competition must be almost a winner-take-all contest. In elephant seals, for example, one dominant male may account for over 80 percent of all copulations with females on a particular beach, and almost as high a proportion of all offspring. (Polygyny does not mean that every male gets to father the offspring of many females—

that would be a mathematical impossibility, given an equal sex ratio. It means rather that a few males mate often and produce many offspring, and most males mate rarely, producing very few offspring.)

If our ancestors were perfectly monogamous, runaway sexual selection could not have favored large brains, or creative intelligence, or anything else. Runaway would never have started. A crucial question is how polygynous our ancestors were. The more polygynous they were, the more potent runaway sexual selection could have been. The modern understanding of human evolution suggests that our ancestors were moderately polygynous—neither as polygynous as elephant seals, gorillas, or peacocks, nor as perfectly monogamous as albatrosses. The evidence comes from many sources, but I shall mention just two: body size differences and anthropological records. Across primates, species where males are much larger than females tend to be highly polygynous. This is because males compete more intensely and violently in more polygynous species where the stakes are higher, and this competition drives up their relative size and strength. Generally, the larger the sex difference in body size, the more polygynous the species. In humans, the average male is about 10 percent taller, 20 percent heavier, 50 percent stronger in the upper body muscles, and 100 percent stronger in the hand's grip strength than the average female. By primate standards, that is a moderate sex difference in body size, implying a moderate degree of polygyny.

Other evidence of polygyny comes from anthropological studies of human cultures and human history. Most human cultures have been overtly polygynous. In hunter-gatherer cultures the men who are the most charming, the most respected, the most intelligent, and the best hunters tend to attract more than their fair share of female sexual attention. They may have two or three times as many offspring as their less attractive competitors. In pastoral cultures the men who have the largest herds of animals attract the most women. In agricultural societies the men who have the most land, wealth, and military power attract the most women. Before the middle ages, in urban civilizations with high population densities, the men at the top of the hierarchy almost always had harems of hundreds of women producing hundreds of babies. The first emperor of China reputedly had a harem of five thousand. King Moulay Ismail of Morocco reputedly produced over six hundred sons by his harem. In European Christian societies from the medieval era onwards, monogamous marriage became the religious and legal norm, though powerful men still tended to attract many mistresses and to re-marry more quickly if their first wife died. For example, anthropologist Laura Betzig showed that throughout American history, presidents tended to mate more polygynously than men of lower political status. (This may be little consolation to politicians of mediocre musical ability, since popular male musicians such as Bob Marley and Mick Jagger allegedly behaved even more polygynously than presidents.)

Those of us brought up in European-derived cultures tend to think of humans as monogamous, but in fact mating in our species has almost always been moderately polygynous. For millions of years, there was enough variation in male reproductive success to potentially drive runaway sexual selection during human evolution.

Runaway Is Unpredictable

The runaway process is very sensitive to initial conditions and random events. Runaway's initial direction depends on the female preferences and male traits that happen to exist in a population. Runaway's progress depends on several kinds of random genetic events such as sexual recombination, which mixes genes randomly every time two parents produce offspring, and the evolutionary process called genetic drift, which eliminates some genes by chance in small populations, as a result of an effect called "sampling error." Because runaway is a positive-feedback process, its sensitivity to initial conditions and random events gets amplified over evolutionary time. These effects make runaway's outcome quite unpredictable. It never happens the same way twice.

Runaway's unpredictability is apparent if you look at the

diversity of sexual ornamentation in closely related species. Of a dozen species of bowerbirds, no two construct the same style of courtship nest. Of three hundred species of primate, no two have the same facial hair color and style. These differences cannot be explained as adaptations to different environments—they are the capricious outcomes of sexual selection.

Computer simulations confirm runaway's unpredictability. In the early 1990s when we were psychology graduate students at Stanford, Peter Todd and I spent months running simulations of runaway sexual selection. We would run the same program, repeatedly, while just changing the initial conditions slightly, or changing the random numbers used by the computer to simulate random events like mutation. The results were quite capricious. Two populations can start out very similar to each other, and evolve slightly different sexual preferences, which lead their sexual ornaments to evolve in slightly different directions, which reinforce their sexual preferences, and so forth. The populations end up in opposite corners of the range of possibilities, sprouting different sexual ornaments, with different sexual preferences. And if you run the same simulation again, with just slightly different random numbers influencing mutations, the populations will evolve in yet another set of directions. A population will often split apart spontaneously into two clusters that are reproductively isolated, creating two distinct species. If you went out for a coffee while running a simulation and came back ten minutes later, the population would usually have moved where you least expected it—not through the physical space of its simulated habitat, but through the abstract space of possible ornament designs.

Suppose you take a dozen species of ape that lived in social groups in Africa about ten million years ago. Think of these species as nearby clusters in the space of all possible sexual ornaments and courtship behaviors. Now turn runaway sexual selection loose in each species. One species might develop a runaway preference for large muscles, and turn into gorillas. Another species might develop a runaway preference for constant sex, and turn into bonobos (previously known as "pygmy chimpanzees"). A third species might develop a runaway preference for creative intelligence, and turn into us.

Depending on your philosophy of science, runaway's unpredictability could be seen as a strength or a weakness. It is a strength if you are looking for an evolutionary process that can explain why two closely related species take dramatically different evolutionary routes. It is a weakness if you expect evolution to be predictable and deterministic, able to explain exactly why one ape species evolved creative intelligence while another did not. Of course, if you think that our mental evolution was driven entirely by natural selection for survival abilities, a fairly deterministic attitude is appropriate. But if you accept that mental evolution could have been influenced by runaway sexual selection, which produces unpredictable divergence, then you can't expect it to be predictable or deterministic.

If our evolution was driven by an unpredictable process like runaway, we should not expect a precise answer to questions like "Why did we, rather than chimpanzees, evolve creative intelligence and language?", or "Why are we the first articulately conscious species on Earth?" It would be like a lottery winner asking why she won. However, we can still ask, "What are the adaptive functions of human creative intelligence, language, and morality?", and "Did these capacities evolve through survival selection, sexual selection, or something else?" Given an adaptation, we can still try to explain why it evolved to have the features and functions it does. We just might not be able to explain why it evolved exactly when and where it did, in the lineage that it did, rather than in other lineages.

The Problems with Runaway in Explaining the Human Mind's Evolution

At first glance, runaway's speed and creative power sound like just what we need to explain the human mind's evolution. Brain size in our lineage tripled in just two million years. From a macroevolutionary viewpoint, that is very fast—much faster than any brain size increase in any other known lineage. Music, art, language, humor, and intelligence all evolved at some time during that explosive growth. On the geological timescale, the human mind's evolution looks faster than the flash from a nuclear strike does on the human timescale.

But evolutionary speed is relative. The human mind's evolution was actually much too slow to be explained by a single runaway event. Two million years is still a pretty long time—about a hundred thousand generations even for a slow-breeding ape like us. During that time, we added two pounds of brain matter—about a hundredth of a gram of brain per generation. A sustained runaway process would have been much more potent. Assuming a modest heritability and a modest amount of variation in brain size, I estimate that runaway could increase brain size by at least one gram per generation. That rough estimate assumes a sexual selection pressure on the low end of pressures that have been measured in other species in the wild. If this estimate is right, a single sustained runaway event would have been at least a hundred times too fast to explain human brain evolution. Brain size would have tripled in 20,000 years, not 2 million years.

Like a ramjet, runaway sexual selection has more of a minimum speed than a maximum speed. It just can't go slow. This is one reason why the simple runaway story makes a poor explanation for human brain evolution. Compared with runaway's hypersonic speed, human brain evolution was like a stroll through the park on a Sunday afternoon. Yet, if this speed objection seems to undermine the runaway brain theory, it undermines every other positive-feedback theory as well. The other processes proposed by E. O. Wilson, Richard Dawkins, Nicholas Humphrey, Andy Whiten, and Richard Alexander would also have run too fast.

This speed problem might be solved by supposing that human brain evolution, like the evolution of almost everything, happened in fits and starts. There were short periods of relatively fast evolution when selection pressures were pushing in some direction, and long periods of stasis when selection just maintained the status quo against mutation. Fossil evidence suggests that brain size increased quickly in a few dramatic bursts. The transition from 450-gram Australopithecine brains to 600-gram *Homo habilis* brains was one such burst (though *Homo habilis* is no longer thought to be our direct ancestor). Another burst produced the early 800-gram *Homo erectus* brain 1.7 million years ago. There were probably several more bursts during the evolution of *Homo erectus* over the next million years. Another burst produced the 1,200-gram archaic *Homo sapiens* brain. A final burst produced the 1,300-gram modern human brain about 100,000 years ago. Each burst looks short in terms of geological time, but lasted for hundreds or thousands of generations, plenty of time for standard selection pressures to mold traits. We do not yet have sufficient fossil evidence to tell whether each burst was driven by a very fast process like runaway or a slower process like ordinary survival selection.

So, where does this leave us? A single runaway event cannot explain two million years of human brain evolution because it would have been too fast and too transient. Instead, we could propose a multi-step runaway process, where each burst in brain size was driven by a separate runaway event. But that would beg the question of why all the runaway events increased rather than decreased brain size. In principle, a species could stumble into runaway sexual selection for the dumbest possible behavior produced by the smallest possible brains. A species of bumbling incompetents could evolve, despite the survival costs of their stupidity, as long as stupidity remained sexually attractive. Runaway is not supposed to be biased in any evolutionary direction, so it should be as likely to decrease a trait's size as to increase it. This makes it a poor candidate for explaining multi-step progressive trends.

Another possible answer to the speed quandary is to forget about fossil brains, and focus on human mental abilities. We do not know when language, art, and creativity evolved. Perhaps they all evolved together when modern *Homo sapiens* emerged about 100,000 years ago. Some archeologists even think that these capacities all evolved in a single burst 35,000 years ago, in an event they call the "Upper Paleolithic revolution." Such rapid evolution might reflect a single runaway process operating over a few thousand generations in a single population, transforming a large-brained but unintelligent hominid into an intelligent, talkative human. The earlier brain-size bursts may have occurred for some other reason. Perhaps the key transition to the human mind was a brain reorganization rather than a simple brain size increase. The reorganization may not be evident in the record of fossil skulls, but may be more psychologically significant than earlier size increases. It may have been driven by a burst of runaway sexual selection relatively late in human evolution.

However, this theory fails to explain why brain size increased in all those bursts before our species evolved. It seems to me that the multi-burst trend toward larger brains should be explained rather than ignored. Pure runaway cannot explain it, because runaway does not have any intrinsic bias toward larger ornament size, higher ornament cost, or greater ornament complexity. The problem with runaway is not just its rocket-like speed. Its more fundamental problem is its neutrality, which makes it weak at explaining multi-step trends that last millions of years. The next chapter examines another sexual selection process that is much better at driving sustained progress in one direction.

Runaway Produces Large Sex Differences

Another problem with the runaway brain theory is that runaway is supposed to produce large sex differences in whatever trait is under sexual selection. Peacock tails are much larger than peahen tails. If the human brain tripled in size because of runaway sexual selection, we might expect that increase to be confined to males. Men would have three-pound brains, and women would still have one-pound brains like other apes. This has not happened. Male human brains average 1,440 grams, while female brains average 1,250 grams. If one measures brain size relative to body size, the sex difference in human brain size shrinks to 100 grams. This 8 percent difference is larger than would be predicted by a sex-blind theory like E. O. Wilson's cultural feedback loop, or the

Machiavellian intelligence hypothesis. But it is much smaller than the runaway brain theory would predict.

Similarly, if creative intelligence evolved through runaway sexual selection, we would expect men to have much higher IQs than women. There are some sex differences in particular cognitive abilities, mostly quite small, with some giving the male advantages, and some the female. However, there appears to be no sex difference whatsoever in the underlying "general intelligence" ability (technically called "the *g* factor") that IQ tests aim to measure. The best analysis has been done by Arthur Jensen in his 1998 book *The g factor*, and he concluded that "The sex difference in psychometric *g* is either totally nonexistent or is of uncertain direction and of inconsequential magnitude." Nor is any sex difference found in average performance on the most reliable IQ tests that tap most directly into the *g* factor, such as an abstract symbolic reasoning test called Raven's Standard Progressive Matrices. Men have a slightly greater variation in IQ, producing more geniuses as well as more idiots, but this greater variation in test scores does not appear to reflect a greater variation in the underlying *g* factor. This absence of a sex difference in general intelligence does not seem consistent with the runaway brain theory that sexual selection on males drove human intelligence.

Sex differences can occur on different levels, however. One could argue that runaway sexual selection did not favor brain size or intelligence directly, but the behavioral manifestations of high creative intelligence. On this view, perhaps runaway sexual selection accounts in part for the greater propensity of males to advertise their creative intelligence through trying to produce works of art, music, and literature, amassing wealth, and attaining political status. A strong version of this theory might suggest that human culture has been dominated by males because human culture is mostly courtship effort, and all male mammals invest more energy in courtship. Male humans paint more pictures, record more jazz albums, write more books, commit more murders, and perform more strange feats to enter the *Guinness Book of Records*. Demographic data shows not only a large sex difference in display rates for such behaviors, but male display rates for most activities peaking between the ages of 20 and 30, when sexual competition and courtship effort are most intense. This effect can be observed from any street corner in the world: if a vehicle approaches from which very loud music is pouring, chances are it is being driven by a young male, using the music as a sexual display.

Certainly there may be many cultural reasons why men behave differently from women. If all sex differences in human behavior are due to sexist socialization, then it may be appropriate to dismiss all cultural and historical evidence concerning a greater male propensity to produce noisy, colorful, costly displays. The runaway brain theory simply suggests that evolved differences in reproductive strategies and display motivations may have been a factor in the historical prominence of male cultural production. Evolution is certainly not the only factor, because the last century has witnessed a rapid increase in women's cultural output, economic productivity, and political influence. Women's ongoing liberation from the nightmare of patriarchy has been due to cultural changes, not genetic evolution. Darwin would probably have been astounded by the political leadership ability of Margaret Thatcher and the musical genius of Tori Amos.

There is a serious problem of scientific method here. The runaway brain theory predicts greater male motivation to display creative intelligence in all sorts of ways, just as male birds are more motivated to sing. Human history reveals that cultural output across many societies was dominated by the behavior of males of reproductively active ages. Yet those societies, and the historical records themselves, were biased by many female-oppressing cultural traditions. (These traditions may have evolutionary roots in male propensities for oppressive mate-guarding, but such propensities would be distinct from any evolved male propensities for creative display.) I honestly do not know how much weight should be given to cultural records that reveal higher male rates of display, and which thereby seem to support the runaway brain theory. We clearly should not accept such records at face value as direct reflections of evolved sex differences. But if we dismiss such records completely, are we doing so because the records are utterly worthless as scientific evidence, or because we find the data politically unpalatable? Should we reject a theory of mental evolution that successfully predicts an observed sex difference, in favor of some other sex-blind theory that predicts a desired sexual equality in culture production that has not yet been observed in any human society?

Male nightingales sing more and male peacocks display more impressive visual ornaments. Male humans sing and talk more in public gatherings, and produce more paintings and architecture. Perhaps we should view the similarities between peacocks and men as a meaningless coincidence, due to sexual selection in the first case and a history of patriarchal oppression that just happened to mimic the effects of runaway sexual selection in the second case. This issue is so scientifically challenging and politically sensitive that it will only be resolved when evolutionary psychologists, cultural historians, and feminist scholars learn to collaborate with mutual respect and an open-minded dedication to seek the truth. Personally, I believe that the current evidence supports two provisional conclusions: sexual selection theory explains many human sex differences (including differences in the motivation to produce creative displays in public), and many pathological traditions have inhibited female creative displays in the last several thousand years. Some people view these two beliefs as mutually exclusive, but I cannot see why they should clash, except at the level of ideological fashion, in the same sense that lime green clashes with electric blue.

In summary, the overall evidence for sex differences is confusingly mixed. At the level of brain size and raw intelligence, human sex differences are too small for the runaway brain theory to work. Although brain size within each sex is correlated about 40 percent with general intelligence, the slightly larger brains of males do not yield a higher general intelligence than those of females. At the level of sexual behavior and cultural output, sex differences are enormous, but they are shrinking rapidly, and are conflated with patriarchal cultural traditions. Overall, this pattern of evidence does not support a strong version of the runaway brain theory, nor does it support any other theory in which male sexual competition through toolmaking, hunting, or group warfare was the driving force behind the human mind's evolution. If sexual selection was important in the mind's evolution, it could not have been a type of sexual selection that produces large sex differences in brain size or general intelligence. At this point, it may help to step back from the runaway brain theory and consider sex differences in a more general evolutionary framework.

Eggs and Sperm

Sexual selection demands sexual reproduction, but it does not demand distinct sexes. If hermaphrodites exercise mate choice, they can evolve sexual ornaments. A small number of animals and a large number of flowering plants are hermaphroditic. Because they still compete to attract mates, they still evolve sexual ornaments. Sexual selection does not require sex differences, and does not always produce sex differences.

However, in most animals, distinct sexes have evolved. They simply specialize in making DNA packets of different sizes. The female sex evolved to make large packets in which their DNA comes with additional nutrients to give offspring a jump-start to their development. The male sex evolved to make the smallest possible packets in which their DNA is almost naked, contributing no nutrients to their offspring. Females make eggs; males make sperm. The fundamental sex difference is that females invest more nutrient energy in offspring than males.

In the early 1970s, biologist Robert Trivers realized that, from this difference in "parental investment," all else follows. Because eggs cost more for females to make than sperm costs for males, females make fewer eggs than males make sperm. But since each offspring requires only one of each, the rarer type of DNA packet, the egg, becomes the limiting resource. Thus, Trivers argued, it makes sense that males should compete more intensely to fertilize eggs than females do to acquire sperm, and that females should be choosier than males. Males compete for quantity of females, and females compete for quality of males. Trivers' supply-and-demand logic explained why, in most species, males court and females choose.

In female mammals the costs of pregnancy and milk production are especially high, amplifying the difference between male competitiveness and female choosiness. For example, the minimum investment human female ancestors could have made in their offspring would have been a nine-month pregnancy followed by at least a couple of years of breast-feeding. The minimum investment our male ancestors could have made in their offspring would have been a few minutes of copulation and a teaspoonful of semen. (For most male primates, that is not only the minimum, but the average.) Females could have produced a child every three years or so. Males could have produced a child every night, if they could find a willing sexual partner. This theoretical difference often plays out as a practical difference. In hunter-gatherer societies, almost no woman bears more than eight children, whereas highly attractive men often sire a couple of dozen children by different women.

Before contraception, a man's reproductive success would have increased with his number of sexual partners, without limit. Every fertile woman he could seduce represented an extra potential child to carry his genes. But a woman's reproductive success reached its limit much more quickly. Conception with one partner was enough to keep her reproductively busy for the next three years. One might think that two children should be enough for each man, because that would sustain the population size. But that implies that evolution is for the good of the species, which it is not. The genes of sexually ambitious men would have quickly replaced the genes of men satisfied with just one sexual partner and two children.

Evolution pays attention to sex differences in reproductive potential because they translate into sex differences in reproductive variation. Males vary much more in the number of children they produce, and this makes sexual reproduction a higher-risk, higher-stakes game for them. Females vary less in their quantity of children, so they care more about quality. So what do the males do with all the extra energy that females are devoting to growing eggs, being pregnant, and producing milk? They use it for reproductive competition and courtship. There is a fundamental tradeoff between courtship effort and parental effort. The more time and energy you devote to growing and raising children, the less time and energy you can devote to driving off sexual competitors and seducing sexual partners.

Jumping Ship

From the point of view of genes in any male body, the body itself is a sinking prison ship. Death comes to all bodies sooner or later. Even if a male devoted all of his energy to surviving, by storing up huge fat reserves and hiding in an armored underground compound, statistics guarantee that an accident would sooner or later kill him. This paranoid survivalist strategy is no way to spread one's genes through a population. The only deliverance for a male's genes is through an escape tube into a female body carrying a fertile egg. Genes can survive in the long term only by jumping ship into offspring. In species that reproduce sexually, the only way to make offspring is to merge one's genes with another individual's. And the only way to do that, for males, is to attract a female of the species through courtship. This is why males of most species evolve to act as if copulation is the whole point of life. For male genes, copulation is the gateway to immortality. This is why males risk their lives for copulation opportunities—and why a male praying mantis continues copulating even after a female has eaten his head.

For a female, too, the body is a sinking ship, but it has almost everything necessary to make more bodies: eggs, womb, milk. The only thing missing is a DNA packet from a male. But there are many willing donors. Finding a partner is usually not the problem. There are often so many willing males that the female can afford to be choosy. Quality becomes the issue. Each of the female's offspring inherits half of its genes from whatever male she chooses. If she chooses an above-average male, her offspring get above-average genes, and are therefore more likely to survive and reproduce. It is for this reason that female mate choice evolved.

Because females can afford to be choosy, and the benefits of sexual choice are large for them, females will typically evolve sexual preferences. As long as the males of a species invest very little in their offspring, they have no reason to refuse to copulate with any female. This is why male mate choice is rarer across species and less discriminating within each species than female mate choice. And as long as males are not sexually choosy, females do not have to bother evolving sexual ornaments. This is why sexual selection produces the sex differences we typically see in most animal species: ardent males with large sexual ornaments courting choosy females without ornaments. (This is sometimes misunderstood by critics as suggesting that males are more "active" and females more "passive." This uselessly simplistic active/passive dichotomy was not prompted by Darwin and is not accepted by modern biologists. Choosy females may be quite active in searching for good mates, comparing males, and soliciting copulations from desired males.)

If the human brain evolved through sexual selection, and followed this typical pattern, we would expect the same sex differences—not only in human behavior, but in human psychology. As far as human sexual behavior goes, the typical biological pattern outlined above seems a pretty good first approximation. Male humans generally invest more time, energy, and risk in sexual courtship, invest less in parenting, are more willing to copulate earlier in relationships with larger numbers of partners, and are less choosy about their sexual partners, at least in the short term. Female humans generally invest less in courtship and much more in parenting, are less willing to copulate early with large numbers of partners, and are more choosy. David Buss, Don Symons, Margo Wilson, Martin Daly, Laura Betzig, and many other evolutionary psychologists have gathered a mountain of data from diverse cultures documenting these sex differences and showing how they can be explained by Darwinian sexual selection. Such studies received a great deal of media attention in the 1990s, and have destroyed the credibility of claims that human sexuality and sex differences are purely a product of culture and socialization.

However, finding the typical sex differences in humans actually makes it harder to argue coherently that sexual selection had a very significant effect on the evolution of the human mind. This is because the typical pattern of male courtship and female choice would have produced much larger sex differences in brain size, intelligence, and psychology than actually exists. Given that we now understand the origins of typical sex differences, how can the human pattern of sexually differentiated courtship result in sexually similar minds?

I do not claim to have a simple answer that explains everything about human sex differences and similarities. I can only ask for you to think through some possibilities with me. Remember, almost every theory of human mental evolution raises the same difficult issues about sex differences, because almost every theory depends on selection pressures that would have affected males and females somewhat differently.

The Sexes Share Genes

There are three factors that could have kept male human minds similar to female human minds despite strong sexual selection. The first factor is called "genetic correlation between the sexes." Males and females in every species share almost the same genes. There is a very high genetic correlation between the sexes. In humans for example, 22 pairs of our chromosomes are shared by both sexes, while only one pair, the X and Y sex chromosomes, are sexually distinct.

The genetic correlation between the sexes inhibits the evolution of sex differences, at least in the short term. Sex differences do not spring up automatically just because sexual selection is at work. Sex differences have to evolve gradually, like everything else. Consider the example of runaway sexual selection for long tails in birds. We assumed that the long tails would be passed on only from father to son. That might happen after many generations, but it is very unlikely to happen that way at first. It is much more likely that a mutation that increases tail length will be passed along to both sexes. Both male and female offspring will inherit longer-than-average tails from their sexually attractive fathers. Initially, tail length will increase with equal speed in both sexes. And both male and female offspring may tend to inherit their mother's sexual preference for longer tails. So, female tail length will ride along on the genetic coattails of male tail length, and male sexual preferences will ride in tandem with female sexual preferences.

Darwin understood the genetic correlation between the sexes in a sketchy way, calling it "the law of equal transmission." In *The Descent of Man* he argued that male human intelligence and imagination evolved mainly through sexual competition, and wrote that "It is, indeed, fortunate that the law of the equal transmission of characters to both sexes has commonly prevailed throughout the whole class of mammals; otherwise it is probable that man would have become as superior in mental endowment to women, as the peacock is in ornamental plumage to the peahen." Basically, Darwin viewed the female brain as riding along on the genetic coattails of sexually selected male brains.

Genetic correlations between the sexes can be measured, and are often fairly strong. Anthropologist Alan Rogers found a very high genetic correlation between male and female height in humans, in a paper he published in 1992. This does not mean that men and women are the same average height. Nor does it just mean that tall fathers have tall daughters, and that tall mothers have tall sons. Technically, it means that a tall parent's opposite-sex offspring are almost as extreme in their height, compared to others of their sex, as their same-sex offspring are, compared to others of their own sex. Rogers saw the implications for sexual selection. If females favored taller-than-average males as sexual partners, then of course male height would increase over evolutionary time because of the sexual selection. But Rogers calculated that female height would also increase, due to the genetic correlation with male height. In fact, female height would increase

98 percent as fast as male height. As you can see, a very unequal sexual selection pressure can produce a very equal outcome.

However, these genetic correlation effects are transient. Eventually, male choosiness should decrease, and the costs of female ornamentation should increase, and these effects will break down the genetic correlation. Male choosiness would probably be eliminated first. Coming back to our long-tailed bird example, any male who rejects a short-tailed female will produce fewer offspring than a male who is less choosy. In most species, the pressures against male choosiness are very strong, causing sex differences in choosiness to evolve very fast. Sex differences in ornamentation might take a bit longer. Females with long tails will be inconvenienced by their cost, and if males do not prefer them to short-tailed females, they should evolve inhibitions against expressing the runaway male ornament. (Typically, this means that they evolve a gene expression mechanism that is sensitive to sex hormones during development, so the genes for long male tails are not turned on in female bodies.)

If genetic correlations between the sexes were not transient, we would never see dramatic sex differences in nature. Peahens would have the same tails as peacocks. Female nightingales would sing like males. The human clitoris would be as large as the human penis. Darwin's coattail theory of female brain evolution doesn't work except in the short term, because sex differences will eventually evolve if the sexes derive different benefits from ornamentation and sexual choice. Genetic correlations between the sexes can explain transient increases in female ornamentation and in male choosiness, but these increases are not evolutionarily stable. Fortunately, there is a second factor that is much more potent over the long term in keeping the sexes similar.

The Mental Capacities for Courtship Overlap with the Capacities for Sexual Choice

The eye of the peahen has very little in common with the tail of the peacock. They are at opposite ends of the body. They are constructed of different materials. They grow under the influence of different genes. During runaway, the genes underlying the sexually selected trait (the tail) may become correlated with the genes underlying the mechanism of sexual choice (the eye), but that is about the limit of their acquaintance.

The same is not true of the mental capacities used in human courtship, such as creative intelligence. There is much more overlap between those aspects of the brain used for producing sexually attractive behavior, and those aspects of the brain used for assessing and judging that behavior. Speaking and listening use many of the same language circuits. The production and appreciation of art probably rely on similar aesthetic capacities. It takes a sense of humor to recognize a sense of humor. Without intelligence, it is hard to appreciate another person's intelligence. The more psychologically refined a courtship display is, the more overlap there may be between the psychology required to produce the display and the psychology required to appreciate it.

This overlap suggests that runaway sexual selection for psychologically refined courtship may produce much smaller sex differences than runaway sexual selection for long bird tails. Consider the case of language. Suppose that human language evolved through a pure runaway process. Let's say males talked, and females listened, and females happened to favor articulate conversationalists over tedious mumblers. Male language abilities would then improve by sexual selection: their vocabularies might grow larger, their syntax more complex, their story plots more intricate, their ideas more imaginative. But for runaway to work, female choosiness would have to increase as well. How could that happen? Female language abilities would have to keep one step ahead of male abilities, to remain discerning. Females would have to be able to judge whether males used words correctly, so their vocabularies would keep pace. They would have to be able to notice grammatical errors, so their syntax abilities would keep pace. Most importantly, the females would have to understand what the males were saying to judge their meaning. Even if males exerted no sexual selection whatsoever on female language abilities, those abilities would have to evolve as part of the female mate choice mechanism.

To a psychologist like me, this is a much more promising sort of overlap than a mere genetic correlation between the sexes. There is a profound functional reason why males and females evolve in psychologically similar ways when courtship turns psychological. They use the same mental machinery to produce displays that they use to judge the displays produced by others.

There are two further reasons for the overlap between display-producers and display-judgers. To produce a really effective display, it helps to anticipate how the display will be judged. One might mentally rehearse a joke before telling it, to see if it will work, and find another joke if it won't. A painter could look at a picture while painting to see if it's beautiful. A musician could listen to the melody being played to see if it's tuneful. When trying to impress someone during courtship, we routinely do this sort of anticipatory filtering and correcting. Even if only males produced courtship displays, they would benefit by evolving psychological access to the same judgment mechanisms that females use.

Conversely, to be a really good judge of something, it helps to be able to do it oneself. For females to judge which male tells the best jokes, they may benefit by evolving joke-telling ability. We shall see later that mental anticipation is closely related to creativity. To be capable of judging someone's creativity, one must develop expectations about their behavior. Without expectations that can be violated, there can be no sexual selection for novelty and creativity. The mental machinery for generating expectations about someone else's stories, jokes, or music may overlap considerably with the mental machinery that is used in producing stories, jokes, and music.

So, even given a pure runaway process based on male courtship and female choice, male minds will tend to internalize the sexual preferences of females in their own courtship equipment, in order to produce better displays. And female minds will tend to internalize the display-production abilities of males in their own sexual choice equipment, in order to be better judges of male displays. This should lead to many mental capacities being shared by both sexes, even if males are more motivated to use their mental capacities to produce loud, public courtship displays. At present this argument is speculative, but it could be supported if neuroscience research found overlap between the brain areas used in producing and judging particular forms of courtship behavior, and if behavior genetic research were to show that the same genes underlie culture-production and culture-judgment abilities in both sexes.

Mutual Choice

Genetic and psychological overlaps between the sexes are fine as far as they go. They may explain some of the mental similarities between men and women, even if the pure runaway brain theory is right. Still, they raise two problems. First, they portray the female mind as riding along on the evolutionary coattails of the male mind, and female intelligence as an evolutionary side-effect of male intelligence. The runaway brain theory does put female brains in the evolutionary driver's seat, since they make the sexual choices that drive runaway sexual selection. But the males are portrayed as doing all the interesting things: the courtship displays, the storytelling, the music-making, the creative idea-work. In short, the runaway brain theory sounds sexist.

In the game of science though, sounding sexist is not a good reason to ban a theory. Science is the one zone of human thought where ideological preferences are not supposed to influence the assessment of ideas and evidence. Human evolution happened somehow. It may not have happened in a way that coincides with our ideological preferences. Usually, I have a very low tolerance when it comes to injecting ideology into discussions about human evolution. However, some objections that are expressed in ideological terms are actually empirical objections that have scientific merit. In this case, the apparently political objection includes a perfectly valid point: the runaway brain theory ignores male mate choice and female sexual competition, which appear to be fairly important in our species. Women are especially good at noticing this, because they are more aware of their own competitive strategies, just as men are more aware of theirs.

The third factor that keeps the sexes similar is the mutuality of human mate choice. Both sexes are choosy when searching for long-term partners. Both compete for sexual status, both make efforts to display their attractiveness and intelligence, and both experience the elation of romantic love and the despair of heartbreak. The pure runaway theory in which males court and females choose just does not reflect the human mating game as we play it.

Evolutionary psychologists sometimes forget this because sexual selection theory is so good at predicting sex differences, and sex differences are so easy to test. As David Buss has emphasized, human sex differences are most apparent in short-term mating. Men are more motivated to have short-term sexual flings with multiple partners than women are. Women are much choosier than men in the short term. Short-term mating is exciting and sexy, but it is not necessarily where sexual selection has the greatest effect. Human females, much more than other great apes, conceal when they are ovulating. This means that a single act of short-term copulation rarely results in pregnancy. Almost all human pregnancies arise in sexual relationships that have lasted at least several months, if not years. Modern contraception has merely reinforced this effect.

Human males are generally not as choosy about short-term affairs as females. There is very little opportunity cost to short-term mating for men. It does not exclude other sexual options. But men get much choosier about medium- and long-term relationships, because their opportunity costs increase dramatically. If they are in a sexual relationship with one woman, it is very difficult to sustain a sexual relationship with another woman. They cannot give both their full attention. They must make choices—sexual choices.

Evolutionary psychologists such as Doug Kenrick have good evidence that when it comes to choosing sexual partners for long-term relationships, men and women increase their choosiness to almost identical levels. They also converge in the features they prefer. Kenrick found that for one-night stands, women care much more about the intelligence of their partner than men do, but for marriage, men and women have equally high standards for intelligence. For almost every sexually desirable trait that has been investigated, men and women get choosier as relationships get "more serious." For most couples, getting serious means having babies. Sexual selection works through the sexual choices that actually result in babies being born, not just the sexual choices that result in a little copulation.

Women quickly learn the difference between male short-term mating and long-term commitment. They know it is generally easy to get a man to have sex, but hard to get him to commit. Male mate choice is usually exercised not when deciding whether to copulate once, but when deciding whether to establish a long-term relationship. This is why sexual competition between women is usually competition to establish long-term relationships with desirable men, not competition to copulate with the largest number of men. Even polygynous men have limited time and energy, and so have high incentives to be choosy about their long-term partners.

It seems reasonable to assume that most human offspring throughout recent human evolution were the products of long-term sexual relationships. (By primate standards, "long term" means at least a few months of regular copulation.) In picking long-term sexual partners, our male and female ancestors both became very choosy. That choosiness is what drove sexual selection, which depends on competition to reproduce, not competition to copulate. Concealed ovulation in our female ancestors undermined the link between single acts of copulation and effective reproduction. If most human reproduction happened in long-term relationships that were formed through mutual choice, then most human sexual selection was driven by mutual choice, not just by female choice.

Mutual choice is good at producing sexual equality in courtship abilities. If men and women became equally choosy in the long-term relationships that produced almost all babies, then men and women would have been subject to an equal degree of sexual selection. Their mental capacities for courtship would have evolved to equally extreme degrees. Their mental capacities for sexual choice would also have evolved equally.

At first glance, mutual choice seems to offer a solution to the problems posed by the runaway brain theory. It accounts for the sexual equality of brain size and human intelligence that the simple runaway model can not explain. The only problem is that mutual choice renders traditional models of runaway sexual selection irrelevant, because runaway depends on intense choosiness by one sex and intense competition by the other. It depends on sexual asymmetry. If human sexual selection has been driven mostly by sexually symmetric mutual choice to form relatively long-term relationships, then runaway is not the right model for human mental evolution.

Assessing the Runaway Brain Theory

If one acknowledges that sexual selection has played a role in the human mind's evolution, it is crucial to understand the runaway process, even if the runaway brain theory itself does not work. The reason is that runaway sexual selection is ubiquitous. Take any population with mate choice that is not totally monogamous, and runaway will occur sooner or later, going off in some direction. Runaway is endemic in sexual selection. Like convection beneath the Sun's surface, it is always bubbling away, mixing up sexual ornaments and sexual preferences, sometimes shooting off in a random direction like a solar flare. Any species that reproduces sexually using mate choice has probably been caught up in the runaway process repeatedly.

The runaway brain theory proposes that most of our unique mental capabilities evolved through ordinary runaway sexual selection. While the theory has a number of strengths, it also, as we have seen, has a couple of crippling weaknesses. Runaway sexual selection is good at explaining traits that are extreme, striking, and costly; that are attractive to the opposite sex; and that have little apparent survival value. Some of the human mind's more puzzling capacities seem to fit this pattern: art, music, poetic language, religious beliefs, political convictions, creativity, and kindness.

Runaway is especially good at explaining the evolutionarily unpredictable—why extreme traits can arise in one species but not in closely related species. Many of the human mind's most interesting capacities do not appear in other apes, and those of most hominids are not discernible from the archeological record. Runaway requires polygyny, and almost every human culture throughout history has been overtly polygynous to some extent. Runaway is extremely fast once it gets going. The fossil record reveals a few rapid increases in brain size punctuated by long periods of relative stasis, which could mark a series of runaway events.

The two major problems with the runaway brain theory are the multi-step progressiveness of brain size evolution, and the minimal sex differences in human mental ability. Pure runaway is not biased in any particular direction, yet for the last two million years human brain evolution has shown a consistent trend towards larger size and higher intelligence. Runaway should not be so consistent. Moreover, pure runaway should have produced large-brained, hyper-intelligent males, and small-brained, ape-minded females. That has not happened. I have reviewed some factors that may have minimized sex differences: genetic correlation between the sexes, the overlap of mental capacities for courtship behavior and for sexual choice, and mutual mate choice. But the most compelling of these factors, mutual mate choice, is not consistent with a pure runaway process.

I think that mutual mate choice in humans is so important that the pure runaway brain theory just cannot be right. This chapter started by praising it, but has ended by burying it. I do not think that female creative intelligence is a genetic side-effect of male creative intelligence, or arose simply as a way of assessing male courtship displays. I think that female creative intelligence evolved through male mate choice as much as male creative intelligence evolved through female mate choice. I shall turn next to a model of sexual selection that works better with mutual mate choice. It emphasizes how sexual ornaments advertise each sex's fitness to the other sex—a function of mate choice that may stretch back to the origins of sexual reproduction itself.

4

A Mind Fit for Mating

Before sexual reproduction evolved, there were several ways for organisms to accomplish the evolutionary task of spreading their DNA around. There was the divide-and-conquer strategy: wrap DNA in single cells that busily eat nutrients until they grow large enough to split in half, leaving each half to grow and split in turn. Bacteria are the masters of this technique, capable of doubling their populations every few minutes, but vulnerable to mass extermination through perils such as toothbrushes and soap.

There was also the cloning-factory strategy: grow a body with billions of cells, and then assign the task of DNA-spreading to a privileged minority of those cells, which bud off to make new, genetically identical bodies. Many fungi reproduce this way, epitomizing the rustic virtues of simplicity and fecundity. Yet this strategy, though successful in the short term, stores up trouble for the long term. Once a harmful mutation arises, as it sooner or later will, there is no means of expunging it. This propensity to accumulate damaging mutations makes such asexual species quite unsuited to evolving much sophistication. This is because bodily and mental sophistication require a great deal of DNA, and the more DNA one has, the more trouble mutations cause.

In the last few hundred million years, an increasing number of species have turned to a third way of spreading their DNA around—the fashionable new method called sexual reproduction, with improved mutation-cleansing powers. One grows a trillion-celled body to produce packets of DNA, makes sure those DNA packets find complementary DNA packets from suitable others, and permits the DNA to combine with that of another individual to produce offspring that bear traits from both parents. Of the 1.7 million known species on our planet, most engage in sexual reproduction. Sexual species include almost all plants larger than a buttercup and almost all animals larger than your thumb. It includes most insects, all birds, and all mammals, including all primates.

Copying Errors

At the beginning, this DNA-combining called sex was probably not very selective. It was simply the most convenient way to make sure that not all of your offspring inherited your mutations. In evolution, mutations are generally a bad thing. Since almost all mutations are harmful, organisms evolve sophisticated DNA repair machinery to correct mutations. Of course, in the long term, mutations are necessary for evolutionary progress, because a tiny minority prove helpful when a species faces new challenges. But organisms don't plan for the long term. To the organism, mutations are simply copying errors—mistakes made when trying to spread DNA by producing offspring.

If you have only one copy of each gene, it is hard to know when certain kinds of copying error have been made. Some errors just won't look right to the DNA repair machinery. They are chemical nonsense, and easily fixed. But other errors look just like ordinary working DNA. These pseudo-normal mutations are the problem. They look like good DNA to the repair machinery, but they do not act like good DNA when you try to grow an organism using them. They undermine the biological efficiency called fitness. Unless there is some way of eliminating them, they will accumulate, generation after generation, gradually eroding the fitness of offspring.

In very recent work, biologists Adam Eyre-Walker and Peter Keightley calculated that the average human has 1.6 harmful new mutations that neither parent had. Our ancestors would have accumulated mutations at the same rate. Geneticist James Crow thinks this estimate too conservative by half, and suggests that we have 3 new harmful mutations per individual every generation.

That doesn't sound too bad, given that we have about 80,000 genes, yet this mutation rate is near the theoretical limit of what selection can cope with. For a species to avoid going extinct as a result of accumulating too many harmful mutations, selection must be able to eliminate mutations at the same average rate that mutations arise, otherwise the species would suffer a "mutational meltdown." For technical reasons, it is very hard to avoid a mutational meltdown when more than one harmful new mutation arises per individual. In fact, it may be impossible without sexual reproduction.

Sexual reproduction probably arose as a way to contain the damage caused by mutations. By mixing up your DNA with that of another individual to make offspring, you make sure that any mutations you have will end up in only half of your offspring. Your sexual partner will have mutations of their own, but they are almost certain to be different mutations on different genes. Because offspring have two copies of each gene, the normal version inherited from one parent often masks the failures of the mutated version inherited from the other parents. Incest is a bad idea because blood relatives often inherit the same mutations, which are not masked by normal genes when close relatives produce offspring. For example, you may need just a little bit of the protein produced by a gene, so one copy of the gene may suffice. The mutated gene's inability to produce a working protein may not matter very much. This masking effect is called genetic dominance. Dominance makes sex very powerful in limiting the damage caused by mutations.

However, dominance is often not perfect, and it is really only a short-term solution. Two normal genes are sometimes still better than one. And hiding the effects of mutations allows them to accumulate over evolutionary time. To keep mutations from accumulating over the longer term, sexual reproduction takes some chances. Consider two parents with average numbers of mutations. Each contributes half of their genes to each offspring. Most of the offspring will inherit nearly the same number of mutations as their parents had. But some may be lucky: they may inherit a below-average number of mutations from their father, and a below-average number from their mother too. They will have much better genes than average, and should survive and reproduce very well. Their relatively mutation-free genes will spread through future generations. Other offspring may be very unlucky: they may inherit an above-average load of mutations from both parents, and may fail to develop at all, or may die in infancy. When they die, they take a large number of mutations with them into evolutionary oblivion.

This effect is extremely important. By endowing the next generation with unequal numbers of mutations, sexual reproduction ensures that at least some offspring will have very good genes. They will preserve the genetic information that keeps the species working. From a selfish gene's point of view, it does not matter that some offspring have very bad genes full of mutations, because those mutations would have died out sooner or later anyway. Better to concentrate them in as few bodies as possible so they do the least damage over the long term. Investment analysts will recognize that sexual selection is a way of implementing a risk-seeking strategy. Since evolution over the long term is a winner-takes-all contest, it is more important to produce a few offspring that have a chance to do very well, than a larger number of mediocre offspring.

Mutations, Fitness, and Sexual Attractiveness

Now, if the goal of sexual reproduction is to keep at least some of your offspring safe from your harmful mutations, it would be foolish to pick your sexual partners at random. Any sex partner will carry his or her own load of mutations. You should pick the partner with the lowest number of harmful mutations: that will give your offspring the highest expected fitness, which means the best chance of surviving and reproducing. If your choice of sexual partner is very good indeed, your genes may hitch a ride to evolutionary stardom on the genetic quality of your mate. Many biologists are coming to the view that mate choice is a strategy for getting the best genes you can for your offspring.

Because of genetic dominance, many mutations are hidden from view. They do not affect body or behavior, so they cannot be used in mate choice. However, dominance is often incomplete, and a lot of genetic variation between individuals does show up in body and behavior. Some traits reveal more genetic information than others. Complex traits such as peacock tails that vary conspicuously between individuals may be especially informative. Their complexity means that their development depends on many genes interacting efficiently. They summarize more genetic information by being more complicated. And their variation at the visible level of body and behavior means that genetic variation can be perceived during mate choice. With sexual selection there is a big incentive to pay very close attention to traits like these.

Such traits are called "fitness indicators." A fitness indicator is a biological trait that evolved specifically to advertise an animal's fitness. Fitness means the propensity to survive and reproduce successfully. It is determined mainly by an individual's genetic quality, which boils down to their mutation load.

There is a close connection between mutations and fitness. If a species has been living in its present environment for many generations, its average genes are probably very well adapted to that environment. Because they have already been tested again and again by natural selection, the average genes in the species are already optimal. If they weren't, they would already have been replaced by different genes. This suggests that any deviation from the genetic norm is a deviation from optimality. Mutations are deviations from the genetic norm. If a set of mutations makes an individual unable to grow an optimal body and unable to produce optimal behavior, then they impair that individual's ability to survive and reproduce. Since fitness means the ability to survive and reproduce, mutations almost always lower fitness; conversely, high fitness implies freedom from harmful mutations. If fitness indicators advertise high fitness, they are also advertising freedom from mutations, which is what mate choice wants. Normal genes are tried and tested, whereas mutations are shots in the dark.

Sexual selection needs some way to connect the sensory abilities of animals to the mutation levels of the potential mates they are choosing between. Fitness indicators are the connection, for they are the traits that make fitness visible. What they make visible can be favored by mate choice, and what is favored by mate choice can evolve through sexual selection. Fitness indicators are the genetic sieve that lets sexual selection sift out harmful mutations. In this mutation-centered view of sex, sexual ornaments and courtship behaviors evolve as fitness indicators.

The Human Mind as a Set of Fitness Indicators

In the previous chapter we met the runaway brain theory. It has problems: it does not explain the trend of hominid brain evolution toward the big and the bright, and it does not work very well with mutual mate choice. However, there is another possible solution. Perhaps the human mind's most distinctive capacities evolved through sexual selection as fitness indicators.

We could call this the "healthy brain theory," in contrast to the runaway brain theory. The healthy brain theory suggests that our brains are different from those of other apes not because extravagantly large brains helped us to survive or to raise offspring, but because such brains are simply better advertisements of how good our genes are. The more complicated the brain, the easier it is to mess up. The human brain's great complexity makes it vulnerable to impairment through mutations, and its great size makes it physiologically costly. By producing behaviors such as language and art that only a costly, complex brain could produce, we may be advertising our fitness to potential mates. If sexual selection favored the minds that seemed fit for mating, our creative intelligence could have evolved not because it gives us any survival advantage, but because it makes us especially vulnerable to revealing our mutations in our behavior.

Extreme vulnerability to mutation sounds like something that natural selection could not possibly favor. Precisely. It is what sexual selection through mate choice favors. Once sexual choice seized upon the brain as a possible fitness indicator, the brain was helpless to resist. Any individuals who did not reveal their fitness through their courtship behavior were not chosen as sexual partners. Their small, efficient, ironclad, risk-averse, mutation-proof brains died out with them. In their place evolved our sort of brain: huge, costly, vulnerable, revealing.

Our species was not the first to stumble upon the fact that complex behaviors make good fitness indicators. Songbirds reveal their fitness by repeating complicated, melodious songs. Fruitflies do little dances in front of one another to reveal their genetic quality. Bowerbirds construct large mating huts ornamented with flowers, fruits, shells, and butterfly wings, presumably to reveal their quality. In fact, many species appear to use their courtship behaviors as fitness indicators. The distinctive thing about humans is that our courtship behavior reveals so much more of our minds. Art reveals our visual aesthetics. Conversation reveals our personality and intelligence. By opening up our brains as advertisements for our fitness, we discovered whole new classes of fitness indicators, like generosity and creativity.

To suggest that a mental capacity like human creative intelligence evolved as a fitness indicator is not just to throw another possible function into the arena of human evolution theories. This is not a function like hunting, toolmaking, or socializing that contributes directly to fitness by promoting survival and reproduction. Instead, fitness indicators serve a sort of meta-function. They sit on top of other adaptations, proclaiming their virtues. Fitness indicators are to ordinary adaptations what literary agents are to authors, or what advertisements are to products. Of course, they are adaptations in their own right, just as literary agents are people too, and just as advertisements are also products—the products of advertising firms. But fitness indicators work differently. They take long vacations. They are social and sales-oriented. They live in the semiotic space of symbolism and strategic deal-making, not in the gritty world of factory production. The healthy brain theory proposes that our minds are clusters of fitness indicators: persuasive salesmen like art, music, and humor, that do their best work in courtship, where the most important deals are made.

We should not expect sexually selected fitness indicators to look very useful if they are evaluated by traditional survival-of-the-fittest criteria. They do not help animals find food or avoid predators. They do not remove parasites or feed offspring. They look costly and useless. They appear luxuriously superfluous, often resembling a pathological side-effect of something more useful and sensible. But these are precisely the features of many human mental abilities that have puzzled scientists. Art and morality look like evolutionary luxuries. Creative intelligence and language seem useful in moderation, but humans do not have them in moderation—we have them in luxuriant excess.

The idea of mental fitness indicators fills an important gap in evolutionary psychology. Physical fitness indicators form a standard part of sexual selection theory, and are covered in every good evolutionary textbook. Researchers such as Randy Thornhill, Steven Gangestad, David Perrett, Anders Moller, and Karl Grammer have analyzed many aspects of the human face and body as fitness indicators that reveal health, fertility, and youth. Most evolutionary psychologists agree that human mate choice is even more focused on mind than on body, concerned as it is with assessing a person's social status, intelligence, kindness, reliability, and other psychological traits. Yet evolutionary psychology has paid very little attention to the possibility that many of our psychological traits may have evolved as fitness indicators too. The idea is not assessed in Steven Pinker's *How the Mind Works*, David Buss's textbook *Evolutionary Psychology*, or any other major work on evolutionary psychology. In most such works natural selection is used to explain most of the mind's adaptations. Where sexual selection is invoked, it is almost always to explain how our mechanisms for mate choice evolved, or how some basic sex differences in sexual strategies evolved. The idea of sexual selection for mental fitness indicators has yet to be adequately explored.

To understand how these parts of the mind may have evolved as fitness indicators, we have to understand a bit more about what fitness means, why fitness varies enough to be worth worrying about in mate choice, and what makes a good fitness indicator. After we have these principles under our belts, we can have another look at the healthy brain theory.

Evolutionary Fitness and Physical Fitness

Fitness indicators are supposed to reveal fitness—but what does "fitness" really mean? For biologists, fitness means an organism's propensity to survive and reproduce in a particular environment. Fitness in this evolutionary sense has three important features: it is relative to competitors in a species, it is relative to an environment, and it is a statistical propensity rather than an achieved outcome.

Evolutionary fitness is always relative to a population of competitors within a species. "High fitness" for a barnacle, a mayfly, an oak tree, and a human depend on very different traits, and suggest very different numbers of offspring. What ties together fitness across species is the link between fitness and evolutionary change. Genes underlying high fitness will tend to spread through a population, replacing genes for low fitness. Evolution increases fitness, by definition. In this sense, evolution is progressive: when sexual selection favors fitness indicators, it necessarily increases fitness and contributes to evolutionary progress.

Evolutionary fitness is also relative to environment. It depends on the fit between an organism's traits and an environment's features, which is why it is called "fitness" rather than "quality" or "perfection." The *Alien* films notwithstanding, there is no such thing as a super-organism that could survive and reproduce in every possible environment. When biologists talk about an organism's fitness, they usually assume that the organism's performance is being measured in an environment similar to that in which the species has been evolving for many generations. An organism that shows high fitness in an ancestrally normal environment will not necessarily show high fitness in a novel environment.

Fitness as a propensity is the most slippery concept to grasp. Fitness as I use the term is a statistical propensity, an expectation that allows us to predict how an individual will probably fare. We attribute propensities all the time to other people: intelligence, kindness, irritability. Like fitness, these traits must be inferred rather than directly perceived. Like fitness, they allow us to make predictions that work on average over the long term, but those predictions are sometimes overridden by situational factors. Fitness is something we attribute to organisms to explain why they survive and reproduce better than their competitors. It is not just a measure of whether they do in fact survive and reproduce, because accidents can happen. A highly fit organism that we expect to thrive may be hit by lightning, or rejected as a sexual partner through some kind of situation-comedy mix-up. These failures to live up to one's fitness do not imply that the concept of fitness is vacuous. Intelligent people sometimes make errors in mental calculations, but that does not invalidate the concept of intelligence. Not all philosophers of biology agree on this propensity idea of fitness, but most do, and so do I.

In other contexts, fitness means something different. "Fitness centers" do not usually contain biologists scribbling down evolutionary equations. Instead, they are frequented by people trying to get fit, to improve their physical fitness. Fitness in the physical sense implies health, youth, athletic ability, and physical attractiveness. When George Bush appointed Arnold Schwarzenegger to head the President's Council on Physical Fitness in the early 1990s, he did not expect Schwarzenegger to improve the quality of the American gene pool. He expected him to get Americans in better shape.

Physical fitness is not relative to a population or an environment, but is relative to a norm of optimal efficiency for a body of a particular species. When we say a man is physically fit, we do not mean he is merely less fat, weak, stiff, and breathless than his peers. A whole population might be physically unfit. To be physically fit is to have a body near the peak of its potential performance, objectively efficient at turning oxygen and food into muscle power and speed. Physical fitness in this sense could even be compared across species. One could say "She is as fit as a champion greyhound." That may be faint praise, but it is not meaningless.

Physical fitness is still environment-relative in the sense that a fit human could not thrive on a neutron star with gravity a billion times stronger than the Earth's. Yet, within the normal operating parameters of a species, physical fitness is useful across a range of situations. An athlete who is fit enough to climb Mount Everest is probably fit enough to scuba-dive, or to fly a rocket to Mars. Physical fitness manifest in one situation usually transfers fairly well to other situations. This is why triathlons and decathlons exist—there are some tradeoffs between the optimal body for distance running and the optimal body for swimming, but some individuals can be better at both than almost anyone else is at either.

Another contrast to evolutionary fitness is that physical fitness is closer to a measurable achievement than a statistical propensity. It is less abstract, and closer to real behavioral outcomes. We expect strength to be manifest in the consistent ability to lift heavy things. We expect aerobic fitness to be manifest in the ability to climb stairs without losing one's breath. Accidents can still keep the fittest athlete from winning a gold medal, but the correlation between physical fitness and physical performance is usually rather high. This is why manifest physical performance is such a good indicator of physical fitness.

Apart from physical fitness, one might also speak of "mental fitness," implying sanity, intelligence, rationality, and communication ability—as when a witness is fit to testify in court. Mental fitness shares most of the important features of physical fitness: it is relative to a norm of optimal psychological efficiency in a particular species, it is fairly general across psychological tasks, and we expect it to be manifest in real behavior. Indeed, what intelligence researchers call "general cognitive ability" or "the *g* factor" could be construed as mental fitness.

Biology students are often taught to make a very clear distinction between evolutionary fitness and physical fitness, to keep them separated by the social Atlantic that keeps professional athletes from mixing with scientists. This distinction is important in teaching biology students to think in flexible, abstract ways about evolutionary fitness. It reminds us that evolutionary fitness is always a matter of trade-offs, or finding the optimal allocation of resources between competing demands. Physical strength is not synonymous with evolutionary fitness, because investing in larger muscles may often produce fewer offspring than investing in larger testicles, fat reserves, or brains. But the distinction makes it hard to develop good intuitions about fitness indicators, which tend to advertise fitness in both the evolutionary and the physical sense.

The Oxford biologist W. D. Hamilton has reminded his colleagues that, within a given species, physical fitness is often rather tightly linked to evolutionary fitness. In his work on sexual selection he has tried to revive a more intuitive concept of fitness in which survival and reproduction do depend on basic physical variables like health, strength, energy, and disease-resistance. Within a species, healthier, stronger animals do tend to survive better, reproduce better, and attract more mates. This correlation between evolutionary fitness and physical (or mental) fitness keeps "the survival of the fittest" from being a tautology.

Evolutionary fitness is linked to physical and mental fitness by something that biologists call "condition." In fact, an animal's "condition" is basically its physical fitness, health, and energy level. A high-fitness animal may be in poor condition due to a temporary injury or food shortage. A low-fitness animal might be in good condition due to a zoo taking very good care of it. In a science laboratory, we can disentangle condition from fitness. We can randomly assign different diets to different animals, or infect an experimental group with a communicable disease and protect a control group from that disease. But in nature, animals largely determine their own condition through their own efforts. The abilities to find food, resist disease, and avoid parasites are major determinants of condition, and major components of fitness. In nature, fitness generally correlates with condition. Good condition is thus a pretty good indicator of high fitness.

Of course, there may be droughts, disasters, food shortages, and epidemics, when all members of a population suffer from poor condition. But even then, higher-fitness animals may suffer less than lower-fitness animals do. The correlation between fitness and condition may remain, despite fluctuations in a population's average condition. In fact, fitness may sometimes be easier to assess under challenging conditions because individual differences in ability may then become more apparent. This is why romantic novels include adventure and risk: emergencies bring out the best in heroes and the worst in pretenders.

As we shall see, many fitness indicators advertise fitness by revealing an animal's condition. They are "condition-dependent" —very sensitive to an animal's general health and well-being ("condition"), and very good at revealing differences in condition between animals. This sets up a chain of relationships that will prove absolutely central to many arguments in this book: genetic mutations influence fitness, fitness influences condition, condition influences the state of fitness indicators, fitness indicators influence mate choice, and mate choice influences evolution.

From the viewpoint of an animal making sexual choices, fitness indicators are just proxies for good genes. But the sexual selection that results from mate choice does not just influence the genes for fitness. It shapes the fitness indicators themselves. These fitness indicators combine evolutionary fitness with physical fitness and mental fitness. That is the key. By trying to get good genes for their offspring, our ancestors unwittingly endowed us with a whole repertoire of very unusual fitness indicators which have come to form an important component of the human mind.

This theory of fitness indicators suggests that much of human courtship consists of advertising our physical fitness and mental fitness to sexual prospects. Physical fitness may be revealed by body shape, facial features, skin condition, energy level, athleticism, fighting ability, and dancing ability. Mental fitness may be revealed by creative story-telling, intelligent problem-solving, skillful socializing, a good sense of humor, empathic kindness, a wide vocabulary, and so forth.

Clearly, many of the traits advertised during courtship also bring non-genetic benefits to a sexual relationship. As David Buss and others have argued, strong mates offer protection, social intelligence brings social benefits, and kindness signals commitment. Fitness-indicator theory does not deny these other benefits, but points out that they are not the only reasons for mate choice. Good genes are important too—indeed, I shall argue that some human mate preferences have been misunderstood as seeking purely non-genetic benefits, when they have actually been focusing on indicators of genetically heritable fitness.

Ms. Fitness USA

Watch enough American cable television, and sooner or later you will find a pretty good analogy for almost any intellectual revolution in evolutionary biology. For me, the revolution in sexual selection ideas in the last twenty years of the 20th century is nicely symbolized by the eclipse of the "Miss America" beauty pageant by newer, more fitness-oriented contests such as "Ms. Fitness USA." In 1980, before the Ms. Fitness contests were invented, biologists thought that most sexual ornaments were arbitrary. Ornaments supposedly evolved through the runaway process or some other arbitrary process. In this picture, the peacock's tail did not reflect any aspect of a peacock's fitness, so was not a very rational basis for sexual choice. Yet a minority of biologists became skeptical about this view that most beauty is arbitrary. Similarly, feminists protested against Miss America pageants, upset by the apparent arbitrariness of the cultural norms of beauty used by the judges. The ability to totter around in high heels and swimsuit did not seem to reflect any very significant aspect of a woman's being.

In response to such criticisms, a promoter named Wally Boyko turned the tables on the beauty contest industry by inventing the "Ms. Fitness USA" contest in 1985. This contest explicitly favors women with the highest physical fitness, not just the greatest beauty. (Indeed, the Ms. Fitness World contest, founded in 1994, is held in conjunction with the annual Arnold Schwarzenegger Fitness Weekend.) The Ms. Fitness contests include three rounds:

an evening gown round (to judge beauty, grooming, poise, and speaking ability), a swimsuit round (to judge muscle tone, body fat, and apparent fitness), and a fitness outfit round (a high-energy, 90-second display of strength, flexibility, endurance, and creativity, set to music). In the third round contestants usually do somersaults, splits, jumps, and one-handed pushups—in such a way as to make the difficult appear effortless. The whole aesthetic shifted from Miss America's soft-bodied, giggly display of femininity to a hard-boiled, active display of health. The judging criteria no longer looked quite so culturally arbitrary. Miss America contestants could improve their chances by dieting, getting silicone breast implants, dyeing their hair, and skillfully applying makeup. But Ms. Fitness contestants, such as the currently top-ranked Monica Brant, can win only by training like professional athletes with aerobics, weightlifting, stretching, sports, and healthy eating. Their physical fitness would be manifest in any culture at any point in history, regardless of minor cultural variations in the norms of beauty.

Some evolutionary biologists responded to the idea of arbitrary sexual ornaments in the same way that Boyko's "International Fitness Sanctioning Body" responded to the Miss America pageant. They rethought the judging criteria. Why should animals choose mates for arbitrary traits, when they can choose mates for traits that reveal their condition and fitness? Certainly, the runaway process can happen in principle, but maybe it is not so important. Maybe it creates transient sexual fashions that come and go, but it does not explain the sexual ornaments that stick around generation after generation. The ornaments that stick around should reveal some information about fitness, about good genes. Most sexual ornaments should be fitness indicators. The debate over this issue has an illuminating history.

Sexual Choice for Fitness

Sir Ronald Fisher first emphasized that animals could choose their sexual partners for high fitness by favoring certain kinds of sexual display. As we saw in Chapter 2, his 1915 paper introduced this idea of fitness indicators. But his 1930 book barely mentioned them, and devoted more space to the idea of runaway. When runaway sank into the quicksand of scientific skepticism, Fisher's even more obscure fitness-indicator idea sank with it. The idea waited thirty-six years for rescue. George Williams revived it in his influential classic, *Adaptation and Natural Selection.* Several decades on, his description of sexual choice for fitness remains unsurpassed.

> It is to the female's advantage to be able to pick the most fit male available for fathering her brood. Unusually fit fathers tend to have unusually fit offspring. One of the functions of courtship would be the advertisement, by a male, of how fit he is. A male whose general health and nutrition enables him to indulge in full development of secondary sexual characters, especially courtship behavior, is likely to be reasonably fit genetically. Other important signs of fitness would be the ability to occupy a choice nesting site and a large territory, and the power to defeat or intimidate other males. In submitting only to a male with such signs of fitness a female would probably be aiding the survival of her own genes.

Since Williams's book became required reading for the new generation of biologists in the 1970s, the indicator idea started to catch on. It received another publicity boost when Richard Dawkins gave it a sympathetic exposition in his 1976 bestseller *The Selfish Gene.*

By the mid-1980s, biologists were seriously assessing the fitness indicator idea. The basic intuition seemed sound, but there were two technical problems so difficult that they took another ten years to resolve. One concerned the supposedly low heritability of fitness, and the other concerned the supposedly low reliability of fitness indicators. To understand how the human mind may have evolved as a set of fitness indicators, we have to understand these problems and their solution.

Why Is Fitness Still Heritable?

Fitness indicators are pointless unless individuals vary in their fitness. If we take fitness to mean the possession of good genes that can be inherited by offspring, then it seems hard to understand how evolution can allow any variation in fitness to remain. Selection is supposed to maximize fitness, driving it ever upwards. It is not supposed to permit fitness variation to persist in species just to provide an incentive for sexual choice.

To follow this argument, it is crucial to understand the difference between "inherited" and "heritable." All traits that depend on genes are inherited. But the term "heritable" is much more restrictive: it refers to the proportion of individual differences in a trait that are due to genetic differences between individuals. The concept of heritability applies only to traits that differ between individuals. If a trait exists in precisely the same form across all individuals, it may be inherited, but it cannot be heritable. It should come as no surprise that fitness is inherited, because fitness clearly depends on genes. The surprising thing is that fitness still varies between individuals in most species, and that the variation often seems to depend on genetic differences.

To see why the heritability of fitness is surprising, consider what happens in species that mate in large aggregations called "leks." *Lek* is Swedish for a playful game or party. Some birds like sage grouse congregate in these leks to choose their sexual partners. The males display as vigorously as they can, dancing, strutting, and cooing. The females wander around inspecting them, remembering them, and coming back to copulate with their favorite after they have seen enough. Leks resemble music festivals where mostly male rock bands compete to attract female groupies. In species that lek, the males usually contribute nothing but their genes. The females may never see them again, and raise their offspring as single mothers. Leks create a situation where sexual selection is extremely strong. The most attractive male sage grouse may mate with thirty females in one morning; average males usually mate with none. It is a winner-takes-all contest, and it should spread the most attractive male's genes very quickly through the population.

If the lekking females choose males for good genes generation after generation, all the males should end up being perfectly fit and identically attractive. Males of lower apparent fitness will have died unmated, their mutations having died with them. After a few generations, all the mutations that show up in fitness indicators should be gone. Only the good genes should be left. If every male has the same high fitness, there is no variation for fitness indicators to reveal. If there is no variation in genetic quality, and if genes are all that females get, there is no longer any incentive for females to be choosy about their mates. Instead of spending time and energy wandering around the lek admiring male displays, the females might as well pick randomly. The reasons for mate choice should disappear as the heritable variation in fitness disappears. According to this evolutionary logic, leks should be temporary phenomena. Yet leks still exist. Presumably, sage grouse have been gathering in leks for thousands of generations. Biologists call this the "lek paradox."

The lek paradox is the most extreme case of a general problem with the heritability of fitness. Any form of sexual selection for fitness indicators should even out genetic variation in fitness. If female choice in our species favored tall males, all males should be equally tall. If male choice favored large breasts, all females should be equally large-breasted. If both sexes favored high intelligence and beautiful faces, all humans should be equally bright and beautiful. Yet we are not. The differences remain, and they are still genetically heritable. So why would selection allow such differences to persist?

Once biologists agreed that the lek paradox was a problem, the hunt was on for evolutionary forces that could maintain variation in fitness. Two major candidates emerged. One emphasized that fitness is environment-relative; the other emphasized the ubiquity of harmful mutations that erode fitness.

Time, Space, and Fitness

We saw earlier that fitness is relative to a particular environment. Environment-relative fitness implies that if a population's environment fluctuates over time or space, then the meaning of fitness will fluctuate too. If the meaning of fitness fluctuates, and the population will not stabilize on any one set of genes that will be good in every environment, then environmental variation could maintain genetic variation.

On evolutionary time-scales, physical environments are changing all the time. The climate gets colder or hotter. Rivers shift course. Mountains rise and fall. Meteorites strike. But such physical changes are usually too slow or rare to maintain variation in fitness. Species adapt fairly quickly to changes in their physical environments, reaching a new equilibrium where all individuals should have optimal traits and high fitness.

More important is the biological environment: the other species that are evolving alongside a given population. Predators may get faster or smarter. New parasites may evolve. Viruses mutate at great speed. In the early 1980s, W. D. Hamilton and John Tooby independently developed the idea that variation in fitness could be maintained over very long periods by populations evolving interactively with their parasites. Every animal large enough for us to see has parasites. Because the parasites are smaller than their hosts, they can grow faster and breed faster—their generation time is shorter. The human generation time is about twenty-five years. For bacteria it can be as little as twenty minutes. For every generation that hosts can evolve to have resistance against parasites, parasites can evolve many generations to exploit their hosts, so parasites can adapt much faster to hosts than vice versa. From a parasite's viewpoint, the host's body is the environment to which it adapts. The host's body determines what counts as fitness for the parasite. But the converse is true as well. From the host's viewpoint, parasites are a major part of the biological environment. The capabilities of parasites determine what counts as fitness for the host. Because parasites are constantly evolving against all large-bodied animals, the biological environment is constantly changing for all such animals. Genes that are good against today's parasites might not be so good tomorrow.

In Hamilton's view, the high-speed evolution of parasites is a major force in moving the goal posts of fitness. No large-bodied species ever reaches the hypothetical equilibrium where every individual has high fitness, because parasites always evolve faster. Hamilton saw the implications for sexual selection. Mate choice should favor fitness indicators that are especially good at revealing how individuals resist parasites like viruses, bacteria, and intestinal and skin-burrowing worms. A large, bright peacock's tail proclaims, "I have conquered my parasites. If I had not, my tail would be small, drab, and diseased-looking. If you mate with me, your offspring will inherit my resistance." In an influential 1982 paper, W. D. Hamilton and Marlene Zuk proposed that many sexual ornaments evolved as fitness indicators that signal freedom from parasites. For example, an uakari monkey's bright red face may have evolved to reveal that it is not infected by blood parasites that would cause pale-faced anemia. As long as there are parasites in the world, the meaning of fitness will vary from one generation to the next. Large-bodied species are thus chasing an optimal fitness that remains always one step ahead of them. That, in Hamilton's view, explains why fitness remains heritable in most species most of the time. Matt Ridley's book *The Red Queen* lucidly describes how arms races between parasites and hosts could maintain the incentives for mate choice.

Our ancestors had plenty of parasites and germs to worry about too: tapeworms, herpes, crab lice, common colds, malaria, stomach flu. Their communicable diseases were probably not as severe as those that arise in urban civilizations, because their population densities were much lower. They did not have plagues like medieval European cities. But every one of our hominid ancestors was probably exposed to dozens of species of fast-breeding, fast-evolving, energy-sapping organisms, from micro-parasites like viruses and bacteria to macro-parasites like head lice. The variable was not whether they had parasites, but how well they maintained their health and energy despite them. The sexual repulsion we may experience toward someone heavily infected with parasites may reflect more than a fear of contamination. It may be showing that Hamilton is right: that resisting parasites is a major part of fitness for any large animal, and advertising that resistance is a major function of sexual ornaments.

Environments fluctuate across space as well as time. Our ancestors lived in small groups spread out over wide areas of Africa. The African continent is not one big flat savannah. Each area has slightly different weather, geology, vegetation, competitors, predators, and parasites. There are many micro-habitats. What is optimal in one area may not be optimal in another. Survival pressures vary across space, so each individual's fitness varies across space. As long as some of our ancestors migrated from one area to another in every generation, they would never evolve to the point where every individual in every area has maximum fitness relative to their local environment. Like variation in selection pressures over time, this variation in space helps explain why fitness remains heritable.

Environmental fluctuations across time and space are best at explaining why physical fitness and health remains heritable. But they are not so useful to us if our interest is in mental fitness indicators. Parasites put evolutionary pressure more on immune systems and bodies than on brains. Variations in climate from one part of Africa to another might maintain heritable variation in physical adaptations, but it is not clear why they should maintain variation in mental adaptations. To explain persistent variation in mental fitness, we need something more.

The Black Rain of Mutation

In science-fiction films and comic books, "mutations" are Faustian bargains that confer superhuman powers while damning their possessors to abnormal appearance and impaired sexual attractiveness. Spiderman was bitten by a "mutated" spider, and acquired wall-clinging powers but became alienated from his girlfriend. Monster Island apparently had high levels of mutagenic radiation, which is how Godzilla acquired his "atomic breath" that incinerates his enemies but keeps him single. This comic-book view of mutations is only half right. Mutations do undermine normal appearance and sexual attractiveness, but they very rarely bring survival or fertility benefits.

Since the late 1980s, many biologists have been coming around to the view that fitness remains heritable mostly because new mutations are constantly arising and causing trouble. As we saw before, mutations almost always lower fitness. The more mutations an individual carries, the lower its expected fitness. To avoid mutational meltdown and extinction, selection had to be potent enough to eliminate those mutations at the same average rate at which they arose. (As we saw, Eyre-Walker and Keightley estimated that at least 1.6 harmful new mutations per individual every generation have been arising in our lineage for the last several million years.)

In most species for most of the time, almost all of the natural selection and sexual selection consists simply of removing harmful new mutations and maintaining the status quo. Selection is mostly conservative and stabilizing. Very rarely does selection favor a new mutant, because only rarely is a mutated gene better than the existing gene at helping an organism survive and reproduce. These rare occasions attract the biologist's attention because they are the times when evolution—genetic change in a species—can occur. But for the rest of the time, there is a tension between selection and mutation. Selection tends to maintain adaptations in their current effective form, while mutation tends to erode them into a chaotic, ineffectual mess.

The Brain as a Target for Mutation

For simple traits that depend on just a few genes, selection is pretty good at eliminating mutations. Each mutation is likely to cause such dramatic change that natural selection rapidly eliminates it. But for very complex traits, like human brains, that grow through the interaction of many genes, mutations are harder for selection to eliminate. There are more genes vulnerable to mutation in the first place, and selection's effects get diluted across more genes. This decreases selection's power to eliminate mutations on any one gene. With mutation stronger and selection weaker, complex traits are less likely to be perched on the peak of perfection.

Genetic variation is more likely to be manifest in complex traits. This makes complex traits like the human brain better fitness indicators.

Imagine all the DNA in our 23 pairs of chromosomes laid end to end in a single strip. The DNA from a single human cell would be about six feet long, and contain about 80,000 genes. Imagine that the genes involved in growing a particular trait are lit up in bright green, and that each gene has a tiny chance of having a mutation that turns the green light red. For a very simple trait like skin color, there might be only half a dozen lights sprinkled along the six-foot length of DNA. It is very unlikely that any of them would be red. For a moderately complex trait like the shape of the human face, there might be several hundred lights. It is likely that a few of them might be red. For a very complex organ like the human brain, there might be tens of thousands of lights. Our DNA would light up like a Christmas tree. Although the proportion of red lights would still be very low, the absolute number would be much higher. The brain would give much better information about mutation load and fitness, because it gives mate choice a wider window on a larger sample of our DNA. (The larger the sample of genes, the more accurate the estimate of mutation load.) This is what biologists mean by the "mutational target size" of a trait: the proportion of the genome that is involved in a trait's development determines the proportion of all mutations that are visible in the trait.

At the moment, nobody knows exactly how many of our genes are involved in growing our brains. Geneticists sometimes estimate that about half of our genes are involved in brain development, and about a third might be active only in the brain. If this guess is about right (and we shall know within a decade or two whether it is), then the mutational target size of the human brain is about half the human genome. The brain probably has a larger mutational target size than any other organ. Of all the new mutations that mess up something during human development, half of them mess up something in the human brain.

If mutations maintain most of the variation in fitness that we see, then the organs with the largest mutational target sizes will make the best fitness indicators. The human brain should make a very good fitness indicator indeed. Its vulnerability to mutation is precisely why sexual choice mechanisms should evolve to pay attention to its performance.

In the rest of this book, I shall take the heritability of fitness for granted. The expectation that fitness should not be heritable was based on theoretical arguments developed in the 1930s. Those arguments are contradicted by the evidence. Wild populations show large amounts of genetic variation. Biologists routinely find individual differences in reproductive success in the wild, differences which are often genetically heritable. Fitness remains heritable in most species for most of the time. It seems likely that a lot of this continuing heritability is due to the continual rain of mutations. Some biologists even wonder how selection can possibly be strong enough to eliminate all these new mutations, and keep the species from falling apart. Fitness-eroding mutations are ubiquitous, and usually stick around for a fairly long time. There is always a tension between mutation and selection. And there are always fluctuations in fitness across time and space which keeps fitness heritable. These are just the facts of life. Mate choice evolves to deal with them.

How to Advertise Fitness

Fitness is like money in a secret Swiss bank account. You may know how much you have, but nobody else can find out directly. If they ask the bank, the bank will not tell them. If they ask you, you might lie. If they are willing to mate with you if your capital exceeds a certain figure, you may be especially tempted to lie. This is what makes mate choice difficult. The supposedly low heritability of fitness was one argument against the importance of fitness indicators in sexual selection. The other problem is the potentially low reliability of fitness indicators. An animal trying to find a high-fitness mate is in the position of an attractive gold digger seeking a millionaire. She has incentives to mate only with a male who offers high genetic or financial capital. But every male has incentives to pretend to be richer than he is, to attract more mates. What is a poor girl to do?

Anita Loos's classic 1925 novel *Gentlemen Prefer Blondes* suggested one good strategy. The blonde protagonist Lorelei Lee forced her suitors to spend vast amounts of money on her, to show how much they really had. Her suitor Gus Eisman may have called himself "the Button King," but who can say whether his business is really profitable? Miss Lee was not the brightest button ever to baffle "Doctor Froyd" in Vienna, but she understood the principle of costly display. If a man can afford to dress as well as a peacock, he is probably not poor. If he gives you a very large diamond, he is likely to be rich. The more they can spend, the more they must have.

Lorelei was not the first to realize this, of course. Thorstein Veblen's *Theory of the Leisure Class* introduced the idea of "conspicuous consumption" in 1899. Veblen argued that in modern urban societies, where strangers come and go, people increasingly advertise their wealth by ornamenting themselves with costly luxuries. Where nobody knows anyone else's true wealth directly, conspicuous consumption is the only reliable signal of wealth. Sociologists and economists understood this logic immediately. Capitalist consumerism evolved in part as a set of wealth indicators.

It took biologists another three-quarters of a century to apply the same principle to sexual selection for fitness indicators. As we saw earlier, in 1975 Israeli biologist Amotz Zahavi argued that many animal signals—including sexual ornaments—evolved as advertisements of the animal's fitness. He suggested that the only reliable way to advertise one's true fitness is to produce a signal that costs a lot of fitness. This explains why sexual ornaments are so often large, extravagant, costly, and complicated. The peacock's tail is not just a cheap, transient advertisement visible only to peahens. It is heavy, encumbering, hard to grow, hard to preen, and highly visible to predators. Peacocks have to drag it around everywhere they go. Unfit peacocks might be able to grow large tails, but they would not be strong enough to carry them while finding food, or fast enough to escape from predators. Only highly fit peacocks can afford very large tails.

Therefore, if a female sees a male sporting a very large tail, she can be confident that he has high fitness, and that his good genes could be passed on to her offspring. Since very fit peacocks tend to have fit sons and daughters that are more likely to survive and reproduce, peahens benefit by choosing big-tailed peacocks. Their preferences for larger-than-average tails can spread. Conversely, peahens that preferred shorter-than-average tails did not leave many descendants to inherit their misguided preference, because their offspring were less fit than average. Sexual selection favors both the preference for costly sexual displays and the displays themselves.

Zahavi suggested that most sexual ornaments are "handicaps": they advertise true fitness by handicapping an individual with a survival cost. He also argued that handicaps should be the only evolutionarily stable kinds of sexual ornament, because they are the only ones that convey the information about fitness that individuals really want when making sexual choices. His paper unleashed a storm of protest. The handicap idea seemed absurd. Throughout the late 1970s the handicap principle was attacked by almost every eminent evolutionary theorist. Surely sexual selection could not have an intrinsic drive to produce wasteful displays that impair survival?

Apparently, most biologists in the 1970s had not read Thorstein Veblen. They did not make the connection between conspicuous consumption to advertise wealth and costly sexual ornaments to advertise fitness. Without that connection it was hard to see how Zahavi's handicap principle could work (or rather, which of the several possible versions of it might work). How could sexual selection favor fitness indicators that impaired an animal's survival prospects? How could mate choice favor a costly, useless ornament over a cheaper, more beneficial ornament? (Why should a man give a woman a useless diamond engagement ring, when he could buy her a nice big potato, which she could at least eat?)

A clever peahen able to read Veblen might propose that, for the good of the species, peacocks should stop this mad waste. Suppose, for example, that each peacock agreed to wear a little hat showing a number between one and ten that revealed his actual fitness (perhaps a composite score of health, strength, fecundity, intelligence, and screeching ability). The problem with this system of quality-signs is that there would be no effective way to police it. Low-fitness peacocks would lie, because they could attract better mates by lying—they would all proclaim a perfect ten. Zahavi realized that the signaling system has to be self-policing. It has to include a range of sexual signals that differ in cost, and thus differ in affordability by individuals of different fitness, by virtue of which they honestly reveal their fitness.

The handicap principle suggests that prodigious waste is a necessary feature of sexual courtship. Peacocks as a species would be much better off if they didn't have to waste so much energy growing big tails. But as individual males and females, they have irresistible incentives to grow the biggest tails they can afford, or to choose sexual partners with the biggest tails they can attract. In nature, showy waste is the only guarantee of truth in advertising.

The handicap principle was also rejected initially because most biologists did not know about economists' research into costly signaling. During the 1960s, game theorists working in economics departments did a lot of work on what makes signals reliable, given incentives to lie. They developed something called signaling theory, which distinguishes two kinds of signal. There are signals that incur a significant cost or commitment, which can therefore be reliable indicators of someone's intentions. And then there are signals that cost nothing, which are called "cheap talk." Economists realized that cheap talk is not to be trusted. It does not commit someone to a course of action. It does not reveal their capabilities. It means nothing, because it costs nothing. If a car company proclaims "We will defend our share of the four-door market at all costs," that is just cheap talk and hot air. But if the company spends a billion dollars building a factory specialized for four-door car production, their proclamation carries some weight.

The factory is not just a capital investment—it is also a strategic signal. It deters competitors from entering the same market niche by reliably revealing the company's financial strength and strategic commitment. In fact, the more (wasteful) excess capacity the factory has, the better a strategic signal it is. Likewise, proclaiming "I have a straight flush" in poker carries less credibility than placing a large bet on one's hand. This costly-signaling principle became so widely accepted among economists in the 1960s that signaling theory withered for lack of controversy.

Advertising Within One's Budget

It took biologists about fifteen years to accept Zahavi's handicap principle. Much of that time was spent clarifying what kinds of handicaps could evolve and what kinds could not. Since handicaps are basically fitness indicators, the debate over handicaps helped lay the foundation for the modern theory of fitness indicators.

A handicap cannot usually evolve if it commits all the males to producing a costly signal regardless of their true fitness. This would be like all men buying a five-carat diamond engagement ring regardless of their salaries. Such a fixed-cost strategy is not sensible for anybody—all the poor men would go bankrupt and starve before their wedding day, while the super-rich men would be indistinguishable from the moderately rich men. The same problems explain why we rarely see sexual ornaments in nature that are produced by all males to an equal degree. A handicap gene that committed all low-fitness males to produce a very costly sexual ornament would simply kill them all. The handicap would help females to recognize high-fitness males, but the females could not tell which of the high-fitness males was best. Mathematical models and simulations suggest that this sort of fixed-cost handicap cannot evolve under reasonable conditions.

Handicaps can evolve much more easily if they are a little more sensitive to an animal's fitness level. A gene that says "spend 50 percent of your disposable energy on courtship dancing" could easily spread through a population if females appreciate dancing.

It would be just like the cultural rule invented by the De Beers diamond cartel that insists, "Spend two months of your salary on your engagement ring." Costly signals that take fitness budgets into account evolve much more easily than do costly signals that ignore budgets. Sensitivity to this budget constraint is called "condition-dependence" by most biologists. It could equally be called "fitness-dependence," to reflect the intuition that fitness indicators should be fitness-dependent. Alan Grafen showed that condition-dependent indicators could evolve, giving Zahavi's handicap principle much more credibility.

This sort of condition-dependence seems intuitive when you think of examples. Better-fed animals can afford to grow larger sexual ornaments. Most energetic animals can afford to exert more effort in courtship. Stronger animals can afford to fight other strong animals in ritualized contests. Faster animals can afford to taunt predators from a closer distance. Animals with better memories can afford to learn a large repertoire of courtship songs. Animals with higher social status can afford to act more confident and relaxed around their peers.

Such condition dependence is one of the most important concepts in sexual selection today. It protects low-fitness animals from incurring the costs of sexual ornamentation and courtship if they do not feel up to it. If you are a really unfit peacock, you are not forced to grow a huge tail that will kill you through exhaustion within a week; instead you can grow a drab little tail and hope for the best. Compared to sexual ornamentation that grows on the body, courtship behavior is even more flexible and condition-dependent. If you are a human feeling really ill, you do not have to go to the Ministry of Sound nightclub with your significant other and dance all night after taking lots of drugs. If you are in poor aerobic condition you do not have to run the Olympic marathon and die of heatstroke. If you are not very bright you do not have to go to Stanford Business School and fail. Condition-dependence lets us choose our battles.

Condition-dependence is equally useful at the high end of the fitness scale, for it enables one to tailor the amount one spends on fitness indicators to one's fitness level. This helps the extremely fit to distinguish themselves from the very fit. It spreads out the apparent differences between individuals so that their fitness is easier to judge. Condition-dependence makes mate choice easier because it lets one infer fitness directly from the apparent costliness of a courtship display.

An Infinite Variety of Waste

Zahavi's handicap principle and the idea of condition-dependence are different perspectives on the same thing. The handicap idea emphasizes that sexual ornaments and courtship behaviors must be costly in order to be reliable fitness indicators. Their cost can take almost any form. They can increase risk from predators by making an animal more conspicuous with bright colors. They can increase risk from germs by impairing an animal's immune system (which many sex hormones do). They can burn up vast amounts of time and energy, like bird song. They can demand a huge effort to obtain a small gift of meat, as in human tribal hunting.

As with Veblen's conspicuous consumption principle, the form of the cost does not matter much. What matters is the prodigious waste. The waste is what keeps the fitness indicators honest. The wastefulness of courtship is what makes it romantic. The wasteful dancing, the wasteful gift-giving, the wasteful conversation, the wasteful laughter, the wasteful foreplay, the wasteful adventures. From the viewpoint of "survival of the fittest," the waste looks mad and pointless and maladaptive. Human courtship even looks wasteful from the viewpoint of sexual selection for non-genetic benefits, because, as we shall see, the acts of love considered most romantic are often those that cost the giver the most, but that bring the smallest material benefits to the receiver. However, from the viewpoint of fitness indicator theory, this waste is the most efficient and reliable way to discover someone's fitness. Where you see conspicuous waste in nature, sexual choice has often been at work.

Every sexual ornament in every sexually reproducing species

could be viewed as a different style of waste. Male humpback whales waste their energies with half-hour-long, hundred-decibel songs that they repeat all day long during the breeding season. Male weaverbirds waste their time constructing ornamental nests. Male stag beetles waste the matter and energy from their food growing huge mandibles. Male elephant seals waste a thousand pounds of their fat per breeding season fighting other elephant seals. Male lions waste countless calories copulating thirty times a day with female lions before the females will conceive. Male humans waste their time and energy getting graduate degrees, writing books, playing sports, fighting other men, painting pictures, playing jazz, and founding religious cults. These may not be conscious sexual strategies, but the underlying motivations for "achievement" and "status"—even in preference to material sources—were probably shaped by sexual selection. (Of course, the wasteful displays that seemed attractive during courtship may no longer be valued if they persist after offspring arrive—there is a trade-off between parental responsibilities and conspicuous display.)

The handicap principle suggests that in each case, sexual selection cares much more about the prodigious magnitude of the waste than about its precise form. Once the decision-making mechanisms of sexual choice get the necessary information about fitness from a sexual display, everything else about the display is just a matter of taste. This interplay between waste and taste gives evolution a lot of elbow room. In fact, every species with sexual ornaments can be viewed as a different variety of sexually selected waste. Without so many varieties of sexual waste, our planet would not be host to so many species.

Evolving Better Indicators

The late 1990s have brought an ever-deeper understanding of fitness indicators in sexual selection theory. Biologists such as Alan Grafen, Andrew Pomiankowski, Anders Moller, Rufus Johnstone, Locke Rowe, and David Houle have pushed the idea of condition-dependence deeper into the heart of sexual selection, relating it to the heritability of fitness arguments and the idea of mutation selection balance. Indicator theory is still developing very quickly, and no one has yet had the final word. However, I am especially intrigued by some ideas that Rowe and Houle developed about condition-dependence in a 1996 paper, because they seem most relevant to the human mind's evolution.

In Rowe and Houle's model all fitness indicators start out as ordinary traits. Each trait has certain costs. Higher-fitness individuals have larger energy budgets, so are better able to bear these costs. Initially, a trait may be favored by sexual choice because of some random runaway effect. But once it is favored, individuals with more extreme, costlier versions of the trait will spread their genes more successfully. This sexual selection increases average fitness in the population, because the trait acts as a weak fitness indicator. But here is the crucial point: the sexual selection also puts pressure on the trait to recruit a larger share of the individual's energy budget for itself. Individuals who allocate a low proportion of their fitness to the sexually favored trait will lose out to those who allocate a lot. As the sexually favored trait grabs a larger share of an organism's resources for itself, it becomes ever more dependent on the organism's total fitness budget. The trait turns from a cheap ordinary trait into a true handicap with large costs—in other words, its condition-dependence increases. And the increasing condition-dependence becomes an ever more valuable source of information about fitness. In this way, sexual selection has turned an ordinary trait into a really good fitness indicator.

The fitness indicator does not just recruit an increased share of an organism's energy: it also makes itself dependent on an increased proportion of an organism's genes. Rowe and Houle call this process "genic capture." The indicator captures a larger amount of information about an individual's genetic quality. Typically, this might work by a trait evolving a little bit more complexity, recruiting some of the genes that influence growth and development processes already evolved for other adaptations. This genic capture process makes the fitness indicator a window on an animal's genome. As the window grows wider through genic capture, the indicator lets an observer see a larger amount of all the genetic variation in fitness with the population, making it easier to choose mates for their good genes. Good fitness indicators give sexual choice a panoramic view of a potential mate's genetic quality.

It is not clear yet exactly how genic capture works, and this feature of Rowe and Houle's model needs further research. If it does work, and if the human brain's complexity evolved in part through genic capture, then there is an interesting implication. It would explain why so many unique human mental abilities look to some biologists like "spandrels," mere side-effects of other adaptations. Stephen Jay Gould has argued that most of our uniquely human capacities did not evolve for specific adaptive functions, but emerged as side-effects of already-existing brain circuits and learning abilities. Like most evolutionary psychologists, I find that argument weak for many reasons—for example, it fails to explain why other large-brained species such as dolphins, whales, and elephants did not invent paleontology or socialism.

However, Gould's argument may have this grain of truth: the human brain's distinctive power is its ability to advertise a lot of the computational abilities that were already latent in the brains of other great apes. This does not mean that music, art, and language came for free just because an ape brain tripled in size. But it might mean that when sexual selection seized upon the ape brain as a set of possible fitness indicators, the genic capture process recruited a lot of pre-existing brain circuitry into human courtship behavior. It made that brain circuitry more manifest in courtship behavior, more condition-dependent, and more subject to sexual choice. Our brains may look like a set of spandrels, but they look that way only because our mental fitness indicators are so efficient at advertising the brain's many abilities. (Of course, fitness indicators are different from spandrels because they evolved through sexual selection to have a specific courtship function, whereas spandrels, by definition, do not have any specific evolved function.)

Mental Traits as Fitness Indicators

Fitness indicator theories like Rowe and Houle's model can help us to understand the evolution of the human mind. Our capacities for music, art, creativity, humor, and poetry do not look like ordinary adaptations are supposed to look. Evolutionary psychologists like John Tooby, Leda Cosmides, David Buss, and Steven Pinker have developed some rules for recognizing mental adaptations. If a human mental trait evolved through natural selection for some specific function, it is supposed to show small differences between people, because selection should have eliminated maladaptive variation long ago. It is supposed to show low heritability, because selection should have eliminated all genes other than the optimal ones long ago. It is supposed to be efficient and low in cost, because natural selection favors efficient problem-solving. And it is supposed to be modular and specialized for solving a particular problem, because modular specialization is the efficient way to engineer things.

Fitness indicators violate all these criteria. If a mental trait evolved through sexual selection as a fitness indicator, it should show large differences between people. It evolved specifically to help sexual choice discriminate in favor of its possessor at the expense of sexual rivals. Fitness indicators can show high heritability because they tap into genetic variation in fitness, and fitness usually remains heritable. For fitness indicators to be reliable, they have to be wasteful, not efficient. They have to have high costs that make them look very inefficient compared with survival adaptations. Finally, fitness indicators cannot be totally modular and separate from other adaptations, because their whole point is to capture general features of an organism's health, fertility, intelligence, and fitness. The peacock's tail appears to fit this profile as a fitness indicator, and many human mental abilities do as well.

To traditional evolutionary psychologists, human abilities like music, humor, and creativity do not look like adaptations because they look too variable, too heritable, too wasteful, and not very modular. But these are precisely the features we should expect of fitness indicators. If a human mental trait shows large individual differences, high heritability, high condition-dependence, high costs, and high correlations with other mental and physical abilities, then it may have evolved through sexual selection as a fitness indicator.

If we make an inventory of what the human brain can do, we find two general themes: very few of the ancient mental abilities that we share with other apes look like fitness indicators, but many mental abilities unique to humans do look like fitness indicators. There are probably thousands of psychological adaptations in the human mind. The vast majority are shared with other species. Some evolved hundreds of millions of years ago and are shared with thousands of species. Some evolved only a few million years ago and are shared only with other great apes. We have exquisitely efficient mechanisms for regulating our breathing, controlling our limbs, keeping our balance, seeing colors, remembering spatial locations, learning foraging skills, being kind to offspring, feeling pain when injured, remembering faces, making friends, punishing cheats, perceiving social status, estimating risks, and so forth. Steven Pinker has explored many of these mechanisms in his book *How the Mind Works*. When I propose a shorthand slogan like "the human mind evolved through sexual selection," I do not mean that sexual selection shaped all of these adaptations that we share with other primates. Of course, about 90 percent of our psychological adaptations evolved through standard natural selection and social selection to solve routine problems of surviving and living in groups. Evolutionary psychology has proven very good at analyzing these adaptations.

My interest is in the psychological adaptations that are uniquely human, the 10 percent or so of the brain's capacities that are not shared with other apes. This is where we find puzzling abilities like creative intelligence and complex language that show these great individual differences, these ridiculously high heritabilities, and these absurd wastes of time, energy, and effort. To accept these abilities as legitimate biological adaptations worthy of study, evolutionary psychology must broaden its view of what an adaptation should look like. At the moment, too many scientists are mis-describing effective fitness indicators like music and art as if they were nothing more than cultural inventions or learned skills. Their expression certainly depends on cultural traditions and years of practice, but other species with different genes cannot learn to do them no matter how hard they might try. If one banishes all these fitness indicators to the realm of "culture," then it does not look as if sexual choice had much impact on the human mind's evolution. But if one accepts fitness indicators as legitimate biological adaptations, then one starts to see the tracks of sexual selection all over our minds.

The Hominid That Wasted Its Brain

To sum up the last few sections, I think that the handicap principle casts a new light on the human brain. Everyone who proposes a theory about the brain's evolution mentions its costs. Our brains are only 2 percent of our body weight, but they consume 15 percent of our oxygen intake, 25 percent of our metabolic energy, and 40 percent of our blood glucose. When we spend several hours thinking really hard, or just conversing with people whose opinion matters to us, we get hungry and tired. Our brains cost a lot of energy and effort to run. Usually, theorists argue that these costs must have been balanced by some really large survival benefits, otherwise the brain could not have evolved to be so large and costly. But that survivalist argument holds only as long as one ignores sexual selection.

If we view the human brain as a set of sexually selected fitness indicators, its high costs are no accident. They are the whole point. The brain's costs are what make it a good fitness indicator. Sexual selection made our brains wasteful, if not wasted: it transformed a small, efficient ape-style brain into a huge, energy-hungry handicap spewing out luxury behaviors like conversation, music, and art. These behaviors may look as if they must be conveying some useful information from one mind to another. But from a biological viewpoint they might signify nothing more than our fitness, to those who might be considering merging their genes with ours.

The better our ancestors become at articulating their thoughts, the deeper the principles of wasteful sexual signaling could reach into their minds. By favoring fitness indicators, sexual choice demanded courtship behavior that stretched the mind's capacities. It demanded that which is difficult. It forced the human brain to evolve ever greater condition-dependence, and ever greater sensitivity to harmful mutations. It asked not what a brain can do for its owner, but what fitness information about the owner a brain can reveal.

Are Fitness Indicators Immoral?

The idea that the human mind evolved as a bundle of fitness indicators does not sit comfortably with contemporary views of human nature and human society. In fact, it violates at least eight core values commonly accepted in modern society. Variation in fitness betrays our belief in human equality. The heritability of fitness violates our assumption that social and family environments shape most of human development. Loudly advertising one's fitness violates our values of humility, decorum, and tact. Sexual status hierarchies based on fitness violate our belief in egalitarian social organization. The idea that people sort themselves into sexual pairs by assessing each other's fitness violates our romantic ideal of personal compatibility. The conspicuous waste demanded by the handicap principle violates our values of frugality, simplicity, and efficiency. The sexual choice mechanisms that judge individuals by their fitness indicators violate our belief that people should be judged by their character, not the quality of their genes. Finally, it seems nihilistic to propose that our capacities for language, art, and music evolved to proclaim just one message that has been repeated loudly and insistently for thousands of generations: "I am fit, my genes are good, mate with me." A mind evolved as a set of fitness indicators can sound like a fascist nightmare.

How is it possible for one biological concept to affront so many of our fundamental values? It seems quite astounding that a scientific idea should so consistently fall on the wrong side of the ideological fence. I think it is no coincidence. Look at it this way: our human norms and values developed as reactions to patterns of natural human behavior that we decided should be discouraged. If a great deal of human behavior consists of advertising one's fitness, and if many ways of doing that impose social costs on others, and if moral norms develop to minimize social costs, then a lot of moral norms should be aimed directly against the irresponsible use of fitness indicators. We value humility precisely because many people are unbearable braggarts who try to flaunt their fitness indicators so relentlessly that we cannot hold a decent conversation. We value frugality because so many people embarrass everyone with their ostentatious displays of luxuries, and waste limited resources that others need. We value egalitarianism because it protects the majority from aspiring despots intent on power and polygyny.

These norms do not just fall randomly from the sky. They emerged as moral instincts and cultural inventions to combat the excesses of sexual self-advertisement and sexual competition. Our moral aversion to fitness indicators may tempt us to reject them as an important part of sexual selection. But if we reject them, then it is hard to see how our moral norms evolved in the first place. It is possible, perhaps even necessary, to admit that much of human behavior evolved to advertise fitness, while simultaneously realizing that the essence of wisdom and morality is not to take our fitness indicators too seriously. This is not to say that our capacities for wisdom and morality are cultural inventions that liberate us from the imperatives of our genes. Our moral instincts may be just another set of evolved adaptations. It is not a question of "us" overriding our genetic predispositions, but of using one set of predispositions to overrule others—just as our evolved desire to preserve our looks can override our evolved taste for fat and sugar.

Another response to such worries is to point out that practically every theory of human mental evolution sounds like a fascist nightmare when we compare it with our comfortable modern lives and our political ideals. According to the Machiavellian intelligence theory, our minds evolved to lie, cheat, steal, and deceive one another, and the most cunning psychopaths became our ancestors by denying food, territory, and sexual partners to kinder, gentler souls. Richard Alexander's group warfare theory suggests that our minds evolved through genocidal violence, with larger-brained ancestors killing off smaller-brained competitors. The theory that human genes and human cultures co-evolved sounds slightly less bloody in the abstract, but it sounds that way only because it fails to specify any selection pressures that could have actually shaped anything. In terms of survival selection, what it boils down to is the view that those with brighter brains learned better technologies to grab resources before those with dimmer brains could, leaving the dimmer brains to starve, die of infectious disease, or be eaten by predators.

No theory of human origins can avoid the fact that evolution depends on reproductive competition, and competition means that some individuals win and some lose. With survival selection, the losers die. With sexual selection, the losers merely get their hearts broken (as their genes die out). If one demands moral guidance from a theory of human evolution, one is free to pick which of these options sounds better. Personally, I think that scientific theories should try to account for facts and inspire new research, rather than trying to conform to contemporary moral values.

5

Ornamental Genius

Sexual choice is mediated by the senses. We cannot use telepathy to pick sexual partners. We have to rely on the evidence of our eyes, ears, noses, tongues, and skin. Since the senses are the first filter for sexual choice, sexual ornaments evolved to play upon the senses. Biologists have started to analyze sexual ornaments as sound and light shows designed for sensory appeal.

Yet sexual choice also runs deeper than the senses. It depends on memory, anticipation, judgment, decision-making, and pleasure. Psychological preferences go beyond sensory preferences. For most species these more sophisticated psychological preferences probably do not matter very much. As far as we know, their sexual ornamentation has no way of activating ideas, concepts, narratives, or philosophies in the minds of other members of their species. Stimulating the senses is about as deep as they can go, because they have no communication system capable of conveying rich ideas. But after our ancestors evolved communication systems such as language, art, and music, psychological preferences may have become crucial in sexual selection.

Those preferences could have gone far beyond the eye's love of bright color and the ear's response to rhythm. They could have included mental quirks that make us prefer novelty to boredom, grace to clumsiness, knowledge to ignorance, logic to inconsistency, or kindness to meanness. If these quirks influenced the sexual choices that shaped the mind's evolution, then the mind could be viewed as an entertainment system that appeals to the psychological preferences of other minds. Just as some books become best-sellers for their contents rather than their covers, our ancestors attracted mates by displaying interesting minds, not just shapely bodies and resonant voices. Our minds may have evolved as sexual ornaments, but ornamentation is not limited to a superficial appeal to the senses. As far as sexual selection is concerned, creativity can be ornamental. Consciousness itself may be ornamental.

As we saw in the previous chapter, many sexual ornaments work as fitness indicators. But almost any trait that varies conspicuously and costs a lot can work as a fitness indicator. One important question is, which fitness indicators will evolve, out of the huge number possible? The runaway process cannot help us here, because it is arbitrary about what kinds of trait it favors. Sensory preferences might be more help in understanding which indicators evolve, because, by definition, they prefer some styles of ornamentation over others. This chapter reviews how biologists have been thinking about sensory preferences, and then generalizes their ideas to consider how psychological preferences may have influenced sexual selection among our ancestors. We shall also see how fruitful interactions occur between all three sexual selection processes we have been considering—runaway processes, fitness indicators, and, in this chapter, ornaments that appeal to the senses and the mind. When I go on to analyze specific human capacities such as art and creativity, I shall draw on all three of these ideas. They are not only complementary processes in evolution, but they offer complementary perspectives on the human mind.

The Senses as Gatekeepers

For an individual making a sexual choice, the senses are trusted advisors for making one of life's most important decisions. But for the individual being chosen, the chooser's senses are simply the gateway to the royal treasury of their reproductive system. The gateway may have heavy security. It may be guarded by decision-making systems that must be charmed or circumvented. It may respond only to secret passwords or badges of office. But it may be vulnerable to flattery, bribery, or threats. Like burglars learning about the security systems of banks, animals evolve courtship strategies to sneak through the senses of other animals, through the antechamber of their decision-making systems, into the vault of their reproductive potential. Every security system has weaknesses, and every sensory system used in mate choice can be stimulated by the right ornamentation.

Since the early 1980s, biologists have paid more attention to the role of the senses in sexual selection. This shift in focus was prompted by a radical paper by Richard Dawkins and John Krebs in 1978. They argued that when animals send each other signals, they are selfishly trying to influence each other's behavior. Signals are for the good of the sender, not the receiver. They are sent to manipulate behavior, not to convey helpful information. If the receiver's genetic interests overlap with the sender's interests, they may cooperate. The receiver may evolve greater sensitivity to the signaler's messages, and the messages may evolve to be quieter, simpler, and cheaper. Cells within a body have almost identical interests and strong incentives to cooperate, so intercellular signaling evolves to be very efficient. On the other hand, if the receiver's interests deviate from the sender's, signals will tend to become exploitatively manipulative. Predators may trap prey by evolving lures that resemble the prey's own favorite food. In defense, receivers may become insensitive to the signal. Prey may evolve the ability to discriminate between the lure and the real food. This may be why lures are so rare in nature.

Dawkins and Krebs realized that courtship is especially complicated because it is sometimes exploitative and sometimes cooperative. Typically, males of most species like sex regardless of their fitness and attractiveness to the females, so they tend to treat female senses as security systems to be cracked. This is why male pigeons strut for hours in front of female pigeon eyes, and why male humans buy fake pheromones and booklets on how to seduce women from the ads of certain magazines. On the other hand, females typically want sex only with very attractive, very fit males, so tend to evolve senses that respond only to signals of high attractiveness and high fitness. When a truly fit male courts a fertile female, they have a shared interest in successful mating. They both benefit. He produces more offspring, and she produces the best offspring she could. But there can also be conflicts of interest. When an unattractive, unfit male courts a female, he would gain a net benefit from copulation (extra offspring at minimal cost to him), but she would not. Her reproductive system would be monopolized producing his inferior offspring when she could have produced better offspring with a better male. So, the female's senses must remain open to courtship by attractive, fit males; but they must resist seduction by inferior males. She must be discriminating.

Sexual discrimination depends on the senses. But the senses may not be perfectly adapted for mate choice, because they must be used in other tasks of survival and reproduction. Primates have just one pair of eyes, which must serve many functions—finding food, detecting predators, avoiding collisions, caring for infants, and grooming friends, as well as discriminating between sexual partners. Visual systems embody design compromises because they fulfill several functions. Eyes for all trades cannot be masters of mate choice.

For example, primate color vision evolved in part to notice brightly colored fruit. The fruit evolved to spread its seeds by advertising its ripeness with bright coloration, to attract fruit-eaters such as primates and birds. Primates benefit from eating the fruit, so they evolve visual systems attracted to bright colors. The fruit's genes can reproduce only by passing through the digestive tract of a primate, so the ripe fruit's coloration is analogous to a sexual display. The fruit competes with the fruit of other trees to attract the primate's attention. Yet the fruit's sexual display can have side-effects on the sexual displays of the primates themselves, as a result of the primates' attraction to bright colors. (Eve's offer of the apple to Adam symbolizes the overlap between the sexual displays of fruit and those of primates.) If a male primate happens to evolve a bright red face, he might prove more attractive to females. He might catch their eyes, because their survival for millions of years has depended on seeking out ripe red fruit. Her senses are biased to notice bright colors, and this "sensory bias" may influence the direction that sexual selection takes.

Sensory Bias

The engineering details of sensory systems can influence the direction of sexual selection. Investigating these sensory details became a hot topic in the 1980s, but the research area has as many names as there are biologists. John Endler called it "sensory drive"; William Eberhard and Michael Ryan called it "sensory exploitation"; Amotz Zahavi called it "signal selection"; Tim Guilford and Marian Stamp Dawkins called it "the influence of receiver psychology on the evolution of animal signals." The most common term for the design of sensory systems driving the direction of sexual selection is "sensory bias," so I'll use that.

Sensory bias theory is a rapidly developing set of ideas that deserves much more research. It tries to ground the evolutionary study of animal signaling in the design of animal senses. It recognizes that there are always design compromises in animal sensory systems, and that these compromises sometimes make it possible to predict the direction in which sexual selection will go. It also suggests that there are many possible ways for a perceptual system to evolve a sensitivity to particular patterns of stimulation. The selection pressures on senses do not determine every detail of sensory system design: there are always contingent details about the responsiveness of senses that could not be predicted from their adaptive functions. These contingencies may influence the direction of sexual selection, by leading senses to respond more strongly to some stimuli than to others. Finally, sensory bias theory recognizes that senses evolve interactively with the signals they favor.

Displays Match Senses

The senses used for mate choice in each species tend to be well matched to the sexual ornaments displayed by that species. This is one piece of evidence consistent with sensory bias theory.

Michael Ryan found that in several Central American species of frog, female ears are most sensitive to the auditory frequency of male courtship calls. If female ears of one species hear best at 800 hertz, then the males of that species tend to produce calls at around 800 hertz. This is reasonable, given that females of these species must use the calls to locate suitable males in the forests of Central America. Male frogs calling at the wrong frequency would be harder to hear and harder to find, so would not produce so many offspring, and their genes for off-pitch calls would die with them.

Where there is a mismatch between frog ears and frog calls, Michael Ryan argued, the ears would exert sexual selection on the calls. Often, the female ears were more sensitive to calls slightly lower in pitch than the average male of their species was capable of producing. Females would find it easier to locate males who produced deeper-than-average calls, because they would be more audible. This should favor males who produce deeper calls. Ryan interpreted this as an example of sensory bias. The female senses are biased towards lower-than-average calls, and that bias appears to drive sexual selection.

However, this may just be an example of females favoring males of higher fitness. Larger frogs produce lower-pitched calls, so any female preference for larger frogs could be manifest as greater auditory sensitivity to lower-pitched calls. It may not be a sensory bias at all, but an adaptive way for females to discriminate between large and small males. Any mate choice mechanism that favors fitness indicators will look "biased" because it will not be most sensitive to the commonest sexual display in the current population. Instead, it will be most sensitive to the sexual display associated with the highest fitness. Nonetheless, it was useful for Michael Ryan to focus attention on call frequency as the relevant variable that connects the female senses to the male displays.

Senses as Engineering Compromises

A more significant claim from sensory bias theory is that animal senses have certain features that evolved just because they efficiently solve the information-processing problems of perception, and these features can drive sexual selection. Eyes have to perceive objects in general, and there may be general principles relevant to this task, principles which may influence mate choice.

Consider the area at the very back of the brain called the primary visual cortex, or "V1." This is the conduit for almost all information that passes between the eyes and the rest of the brain. Each V1 brain cell covers a tiny area of the visual world, and fires most actively when the local pattern of light in that area corresponds to the edge of an object. V1 seems to be a set of edge-detectors. Vision researchers believe that this is simply an efficient way to process visual information about the world, since vision is about seeing objects, and objects tend to have edges. This edge-detection principle has been used in most successful robot vision systems designed by humans.

Now consider how a male could grab the attention of a female's V1 system. He has to activate her edge-detectors. He could evolve a body that has many more real edges than average, perhaps a sort of fractal design. But the more real edges he has per unit of body volume, the more fragile his body would be and the more heat he would lose. Better to evolve sexual ornaments that display lots of fake edges. Dots would work, but thin parallel stripes would be even better, displaying more edge information per unit area. Perhaps stripes became popular sexual ornaments across many species because stripes are optimal stimuli for activating the visual cortex.

A similar explanation might account for the popularity of sexual ornaments with bilateral and radial symmetry. Biologist Magnus Enquist suggested that symmetric patterns might be the most exciting way to stimulate animal visual systems. He argued that any visual system capable of recognizing objects when they are rotated will tend to be "wired" in such a way that it is optimally excited by radically symmetric patterns. Enquist and his collaborator Arak did some evolutionary simulations in support of their claim that any neural network capable of recognizing rotated objects would be optimally excited by radially symmetric patterns. Supposedly, this explains the popularity of sexual ornaments that resemble stars, sunbursts, and eyespots.

In addition to fulfilling general engineering principles, the senses of each species must also adapt to its particular habitat and ecological niche. Sensory bias theorists such as John Endler have investigated how different lighting conditions influence the sensitivities of different animal visual systems. This sort of research promises to help biologists predict which animal lineages are more likely to evolve particular kinds of sexual ornament that play upon particular sensitivities. This application of sensory bias theory might help biologists to discern more patterns beneath the apparently chaotic proliferation of sexual ornaments in different species.

Yet a different view of sensory biases may explain why ornamentation evolves so unpredictably. For example, given the same problem of categorizing visual shapes, two different species may evolve two rather different solutions. One may evolve to represent visual shapes as variations on some sort of generalized cylinder, while the other may represent visual shapes as sets of facets and angles. Both ways of mentally representing shape may work perfectly well, but they might respond very differently to a novel sexual ornament that has a particular shape. The ornament might make an aesthetically pleasing generalized cylinder, but a very unappealing set of facets, or vice versa. The ornament may prove a sexual success in one species but not in the other.

One of the deepest insights from sensory bias theory is that there is always some evolutionary contingency in the design of perceptual systems. These contingencies make it impossible to predict all possible responses to all possible stimuli just from knowing what a perceptual system evolved to do. Therefore, if a new sexual ornament evolves that excites a perceptual system in a novel way, it may be favored by sexual selection in a way that could never have been anticipated. For example, biologist Nancy Burley found that female zebra finches just happen to be attracted to males that have tall white plumes glued on top of their heads. Their white-plume preference probably did not evolve as an adaptation, because as far as we know, ancestral finches never had white plumes on their heads. The preference just happened to be a latent possibility in a visual system that evolved for other purposes. I think this idea of evolutionary contingency in perceptual systems is one of the most intriguing ideas to come out of sensory bias theory. It might even work better than runaway sexual selection as a general explanation of why sexual ornaments diversify so unpredictably in different species.

From Sensory Appeal to Sexual Appeal

My main worry about sensory bias theory is that stimulating a sensory system is only the first step in influencing a mate choice decision. Grabbing a potential mate's attention is a long way from winning his or her heart. Granted, for animals that live widely separated from one another, it may take a lot of effort to find anyone of the opposite sex during the mating season. Under these conditions, making a strong sensory impression would give an animal a reproductive advantage. A whale song audible from hundreds of miles away can help two lonely whales to find each other. For many species, locating a mate—any mate—is a big problem. The sensitivity of their senses may be crucial to finding a mate, so may have a significant impact on sexual selection.

For highly social animals like most primates, finding potential mates is not the problem. Many primates already live in large groups, and interact regularly with other groups. They are spoiled for choice. When mate choice depends more on comparing mates than locating mates, the sensory engineering argument seems weaker. Why should an individual be perceived as a more attractive sexual partner just because its ornamentation happens to excite some brain cells in the lowest level of one's sensory systems? If it were that easy to make animals come running, predators would more often evolve lures to dupe prey into approaching them.

Our intuition may tell us that strong sensory effects are sexually attractive, but I doubt this attractiveness is explained entirely by sensory bias arguments. There are good adaptive reasons why ornaments that produce strong sensory effects make good fitness indicators. Consider the list of sensory bias effects that Michael Ryan and A. Keddy-Hector compiled in an important review paper of 1992. They noted that animals usually respond more strongly to visual ornaments that are large, brightly colored, and symmetrical, and to auditory ornaments (e.g. songs) that are loud, low in pitch, frequently repeated, and sampled from a large repertoire. These responses could be attributed to sensory engineering effects. But that begs the question of whether the sensory engineering evolved to help animals choose good sexual partners. Large, healthy, well-fed, intelligent animals can produce larger, brighter, and more symmetric visual ornaments, and louder, deeper, more frequent, and more varied songs. As far as I know, there is no example of a sensory bias that leads animals to favor sexual partners that are smaller, less healthy, less energetic, and less intelligent than average. Most sensory biases are consistent with what we would expect from adaptive decision-making machinery that evolved for mate choice. It may not have evolved specifically for mate choice, but it might as well have.

Many sexual ornaments may look as if they are merely playing on the senses. They may appear to be nothing but fireworks, sweet talk, eye candy, special effects, and manipulative advertising. But maybe we should give the viewers more credit. What look like sensory biases to outsiders may have a hidden adaptive logic for the animal with the senses.

Tickling Senses Versus Advertising Fitness

If sensory biases led animals to choose lower-fitness animals over higher-fitness animals, I suspect that the biases would be eliminated rather quickly. It seems unlikely that an ornament could persist as a pure sensory bias effect that does not convey any fitness information. That grants too much evolutionary power to males evolving ornaments and not enough to females evolving sensory discrimination abilities. Animals choosing mates do not want their senses subverted by meaningless ornaments. They may like fitness indicators that have a lot of sensory appeal, but they should not be favoring sensory appeal over fitness information.

Often there may be no conflict between sensory bias theory and fitness indicator theory. They are complementary perspectives on sexual selection. Sensory bias theory reminds us that mate choice is mediated by perceptual abilities, and that as new perceptual abilities evolve, the way is opened for new kinds of sexual ornaments to evolve. With the evolution of eyes came the possibility of visual ornaments. With the evolution of bird ears came the possibility of bird song. And perhaps, with the evolution of language comprehension abilities in our ancestors, came the possibility of sexual selection for much more complicated thoughts and feelings expressed through language.

Pleasure-Seekers

Biologists Tim Guilford and Marion Stamp Dawkins have argued that sensory bias theory can be generalized to deal with all sorts of psychological biases, which may also affect the evolution of animal signals and sexual ornaments. Any aspect of an animal's nervous system that influences how it reacts to a signal can influence how signals evolve. Apart from sensory biases, there can be attentional, cognitive, memory, judgment, emotional, and hedonic biases. These may be even more important in accounting for complex courtship behaviors of the sort that our species has evolved.

For example, maybe we can understand the mind as a sexually selected entertainment system that plays not just upon our sensory biases, but upon our thirst for pleasure. Consider two hypothetical kinds of animal. One has evolved some hard-wired brain circuits to do mate choice. It searches through several potential mates, remembers their ornaments and courtship behaviors, compares them using some decision algorithm, and picks one for copulation. It derives no pleasure from impressive ornaments to which it attaches a high value. It simply registers the value in an automatic, businesslike way. It has no hedonic experience. A good mate brings it no pleasure, only good genes.

It could be called the "cold chooser." I suspect that most insects work in this way.

The other animal is a "hot chooser." Its behavior may look similar, but its experience is very different. Its mate choices are influenced by subjective feelings of pleasure. When an attractive individual performs a charming courtship dance in front of the hot chooser, the hot chooser experiences some combination of aesthetic rapture, curiosity, warmth, happiness, awe, lust, and adoration. These feelings play a direct causal role in the mate choice process. The more pleasure a potential mate arouses in the hot chooser, the more likely that individual is to be chosen.

Given this description of cold and hot choosers, there is no way an external observer could tell them apart. Now I shall add the crucial feature that makes an observable difference, a difference that could influence evolution through sexual selection. Suppose that the pleasure system the hot chooser uses for mate choice is the same pleasure system it uses for all other domains of survival and reproduction. The hot chooser has a big pleasure-meter in its brain—it may be something like the level of endorphins floating around its nervous system. Its pleasure in watching an attractive male is subjectively similar to its pleasure in eating good food, escaping a dangerous predator, viewing a propitious landscape, watching its children thrive, or doing anything else that contributes to survival or reproduction. All of its decisions are mediated by this pleasure-meter.

Over the short term, the cold chooser and the hot chooser will behave in the same way. They will make the same mate choices. But over the long term, they can evolve in different directions because they will react differently to new courtship behaviors. Suppose that a male happens to have a mutation that leads him to give good food to a female. A cold-choosing female may eat the food, but the food might not influence her mate choice, because her eating system is separate from her mate choice system. Her systems do not share the common language of pleasure. The mutant may have no reproductive advantage, and his food-giving tendencies will probably die out. (Females of many species have evolved preferences for food gifts during courtship; my point here is that they may not have automatically wanted to mate with the first generation of males that offered food.)

If the same food-giving mutation arises in the hot chooser species, the female's reaction would be much more positive. The mutant gives her food, which increases her pleasure in his company. Since her pleasure-meter is what determines her mate choice decisions, she favors the food-giving mutant. He gains an immediate reproductive advantage over his competitors. The gene for food-sharing spreads through the population because it brings pleasure, and pleasure influences mate choice. The hot choosers would equally favor any novel courtship behavior that saved them from predators, or led them to a rich new habitat, or helped their existing children thrive, or brought them any other kind of pleasure.

Why would any animal evolve a pleasure-meter? I think that the main benefit of a unified pleasure system is that it simplifies learning by allowing the hot chooser to use similar kinds of reinforcement learning in many different contexts. If it feels pleasure when eating, it can use that pleasure as a reinforcement signal to tell it to do more of the foraging strategy that was just successful. If it feels pleasure when copulating, it can use that pleasure as a reinforcement signal to make more use of the mate choice strategy that was just successful. Designers of robot control systems have realized that smart robots need reinforcement learning abilities. Moreover, artificial intelligence researcher Pattie Maes has argued that when robots need to juggle many priorities, a central pleasure system can help them rank those priorities. Pleasure helps solve the problems of reinforcement learning and prioritizing behaviors.

The stern sensory bias theorist might warn that this sort of pleasure system makes the hot choosers vulnerable to sexual exploitation. Courtship behaviors would evolve that simply activate the pleasure centers, influencing the hot choosers to mate with their manipulators. That sounds bad. But is it? In terms of the subjective experience of the hot choosers it cannot be bad, because activation of their pleasure centers is, by definition, pleasurable. As long as pleasure is defined broadly enough, to encompass everything from a full belly to a fulfilled life, an individual cannot wish for any subjective experience beyond pleasure. For utilitarians who value the greatest happiness for the greatest number, sexual selection driven by pleasure is a dream come true.

The real question is whether pleasure-giving courtship imposes any evolutionary costs on the hot choosers. If it did, the hot choosers would evolve a barrier between that form of pleasure and their mate choice system. However, pleasure is not arbitrary in the way that some sensory biases may be arbitrary. Pleasure systems evolve for a reason: they encourage animals to do things that improve their survival and reproduction prospects. Food brings pleasure because our bodies require energy. Predators bring displeasure because they want to kill us. If a hot chooser's pleasure systems are well calibrated, any courtship behavior that brings it pleasure will increase its fitness somehow. The behavior brings evolutionary benefits, not evolutionary costs.

The only remaining worry is that pleasure-giving courtship might not be a very good indicator of an individual's fitness. A hot chooser might favor pleasure at the expenses of good genes. If good genes are very important, and if pleasurable courtship does not correlate with good genes, then the hot choosers should evolve a defensive barrier between their pleasure system and their mate choice system. But I don't think that such a defense would usually be necessary. Remember the basic requirements for a fitness indicator: it should vary perceptibly, and it should be sufficiently costly that low-fitness pretenders cannot fake it. Pleasurable systems evolved in the first place as discriminatory systems very sensitive to variation between situations, so noticing individual variation between sexual prospects should not be a problem.

So how costly is it to give pleasure? If the pleasure comes from gaining a significant fitness benefit such as food, shelter, protection, or access to good territory, then the pleasure-giver probably incurred significant costs to acquire such a gift. If the pleasure comes from dextrous grooming, brilliant conversation, attentive foreplay, or prolonged copulation, there are time, energy, and skill costs. Giving pleasure is generally harder than exploiting sensory biases, because pleasure has to reach much deeper into the receiver's brain. For this reason, pleasure-giving courtship behavior is probably a better fitness indicator than courtship that merely activates sensations.

Pleasure-giving is rather different from sensory exploitation. It feels better, it is better at tracking fitness benefits given to oneself, and it works better as a fitness indicator. Hot choosers that use pleasure to mediate mate choice are not more evolutionarily vulnerable than cold choosers. On the contrary, they are better positioned to let sexual selection take them off in new evolutionary directions where unknown pleasures await.

The Ornamental Mind

As discussed in Chapter 1, traditional theories viewed the human mind as a set of survival abilities. The dominant metaphors for mental adaptations were drawn from military and technical domains. Cognitive science views the mind as a computer for processing information. Many evolutionary psychologists view the mind as a Swiss army knife, with distinct mental tools for solving different adaptive problems. Some primatologists view the mind as a Machiavellian intelligence center devoted to covert operations.

Our discussion of sensory bias theory and pleasure leads to a different view. Perhaps we can do better by picturing the human brain as an entertainment system that evolved to stimulate other brains—brains that happened to have certain sensory biases and pleasure systems. At the psychological level, we could view the human mind as evolved to embody the set of psychological preferences our ancestors had. Those preferences were not restricted to the surface details of courtship like the iridescence of a peacock's tail; they could have included any preferences that lead us to like one person's company more than another's. The preferences could have been social, intellectual, and moral, not just sensory.

This "ornamental mind" theory leads to some quite different metaphors drawn from the entertainment industry rather than the military-industrial complex. The mind as amusement park. The mind as a special-effects science-fiction action film, or romantic comedy. The mind as a Las Vegas honeymoon suite. The mind as a dance club, cabinet of curiosities, mystery novel, computer strategy game, Baroque cathedral, or luxury cruise ship. You get the idea.

Psychologists who pride themselves on their seriousness may consider these metaphors trivial. To them, the mind is obviously a computer that evolved to process information. Well, that seems obvious now, but in 1970 the mind as a computer was just another metaphor. It was just slightly better than Sigmund Freud's metaphor of the mind as a hydraulic system of liquid libido, or John Locke's metaphor of the mind as a blank slate. The mind-as-computer helped to focus attention on questions of how the mind accomplishes various perceptual and cognitive tasks. The field of cognitive science grew up around such questions.

However, the mind-as-computer metaphor drew attention away from questions of evolution, individual differences, motivation, emotion, creativity, social interaction, sexuality, family life, culture, status, money, power, birth, growth, disease, insanity, and death. As long as you ignore most of human life, the computer metaphor is terrific. Computers are human artifacts designed to fulfill human needs, such as increasing the value of Microsoft stock. They are not autonomous entities that evolved to survive and reproduce. This makes the computer metaphor very poor at helping psychologists to identify mental adaptations that evolved through natural and sexual selection. "Processing information" is not a proper biological function—it is just a shadow of a hint of an abstraction across a vast set of possible biological functions. The mind-as-computer metaphor is evolutionarily agnostic, which makes it nearly useless as a foundation for evolutionary psychology. At the very least, the metaphor of the mind as a sexually selected entertainment system identifies some selection pressures that may have shaped the mind during evolution.

This entertainment metaphor suggests that the human mind shares some features with the entertainment industry. The mind has to be open for business, with a clean, safe, welcoming interior. It needs good public access routes and good advertising. It must provide a world of stimulation, ideas, adventure, interaction, and novelty set apart from the ordinary world of tedium, toil, and threatening uncertainty. It must capture the right market niche, and respond to changing consumer tastes. The mind hides the appalling working conditions of its employees (the energy-hungry brain circuits) to provide attentive, smiling service for visitors. Like the future dystopia in H. G. Wells's *The Time Machine*, the Eloi of leisured ideas appear on the surface of consciousness, while the Morlocks of cognitive effort are imprisoned underground.

If the ornamental mind theory has any merit, then the functional demands that evolution has placed on the human mind have been misunderstood. The entertainment industry does not operate like a military campaign. As Darwin realized, sexual selection does not work like survival of the fittest. All of the criteria of success, the strategies, the resources, and the modes of competition are different.

Viewed from a military point of view, Hollywood is a failure. It hasn't even managed to annex the San Fernando valley, or invade Santa Monica, or bomb Santa Barbara, or establish a secret alliance with Tijuana. Its standing army is just a few hundred studio security guards, and it has no navy or air force. Its people are undisciplined, vain, soft, and prone to fantasy. They live on salad. They would be no match for the Spartans, the Mongols, or the British SAS. This is all true, but rather misses the point. If the human mind evolved as an entertainment system like Hollywood, those of its features that look like military-competitive weaknesses may actually be its greatest strengths. Its propensity for wild fantasy does not undermine its competitive edge, but attracts enormous interest from adoring fans. Its avoidance of physical conflict allows it to amass, quietly and discreetly, enormous resources and expertise to produce ever more impressive shows. Its emphasis on beauty over strength, fiction over fact, and dramatic experience over plot coherence, reflects popular taste, and popular tastes are what it lives on. Its huge promotional budgets, costly award shows, and conspicuously luxurious lifestyle are not just wasteful vanity—they are part of the show. Its obsession with fads and fashion do not reflect victimization by exploitative memes, but the strategic appropriation of cultural ideas to promote its own products.

Profit is Hollywood's bottom line, and everything about it that would look baffling to Genghis Khan makes perfect sense to entertainment industry analysts who understand what produces profit. To understand the human mind's evolution, we have to remember that reproductive success is evolution's bottom line. The mind makes very little sense as a Swiss army knife or a military command center. It makes more sense as an entertainment system designed to stimulate other brains, and the ornamental mind theory captures that intuition.

The Space of All Possible Stimulation

The entertainment industry can be viewed as an attempt to explore the space of all possible stimulation that can excite the modern human brain. Every movie, every book, every painting, every music CD, and every computer game is a set of potential stimuli that may or may not work. The human brain is fickle: it responds much more positively to some stimulation than to other stimulation. Nobody knows in advance what stimulation will work, though some can make some good guesses. If evolutionary psychologists like me could make solid predictions about exactly what stimulation patterns would optimally excite the human brain, we could just move to Hollywood and become highly paid entertainment industry consultants. But we cannot do much better than ordinary film producers, because a general understanding of typical human reactions to ancestrally normal events does not allow us to predict the human brain's exact reactions to any possible novel stimulation. Modern human culture is a vast, collaborative attempt to chart out this space of all possible stimulation, to discover how to tweak our brains in pleasurable ways.

The ornamental mind theory suggests that human evolution, like the entertainment industry, pursues promising lines of stimulation that might bring rewards for the producer. Sexual selection explores this space of all possible stimulation, reaching into the perceiver's brain and gauging what excites a positive reaction. Sexual evolution navigates through the brain-space of each species, in search of mutual pleasure and reproductive profit.

Imagine a species that stumbles into an evolutionary utopia in which sexual selection is no longer driven by male competition for dominance and display, but by mutual choice for mutual pleasure. The males who deliver the greatest rapture to females are sexually favored, passing on the pleasure-giving abilities to both sons and daughters. Equally, those females who deliver the greatest bliss and contentment to males are favored, passing on their pleasure-giving abilities to their offspring. Each generation provides more pleasure than the last, and receives more. The species spirals upward into rapture, leaving behind all the genes for unpleasantness, unkindness, inattentiveness, and poor foreplay.

If only. The trouble with mutual choice for mutual pleasure is that all the genes for unpleasantness come aboard as stowaways. Mutual choice implies that individuals sort themselves out in a mating market. As a thought experiment, imagine for the moment that mating is perfectly monogamous. The best pleasure-giving female pairs up with the best pleasure-giving male. Both have their sexual preferences fulfilled, and they live in bliss and produce pleasure-giving children. But their competitors do not just give up and die of embarrassment at the inferiority of their foreplay. Moderately pleasant females mate with moderately pleasant males, because neither can do any better in the mating market. And the most unpleasant females mate with the most unpleasant males, because their only alternative would be to remain single. All else being equal, they will all have children too. In fact, assuming monogamy, the genes for pleasure-giving will not have any reproductive advantage whatsoever over the genes for imposing unspeakable misery on one's sexual partner.

Mutual choice for mutual pleasure will determine which sexual relationships form, but will not increase pleasure from one generation to the next. The sexual choice would not result in any real sexual selection. It would reshuffle genes but would not change which genes persist in the population. It would not make evolution happen. Given monogamy, mutual choice for pleasure is only pseudo-selection. It looks like sexual selection, but it doesn't change genes like sexual selection.

Pleasure alone is not enough. We need either more sexual competition than monogamy provides, or some interaction between sexual selection for entertainment and other sexual selection processes. The ornamental mind theory tends to overlook the interactions between brains as entertainment producers and brains as entertainment consumers. We must remember the possibility of runaway effects, where entertainment consumers become more and more demanding. The ornamental mind theory also ignores the problem of consumer boredom. On evolutionary time-scales, consumers may simply lose interest in useless stimulation. They may simply walk out of sexual selection's amusement park if their sexual choices are not delivering good genetic value. In modern human culture, consumers can be treated as passive systems with stable tastes that can be exploited. But in evolution, entertainment-consumers can evolve as fast as entertainment-producers can. Neither has the upper hand. We have to put the ornamental mind theory together with the fitness indicator theory to explain why some sexual ornaments stick around.

Putting the Pieces Together

On its own, the idea of ornamental evolution through sensory biases has about the same number of strengths and weaknesses as the runaway brain theory and the healthy brain theory do. We probably need to combine all three perspectives to understand human evolution. I would not have spent a whole chapter on the runaway process if I did not think it was important in explaining the capricious divergence of courtship behavior between different ape and hominid species. I would not have spent a chapter on fitness indicators if I did not think that the pressure to advertise good genes was important in mental evolution. And I would not have discussed sensory biases, pleasure, and entertainment if I did not think that the psychological quirks of our ancestors had influenced our psychological capacities through the sexual choices they made.

Later, when I come to discuss particular human abilities like language and creativity, I shall draw on all three viewpoints. Biologists sometimes compare runaway theory, indicator theory, and sensory bias theory as if they were competing models of sexual selection. Such debates helped revive sexual selection theory, but I think that each of the theories now has enough support for them to be considered as overlapping sexual selection processes, not competing models. They all really happen in nature.

Runaway happens because sexual preferences really do become genetically correlated with the sexual ornaments they favor. It helps to explain human mental traits that are extreme, unusual, attractive, and useless for survival, and why such traits evolved in our lineage and not in other ape species. Runaway is endemic to sexual selection, always happening, or just finished, or just about to happen. It explains much of sexual selection's power, speed, and unpredictability.

Sexual ornaments really do evolve higher costs and higher condition-dependence in order to work better as fitness indicators. Indicator theory explains why some sexual ornaments stick around for many generations rather than disappearing as transient runaway effects. It gives sexual selection much of its direction, explaining why individuals usually prefer large tails to small, loud calls to whispers, good territories to bad, winners to losers, health to sickness, and intelligence to stupidity.

Sensory biases really do influence in which direction runaway is most likely to go, and which indicators are most likely to evolve. Sexual selection for pleasure and entertainment explains why so many sexual ornaments like the human mind are pleasing and entertaining. It draws attention to the role of sensation, perception, cognition, and emotion in sexual choice.

How Ornaments and Indicators Interact

Any particular trait that evolved through sexual selection was probably influenced by some combination of runaway processes, pressures to advertise fitness, and psychological preferences. Most sexually selected traits probably work as both ornaments and indicators. Some elements of their design evolved to provide hard-to-fake information about fitness; others evolved just because they happened to be exciting and entertaining. To understand the human mind as a set of sexually selected traits, we have to envision how ornamental and indicator functions can exist side by side in the same trait.

An indicator must accurately indicate a particular quantity. But this requirement does not determine every aspect of an indicator's design: there are always many design elements that are free to vary in ornamental ways. Almost all car speedometers can successfully indicate the car's speed, but there are hundreds of different speedometer designs used in different makes and models of car. All wristwatches indicate the time, but different watch designs may vary in every possible detail according to the aesthetic tastes of manufacturers and consumers. As long as speed, time, or some other indicated quantity is more or less intelligible, the indicator's design is free to vary according to aesthetic whimsy, exploring the fringes of ornamental style.

Actually, the handicap principle makes sexually selected traits a bit more constrained than watch designs. The Rolex Corporation has no incentive to mislead its customers about the time. Animals do have incentives to mislead potential mates about their fitness. Coins make a better analogy for sexually selected traits than do watch-faces. Numismatists are familiar with the two criteria of successful coins: they are hard to counterfeit (a requirement that increases with their monetary value), and they are attractive to the eye and the hand. Coins indicate value just as watches indicate time. But with coins there is a much greater incentive for fakery.

Counterfeiting has been a concern ever since 560 B.C., when King Croesus of Lydia invented true official coinage (government-issued cast disks of standard weight, composition, and guaranteed value). To guard against counterfeiting, authorities produce coins according to the handicap principle. They endow coins with features that would be prohibitively expensive for a counterfeiter with low capital to imitate. In the ancient era, it was usually sufficient to produce coins with hard-to-make iron coining dies. By the 17th century, authorities had to invest in expensive rolling mills, sizing dies, and blanking presses to deter counterfeiting. The modern principles of coinage—accuracy of dimension, perfect reproduction of design, standard weight of an easily tested alloy—all evolved to make coins accurate indicators of monetary value.

And yet there has been enormous scope for coins to vary in ornamental ways. This ornamental elbow room is what gives numismatics its interest, just as sexual selection gives biodiversity its fascination. Ancient Greek coins, though commonly made of precious-metal alloys to a common basic design, were ornamented in different ways depending on the city-state of origin: owls for Athens, bees for Ephesus, the griffin for Abdera, the eagle of Zeus for Olympia, the lion of Leontini, the minotaur of Knossos, the quince of Melos, the silver-miner's pick at Damastium, the grapes of Naxos. The requirement that the famous Sicilian decadrachm of 480 B.C. must properly indicate its value did not determine its beautiful ornamentation, with triumphal chariots above a fleeing lion (symbolizing the recently conquered Carthage) on one side, and, on the other side, Arethusa. (Arethusa was a water nymph who escaped unwanted sexual attention from the river-god Alpheios by asking Artemis to transform her into a freshwater spring—an evolutionarily counterproductive way to exercise female mate choice.) Within a few years of the invention of coinage, Greek city-states were not just worrying about overcoming counterfeiting; they were competing to make coins beautiful. While there were just a few principles to guarantee a coin's value, there were an infinite number of ways to ornament it with a pleasing design.

The principles of coinage, like those of sexual selection, are not just economic but aesthetic. While the economic principles of value-indication tend to produce similarities between coins, the aesthetic principles are more creatively protean, producing endless diversity. To understand the features of any given coin, it is not enough to appreciate the general requirements of money (durability, divisibility, portability), or the particular anti-counterfeiting principles of coinage (standard size, weight, composition, and design). One must also appreciate the aesthetic imperatives, from the universal sensory demands of the human hand and eye, to the historically contingent symbolism of a particular culture. Likewise for a sexually selected trait—one must understand how certain features indicate an animal's fitness, and how other features evolved as aesthetically pleasing ornaments, just because they happened to excite the senses and brains of the opposite sex. As anti-counterfeiting principles rarely suffice to explain every detail of a coin, in almost no case of a sexually selected trait does the handicap principle alone suffice to explain every detail. There is always some aesthetic slack.

In sexual selection, traits that began as indicators tend to grow more complexly ornamental because the sensory preferences of the opposite sex partially impose their own aesthetic agenda on the indicator. Conversely, traits that originate as pure runaway ornaments tend to acquire value as fitness indicators because aesthetically impressive ornaments tend to be costly and difficult to produce. Almost all sexually selected traits that last more than a few hundred generations probably function both as indicators and as ornaments. They may have originated mainly as one or the other, but soon imposed sufficient costs that they indicated fitness accurately, and soon acquired enough aesthetic complexity that they stimulated the senses of the opposite sex in ways that could not be reduced to indicating fitness.

The messy overlap between indicators and ornaments does not mean that we can afford to get messy about sexual selection theory. Zahavi's handicap principle is quite distinct from Fisher's runaway process. But they frequently work together, so we should not worry too much about trying to categorize every sexual trait as either an indicator or an ornament. Instead, we should use different models of sexual selection as lenses to view a given trait from different angles and different distances, to answer different evolutionary questions. The fitness-indicator principles are good at explaining why animals of a given species have such a strong consensus about what they like in a sexual trait: why all peahens like the peacocks to have large, symmetric, bright, many-eyed tails. The fitness-indicator perspective explains the perfectionism and conservatism of sexual tastes within each species. It also explains why large, long-lived animals have not degenerated to extinction under the pressure of harmful mutations. On the other hand, the ornamental principles are good at explaining why animals of different species develop such different tastes: the tails that attract peahens, for example, are not turn-ons for female turkeys or female albatrosses. The ornamental perspective explains the protean divergence of sexual tastes across species over macro-evolutionary time. It also explains why sexually reproducing life on our planet has split apart into millions of different species.

The ornamental view is especially important for appreciating the role of evolutionary contingency in shaping sexual traits, just as it is in appreciating the role of historical contingency in shaping coins. Once King Croesus invented official coinage, we could have predicted that most city-states of the ancient Mediterranean world would adopt coins, would make them hard to counterfeit, and would ornament them with some pleasing designs. However, we could not have predicted that the coin-engraver's art would reach its peak in 5th-century B.C. Syracuse, on the island of Sicily. It could have happened at some other time in Carthage, Crete, or Athens, but it didn't.

Likewise for the products of sexual selection. We can see that, once sexually reproducing animals evolved the capacity for mate choice, every animal species would then evolve some sort of fitness indicator; and that some indicators might be costly, exaggerated body parts, and others would be costly, ritualized courtship behaviors. But we could not have predicted that courtship behavior would reach an especially high degree of sophistication exactly 535 million years after the Cambrian explosion (when multicellular animals proliferated) in our particular species of bipedal ape. Nor could we have predicted that the courtship behavior would take the precise form of interactive conversations using arbitrary acoustic signals (words) arranged in three-second bursts (sentences) according to recursive syntactic rules. Perhaps it could have happened in an octopus, a dinosaur, or a dolphin. Perhaps it was likely that it would happen sometime, in some species of large-brained social animal. Rewind the tape of evolution, and the human mind would probably not have evolved, because sexual selection would have taken a different contingent route in our lineage of primates. But I suspect that in any replay of evolution on Earth, sexual selection would sooner or later have discovered that intelligent minds similar to ours make good courtship ornaments and good fitness indicators.

Sexual Selection, Natural Selection, and Innovations

The interaction of the three major sexual selection processes can explain sexual ornaments. Less often appreciated is how they can interact with natural selection for survival to produce evolutionary innovations. To understand any specific innovation such as the human mind, it may help us to look at what role sexual choice might play in the evolution of innovations in general.

The history of life on Earth is marked by major evolutionary innovations such as the evolution of DNA, chromosomes, cell nuclei, multicellular bodies, and brains. Classic examples of moderately important innovations include legs, eyes, feathers, eggs, placentas, and flowers. Much more frequent are the minor innovations that distinguish one species from another. These micro-innovations are often no more significant than a different mating call or an unusually shaped penis.

The major innovations give their lineages such an advantage in exploring new niches that they result in a burst of biodiversity called an "adaptive radiation." The first species that suckled its young with milk ended up being the ancestor of all 4,000 species of mammal. The first ape that walked upright became the ancestor of a dozen or so species of hominid, including us. Every major group of organisms (such as a phylum or an order) has a major innovation at its root. Every medium-sized group (such as a class or a family) has a moderately important innovation at its root. Every species is distinguished by some micro-innovation. The tree of life is a tree of evolutionary innovations.

It remains to be seen how important the human brain is as an evolutionary innovation. If we became extinct tomorrow, it would count as a micro-innovation characteristic of just one species. If our descendants succeeded in colonizing the galaxy and splitting apart into a hundred thousand species millions of years from now, it would count as a macro-innovation. But an innovation's ability to trigger an adaptive radiation millions of years after its origin cannot explain why it evolved. This raises a serious problem that has remained unsolved ever since Darwin: how can innovations emerge through a gradual process like natural selection? This question has three variants of increasing difficulty.

The easy, most general problem is: how can a qualitatively novel structure arise through gradual, quantitative changes? The answer, of course, is that the whole universe unfolds by processes that turn quantitative change into qualitative novelty. The incremental process of gravitational attraction turns interstellar dust clouds into star systems. The incremental processes of capital investment and education turns poor villages into prosperous cities. The incremental process of growth turns a fertilized egg into a human baby. There is nothing special about evolution in this respect. Every thing in the world that we bother to name is a bundle of qualitatively novel properties emerging from an accumulation of quantitative stuff.

The moderately hard problem is: how can a complex innovation emerge that depends on many parts functioning together? Assuming that natural selection can tinker with only one part at a time, it seems difficult for natural selection to construct multipart innovations. What good is the retina of an eye without the lens, or vice versa? This sounds like a lethal argument against incremental Darwinian evolution, but it isn't. If it were, the existence of Microsoft would force one to be a Creationist. The Microsoft Corporation is composed of thousands of employees who must all work together for the corporation to function: management, accounting, personnel, marketing, finance, programming, and so forth. Could Microsoft have arisen through the incremental accumulation of employees, hiring them one by one? It seems logically impossible. If employee number one was a programmer, the corporation couldn't survive, because there would be no one in marketing to sell her product, no one in personnel to pay her, and no one in the legal department to sue software pirates. But if employee number one was in marketing, she wouldn't have any product to sell. And so on. How could a corporation that includes dozens of different kinds of employee possibly have emerged in just twenty years through incremental hiring? The answer is that the early employees were less specialized, and each filled many roles. When Microsoft consisted of just the teenaged Bill Gates and Paul Allen, they split all the corporate responsibilities between them. As more employees were hired, responsibilities were delegated and became more specialized. If one accepts the possibility of growing large, multi-part corporations by hiring one person at a time, perhaps one should not be too bothered by evolution's ability to produce innovations by compiling one genetic mutation after another. As far back as the 1850s, Herbert Spencer was pointing out that gradual growth through progressive differentiation and specialization is the way that both social organizations and biological adaptations must evolve.

The Threshold of Innovation

The really difficult problem is: how can natural selection favor the initial stages of evolutionary innovations when they are accumulating costs but not yet offering any net survival benefits? Darwin worried a lot about this problem. How could natural selection favor proto-eyes or proto-wings before they grow sufficiently large and complex to yield their survival benefits?

Selection is frugal: it penalizes traits that impose costs without offering benefits. If most innovations give net survival benefits only once they have passed some threshold of complexity and efficiency, it is hard to see how evolution could favor them before they reached that threshold. This has always been the single most serious objection to Darwin's theory of evolution by natural selection. It was argued most forcefully by the zoologist St. George Mivart just after *The Origin of Species* was published, and it has been a stumbling block ever since.

Some minor innovations do not suffer from this threshold effect. A giraffe's neck could have evolved to its present length gradually, each increment of length giving an immediate improvement in reaching higher acacia tree leaves. An insect's camouflage could evolve gradually, each step further reducing a predator's chance of noticing the insect. Neck-stretching and color-changing could provide net survival benefits continually throughout their evolution.

Some evolutionary theorists such as Richard Dawkins and Manfred Eigen suggest that the threshold effect is overstated for many major innovations. They think that there are often ways to evolve dramatic innovations along a continuous path where every step right from the beginning yields a new survival benefit. They might be correct. We do not know enough about the evolutionary dynamics of complex traits to know how common the threshold problem is. Most biologists still believe this to be the most significant problem that theories of evolutionary innovation must address. I agree. In my experience with running genetic algorithm simulations on computers, the threshold problem is a very serious obstacle to evolving innovations. If you actually try to evolve something complicated and useful inside a computer using simulated natural selection, you are likely to be frustrated. Simulated evolution often stalls for no apparent reason, gets stuck in a rut for thousands of generations, and shows a perverse tendency to avoid interesting innovation whenever possible. This frustration with simulated evolution's limited innovation ability is fairly common among genetic algorithm researchers.

The threshold effect boils down to this: the evolutionary costs and benefits of innovations work like the economics of pharmaceutical research. The Pfizer Corporation spent over $100 million and many years developing the drug Viagra before the drug made a single cent of profit. The costs accumulated early, and the benefits came only later. Drug companies can cope with this delayed gratification, and have the foresight to undertake the research that leads to such profitable innovations. But evolution has no foresight. It lacks the long-term vision of drug company management. A species can't raise venture capital to pay its bills while its research team tries to turn an innovative idea into a market-dominating biological product. Each species has to stay biologically profitable every generation, or else it goes extinct. Species always have cash-flow problems that prohibit speculative investment in their future. More to the point, every gene underlying every potential innovation has to yield higher evolutionary payoffs than competing genes, or it will disappear before the innovation evolves any further. This makes it hard to explain innovations.

Sexual Selection and Venture Capital

Let's go back to the Microsoft example. We saw that large corporations could grow from a couple of entrepreneurs by hiring employees one at a time. Evolution's threshold problem is more of a finance problem than a personnel problem. How did Microsoft grow large enough to reach the threshold of profitability? Like most companies, it survived in the early days through bank loans, venture capital, and stock issues. It didn't grow just from the profits it made. It grew because people were willing to lend it money in the hope that they would get paid back in the future. The problem in growing large corporations is not that you have to hire people one by one—that's the easy part. The problem is that most corporations can't break even until they reach a certain critical mass, and they can reach that mass only by borrowing money against their future profits.

Evolution seems to offer no mechanism to do this when there is a potential to develop a major innovation. Capitalism depends on foresight, and evolution has no foresight. The problem of evolutionary innovation boils down to this: evolution needs something like a venture capitalist. It needs something that can protect the very early stages of an innovation against the ravages of the competitive market and the laws of bankruptcy, by granting it some line of credit.

Sexual selection works, I think, as evolution's venture capitalist. It can favor innovations just because they look sexy, long before they show any profitability in the struggle for survival. It can protect the early stages of innovations by giving them a reproductive advantage that can compensate for their survival costs. Of course, this is a risky business. Most innovations may never show any profit, and may never yield any survival advantages. But they don't have to. Venture capitalists can make money when a company floats stock on the stock market, even if the company never sells a single product. Runaway sexual selection can favor evolutionary innovations that never offer a single survival benefit. Both processes work through the magic of runaway popularity. Desire reinforces desire. A confidence bubble grows.

Sometimes the bubble bursts. For every courtship ornament like the peacock's tail that persists, perhaps dozens of ornaments come and go. These ornaments may originate in humble form, become popular for a while, grow a little in complexity and size, and then become unfashionable through various random evolutionary effects, sinking back into evolutionary oblivion. These ornamental fashion cycles may not be good for the species as a whole, but evolution cares no more about the species as a whole than capital markets care about entrepreneurs.

Why Is Evolutionary Innovation Obsessed with Male Genitals?

If many innovations originate through sexual selection, we would expect most micro-innovations that distinguish one species from another to be sexual ornaments. This contradicts some traditional views of how species split apart, but, surprisingly, this is pretty much what biologists see. The vast majority of species-defining innovations seem inconsequential for survival. Francis Bacon, father of the scientific method, disparaged the seemingly pointless variety of plants and animals, calling them "the mere Sport of Nature." Darwin was equally perplexed, often wondering why there was so much variety but so little real novelty. If innovations spread through populations because of their survival benefits, why do so few innovations show the survival improvements associated with major innovations and adaptive radiations?

One clue comes from the criteria that taxonomists use to classify specimens into species. Male sexual ornaments and male genitals are the most useful traits for distinguishing most animal species from closely related members of the same genus. If you can't tell whether a beetle is one species or another, look at its color pattern, its weaponry, and its genitals. In his book *Sexual Selection and Animal Genitalia*, William Eberhard emphasized that male genitals are often the first things to diverge when one species splits off from another. Evolutionary innovation seems focused on the details of penis shape. In Eberhard's view, this is because female choice focuses on the details of penis shape, and female choice apparently drives most micro-innovation. In plant taxonomy, the analogous sexually selected traits are the flowers, and they are often most useful in making species identifications. It is often harder to tell what species a female animal is, because the appearance of females diverges much less between species. Bird watchers know this: given a female, you can often only identify the genus, but given a male, you can zero in on the exact species.

The micro-innovations that distinguish species often evolve through sexual selection, as sexual ornaments (or genitals) shaped by mate choice. At one level, this fact simply restates the modern definition of a biological species: a reproductively isolated group of individuals. The commonest kinds of traits that distinguish species must be traits that can work as sexual isolators to keep one group from interbreeding with other groups. Sexual choice is a very efficient sexual isolator for keeping species distinct. As the biologist Hugh Paterson pointed out in the 1970s, species are basically consensual systems of mate choice. The result is that human taxonomists end up using the same traits to distinguish species that species members themselves use: sexual ornaments. This is why most micro-innovations are concentrated in genitals, ornaments, and courtship behaviors.

Innovation Through Sexual Choice

Sexually selected novelties of this sort could be called "courtship innovations." Most will be nothing more than a slightly novel design for a penis, a minor variation in mating coloration, or a different style of courtship dance. But from these humble origins, a small proportion of courtship innovations and their side-effects may turn out to have some survival benefits in addition to their courtship benefits. They may then become favored by natural as well as sexual selection. Of these survival adaptations, a small proportion may prove significant enough to allow a species to invade many new environmental niches. They produce adaptive radiations, proving themselves over time as major innovations. The ecological success of major innovations may hide the fact that many of them originated as courtship innovations.

The feathered wing may be a good example of a courtship innovation that proved to have large survival advantages in the long term. *Archaeopteryx* fossils from 150 million years ago were first found over a century ago, and paleontologist John Ostrom's 1969 theory that birds evolved from small, fast-running theropod dinosaurs has held up fairly well. However, biologists are still not sure how or why feathered wings evolved on dinosaur-type bodies. Many biologists propose that wings always had an aerodynamic function, even in their early stages of evolution. There is the ground-up theory that wings evolved to help small dinosaurs jump and turn quickly to catch prey, and the trees-down theory that wings helped to break their falls (progressing from parachuting to gliding to powered flight). Other biologists point out that the earliest proto-birds (such as the *Protarchaeopteryx* unearthed in China in the early 1990s) had well-developed wings, but no sign of the lighter skeleton associated with flying, and no sign of the top/bottom asymmetry that gives wings lift. Some have even proposed that feathers originated for insulation, feathered wings helping females to incubate their eggs, as in ostriches. But perhaps wings originated as sexual ornaments, along the following lines. Take a fairly useless dinosaur forelimb. Add a bit of color or an extra skin-flap with a novel mutation. Apply sexual choice and the runaway process. Result: a large surface area ornamented with color, available for display to the opposite sex. Feathers make excellent sexual ornaments—they are light, flexible, and movable. They are still used in courtship displays by male rifle-birds, who snap them open and shut in front of awestruck females. If the male protobirds happened to combine their forelimb displays with energetic jumps during courtship, and if females selected for the best jumpers, then the transition from a display function to an aerodynamic function would be relatively smooth. Once wings proved useful in other contexts such as escaping predators, then survival selection would start shaping them for flight instead of just sexual ornamentation. This would have led to the well-documented proliferation of bird species well before the extinction of their dinosaur cousins 65 million years ago, and continuing to the present.

Of course, this scenario for wing evolution is just one hypothesis, and it is by no means clear whether it is right. At least this speculative example illustrates the general point that courtship innovations can potentially lead to unanticipated survival advantages. If we want to overcome the threshold problem of how evolution can favor the initial stages of innovations before they show net survival benefits, sexual selection seems to be a very strong candidate.

The human mind can be seen as one of these courtship innovations that happened to show some large survival advantages long after it first evolved. Modern *Homo sapiens* evolved about 100,000 years ago in Africa. By that time, our ancestors had brains the same size as ours. Yet almost all of the technological process in toolmaking came tens of thousands of years later. Agriculture took another 90,000 years to invent, and only after that did the global human population climb above a few million. More than 95,000 years after human language probably evolved, we invented writing and reading, allowing useful information to be transmitted down through generations and across great distances.

Neanderthals had already evolved quite large brains 200,000 years ago, yet showed very limited technological progress and very modest abilities to spread into new habitats. Neanderthals may have had most of the courtship innovations that we call the human mind, yet they did not stumble upon the potential survival advantages conferred by our sort of creative intelligence. Our lineage did, so we imagine those survival advantages as projecting all the way back to the mind's origins.

Every inventor knows that innovation depends a lot on serendipity. A novelty may be invented for one purpose, only to prove its value years later for a completely different purpose. The Chinese invented gunpowder for firework displays, and the Europeans adopted it for warfare. The dinosaurs may have evolved proto-wings as sexual ornaments, and evolved into birds that use them for flight. The human mind may have evolved as a set of fitness indicators and sexual ornaments, and now we use it to make movies, give venture capital to start-up companies, and read books on mental evolution. Each species is free to use its sexually selected adaptations for any non-sexual purpose that it can invent—and as long as that purpose contributes somehow to survival or reproduction, selection can favor such use.

Sexual selection thus works as a natural source of serendipity in evolution. It gives evolution the slack it needs to play around without demanding that every cost incurred now must yield some future economic benefit. As all scientists know and most governments forget, this is the only way that productive research and development happens.

From a Production Orientation to a Marketing Orientation

The traditional view of sexual selection in biology is similar to the traditional view of advertising in a production-oriented corporation. Until the 1950s, corporate management usually focused on making production more efficient. The goal was to transform raw materials into physical products as cheaply and reliably as possible. Henry Ford's production line was the icon of good management, even though it made Model-T cars in only one color. Advertising was an afterthought—just a way to get rid of the product once the hard job of making it had been accomplished. This is how many biologists still view evolution. Natural selection does the hard work of creating efficient organisms that transform food into growth, and into more organisms. Sexual selection does a little advertising as an afterthought, once the product—the organism—is available for purchase in the sexual marketplace.

Then, in the 1950s and 1960s, a revolution swept through the business world. Beginning with innovative consumer-oriented companies like Procter & Gamble, the "marketing orientation" took over from the old "production orientation." According to the marketing orientation, a company's goal should not be to manufacture physical objects, but to make profits by fulfilling consumers' needs, wants, and preferences. Production matters only insofar as it contributes to consumer satisfaction. If nobody wants a product, there is no point in making it. If everybody wants something different from what is being made, a company would do better to change what it makes.

The marketing-oriented company works backwards from consumer preferences, not forwards from raw materials. Advertising is not some mysterious luxury hovering above the factory, but the only way to connect consumer preferences to the products on offer, and hence to profits. Indeed, advertising and packaging becomes a major part of the product. A marketing orientation does not just mean more sophisticated advertising. It means reshaping everything a company does so that it contributes to satisfying some consumer preference in a profitable way. (This may, of course, include crafting a culturally learned preference out of the human instincts for acquiring status, displaying wealth, and attracting mates.) The marketing revolution was probably the most significant change in business thinking since the invention of money. It puts consumer psychology at the heart of practical economics. It is responsible for the dazzling proliferation of products and services in modern economies. Not all corporations have shifted from the production to the marketing orientation, but the most successful ones have.

By suggesting that sexual selection plays a major but neglected role in evolutionary innovation in general and the human mind's evolution in particular, I am proposing a sort of marketing revolution in biology. Survival is like production, and courtship is like marketing. Organisms are like products, and the sexual preferences of the opposite sex are like consumer preferences. Courtship displays are not a mysterious luxury soaking up excess energy after the business of survival is accomplished. Rather, they are the only way to get one's genes into the next generation, by fulfilling the sexual preferences of the opposite sex. Survival matters only insofar as it contributes to courtship. If nobody wants to mate with an animal, there is no evolutionary point in the animal surviving.

A marketing orientation does not imply shoddy production. On the contrary, greater sensitivity to consumer demands for high-quality products may force companies to improve production standards. Likewise, mate choice for fitness indicators may drive very fast improvements in fitness. Through fitness indicators, sexual selection preserves the near-perfection of biological adaptations, and protects them against erosion by mutations.

A marketing orientation may result in a seemingly irrational diversification of products and species. Procter & Gamble filled supermarket shelves with dozens of nearly identical detergents and soaps, each aimed at a different market niche. This may seem wasteful, but evolution does the same thing. It fills ecosystems with dozens of nearly identical species, each with slightly different courtship behaviors and displays. This is how sexual selection splits species apart. It may explain the biodiversity of sexually reproducing animals and flowering plants.

Most importantly, a marketing orientation does not imply that advertising crowds out innovation. Quite the opposite: the market's hunger for novelty drives greater investment in research and development, and the efficiency of advertising makes corporations confident that the benefits of innovations will exceed their research costs. Sometimes, by trying to find a superficial variant that attracts consumer attention, a company will stumble upon a major invention that becomes the industry standard after a few years. Likewise, sexual selection rewards the novel and the ornamental, but this does not rule out the useful. A courtship innovation may later prove its worth as a survival advantage.

A marketing orientation in evolution does not just mean paying a little more attention to courtship as a form of advertising. It means that every aspect of an organism's growth, structure, and behavior has been shaped to fulfill the sexual preferences of the opposite sex. It puts courtship at the heart of modern biology, as marketing is at the heart of modern business. This marketing revolution swept through the organic world half a billion years ago, just after the Cambrian explosion produced the first complex, sexually reproducing animals. Any animal that persisted in a production orientation, an obsession with food and survival, lost out to competitors that adopted a marketing orientation, an obsession with profiting genetically by pleasing the opposite sex. The explosion of organic complexity and diversity in the last half billion years is just what we would expect if evolution underwent a marketing revolution.

Animal minds are not uniformly black Model-T cars churned out by the assembly line of natural selection. They are self-advertising, self-promoting, self-packaging products adapted from the bottom up, from the inside out, from birth to death, to the demands of their consumers: the opposite sex. In modern society, we may feel ambivalence about the marketing orientation of the businesses that shape our lives. Their marketing departments take an interest in our attitudes that is both flattering and alarming. But it would be hypocritical to pretend that we are in this marketing world but not of it. I believe that our minds evolved through a million years of market research called sexual selection. From this perspective, we are walking, talking advertisements for our genes.

This marketing perspective has implications not only for evolutionary biology, but also for evolutionary psychology. If species evolve to adopt this marketing orientation dictated by sexual selection, then perhaps natural selection's status in evolution has been overestimated. If mate choice promotes speciation and innovation, then sexual selection may be to macro-evolution what genetic mutation is to micro-evolution: the prime source of potentially adaptive variation, at both the individual level and the species level. Like mutations, most courtship innovations could be viewed as costly wastes. But, also like mutations, a few courtship innovations like the human brain may prove spectacularly useful.

It may be no accident that sexual life forms dominate our planet. True, bacteria account for the largest number of individuals, and the greatest biomass. But by any reasonable measures of species diversity, or individual complexity, size, or intelligence, sexual species are paramount. And of the life forms that reproduce sexually, the ones whose reproduction is mediated by mate choice show the greatest biodiversity and the greatest complexity. Out of the million or so known animal species, the vast majority reproduce sexually, including the majority of insects. Almost all animals larger than a couple of millimeters are sexual reproducers capable of sexual choice: all mammals, all birds, all reptiles. The situation is similar with plants. Of some 300,000 known plant species, about 250,000 reproduce through flowers that attract pollinators. Without sexual selection, evolution seems limited to the very small, the transient, the parasitic, the bacterial, and the brainless. For this reason, I think that sexual selection may be evolution's most creative force. It combines an inventor's playful love of discovery with the venture capitalist's willingness to invest enough in innovations to bring them to the market where they may prove useful. We shall see next how the mating market may have operated among our ancestors, and how courtship and mate choice may have generated the evolutionary innovations that constitute human nature.

6

Courtship in the Pleistocene

To judge a new theory of human evolution, it can be more important to forget one's preconceptions than to learn a set of new facts and ideas. Most of our images of human evolution come from popular culture. Film, television, cartoons, and advertising have filled our heads with a lot of colorful nonsense about prehistory. If the image in your mind is of cave-men clubbing cave-women unconscious and dragging them off, you may not grant sexual choice much significance in human evolution. This chapter aims to confront these preconceptions, inquiring how our ancestors did and did not form sexual relationships.

Popular culture images of prehistory are divided by market segmentation according to consumer age group, and by sexual content ratings. There is a children's G-rated version of prehistory that eliminates all sex and most violence, where neither sexual selection nor natural selection have much force. Playmobil toy sets include multi-ethnic cave-men happily living alongside dinosaurs, hunting lions, and living in jungles. The *Flintstones* cartoons depicted a prehistory of capitalist affluence, suburban family values, and chaste monogamy. In these Gardens of Eden there is no hint of reproductive competition, the engine of evolution.

Then there is a "Parental Guidance" prehistory, with a bit more violence and a few coy allusions to romance. Our PG version of prehistory is usually compiled from *Planet of the Apes* films, television cartoons about time-traveling teenagers, school trips to natural history museums, and summer camp experiences with the odd broken bone or stinging insect. Since this version emphasizes adventure, danger, and survival, it makes more plausible the idea that our minds evolved for toolmaking, hunting, and warfare. The resulting theory of human evolution resembles the opening sequence of Stanley Kubrick's *2001: A Space Odyssey*, in which proto-human apes conquer their rivals by inventing bone clubs, which put us straight on the technological path to moon-going spacecraft. The PG version never shows how the proto-humans produced any offspring, so sexual selection remains invisible.

Adult versions of prehistory include sexual content, but almost always in the form of a prurient male fantasy where female choice is irrelevant. Please, forget the sexual favors Raquel Welch bestowed on the dinosaur-slaying cave-man in the film *One Million Years B.C.* Do not take seriously the scene in *Quest for Fire* in which a rough stranger visiting a more sophisticated tribe is invited to copulate with all of the tribe's fertile women. Erase the memory of Daryl Hannah's rape by Neanderthals in *Clan of the Cave Bear.* The torrid paleolithic romances of Jean Auel are good entertainment, as are the erotic daydreams that may float through the minds of college students during springtime physical anthropology courses. However, they are not good touchstones for judging a theory of mental evolution through sexual choice.

Most media portrayals of prehistory follow one of three strategies: eliminate sexual content entirely, show cave-women falling for adventure heroes who rescue them from peril, or offer a narcissistic sexual fantasy in which only the protagonist (usually male) exercises sexual choice. There seems to be no market for portrayals of our early ancestors exerting mutual choice. If we are to see all the genuine tensions and difficulties between the sexes, media producers assume we must be rewarded with a proper costume drama set in Imperial Rome or Regency England. After all, could Alan Rickman and Sigourney Weaver keep a straight face playing an intense romantic psychodrama set in Pleistocene Zaire, while wearing mangy furs, with ochre-smeared hair, and covered in ticks?

Maybe not, but a romantic psychodrama is just what we need to envision how sexual choice may have worked during human evolution. This is not a vain hope. In some ways we are better positioned to understand sexual selection than survival selection. The sexual challenges our ancestors faced were created by other members of their own species. Likewise today. If our thoughts and feelings about sexual relationships are not too different from those of our ancestors, then our sexual challenges must not be too different. We get infatuated, we fall in love, we feel ecstatic, jealous, or heartbroken, we grow bored with some partners, and, if lucky, we develop a companionable attachment to the sexual partners with whom we raise children. We are attracted to beautiful faces and bodies, but also to a good sense of humor, a kind personality, a keen intelligence, and a high social status. If these sexual tastes are part of human nature that evolved gradually, our ancestors must have felt similarly to some degree. We should not automatically project modern social arrangements back into prehistory, but it is probably valid to project our individual emotions on to our ancestors.

By contrast, it can be difficult to appreciate the survival challenges that shaped our mental adaptations. In the developed world, we drive around in cars, live in the same house for years, use money to buy food, work hard at specialized jobs, and go to hospitals when ill. Our ancestors had to walk everywhere, lived in makeshift shelters in dozens of different places every year, did little work other than foraging for food, and when they fell ill, they either recovered spontaneously or died. The economics of surviving have changed dramatically, while the romantic challenges of mating have remained rather similar.

Pleistocene and Holocene

Why are evolutionary psychologists so preoccupied with the Pleistocene? The Pleistocene was a geological epoch uniquely important in human evolution, because it included the evolution of all that is distinctively human. At the beginning of the Pleistocene, 1.6 million years ago, our ancestors were still relatively small-brained apes who walked upright and made just a few crude stone tools. They were almost certainly without language, music, art, or much creative intelligence. At the end of the Pleistocene, just 10,000 years ago, our ancestors were already modern humans, identical to us in bodily appearance, brain structure, and psychology. The evolution that shaped human nature all took place in the Pleistocene.

After the Pleistocene came the Holocene, occupying the last 10,000 years. The Holocene includes all of recorded history. During the Holocene, humans spread around the planet, invented agriculture, money, and civilization, and grew from populations of a few million to a few billion. The Holocene has been historically crucial but evolutionarily unimportant. Ten thousand years is only four hundred human generations, probably not enough time to evolve many new psychological adaptations. But it is plenty of time for runaway sexual selection to make populations diverge a bit in some aspects of body shape, facial appearance, and psychological traits. However, this book is not concerned with such relatively minor differences between populations. It is concerned with universal human mental abilities that our closest ape relatives do not share.

The Holocene changed patterns of human mating and reproduction dramatically. It saw the emergence of inherited wealth, arranged marriages, hierarchical societies, patriarchy, feminism, money, prostitution, monogamous marriage, harems, personal ads, telephones, contraception, and abortion. These make modern courtship rather different from Pleistocene courtship. But Pleistocene courtship is what drove sexual selection during the relevant period of human evolution, and human behavior in the Holocene still reflects our Pleistocene legacy.

Pleistocene Life

Knowing that the human mind's distinctive abilities evolved in the Pleistocene makes evolutionary psychology much easier. It means that all the ancestral environments that shaped the basic mental capacities of our species were physically contained within the African continent, since all pre-human ancestors lived in Africa, and humans spread out of Africa only towards the end of the Pleistocene. Our ancestors lived in areas of sub-Saharan Africa that contained mixtures of open savanna, scrub, and forest. Instead of caves or jungles, picture Africa's broad, flat plains, with their fever trees and acacias, their wet and dry seasons, their hot days and cool nights, their plentiful hoofed herds and rare, emaciated predators, the incandescent sun, and millions of scrabbling insects.

A fairly coherent picture of Pleistocene life has emerged from anthropology, archeology, paleontology, primatology, and evolutionary psychology. Like other social primates, our hominid ancestors lived in small, mobile groups. Females and their children distributed themselves in relation to where the wild plant food grew, and clustered in groups for mutual protection against predators. Males distributed themselves in relation to where the females were. Many members of each group would have been blood relatives. Group membership may have varied daily and seasonally, according to opportunities for finding food and exploiting water sources.

Our ancestors would have known at least a hundred individuals very well by face and by personality. During their lifetimes they would have come into contact with several hundred or thousand members of the same local population. Almost all sexual partners would have been drawn from this larger tribal group, which, after language evolved, would probably have been identified by their shared dialect.

During the days, women would have gathered fruits, vegetables, tubers, berries, and nuts to feed themselves and their children. Men would have tried to show off by hunting game, usually unsuccessfully, returning home empty-handed to beg some yams from the more pragmatic womenfolk. Our ancestors probably did not have to work more than twenty or thirty hours a week to gather enough food to live. They did not have weekends or paid vacation time, but they probably had much more leisure time than we do.

There was intermittent danger from predators, parasites, and germs, but our ancestors would have become as accustomed to coping with those dangers as we are to crossing roads. Nature was not red in tooth and claw. Usually, it was really boring. Predators would have tended to kill the very young, the very ill, the very old, and the very foolish. Most illnesses would have been due to poor condition brought on by starvation or injury. Our ancestors did not spend all their time worrying about survival problems. They were among the longest-lived species on the planet, which implies that their daily risk of death was minuscule. Like most great apes, they probably spent their time worrying about social and sexual problems.

For most of evolution, our ancestors ranged across wide areas without being tied to a single home base or territory. They owned no more than they could carry, had no money, inherited no wealth, and could not store food today to insure against starvation next month. If individuals consistently appeared healthy, energetic, and well-fed, it was not because they were born rich. It must have been because they were good at foraging and good at making friends who took care of them during rough patches.

To understand how sexual selection may have operated in the Pleistocene, we have to ask how sexual relationships and sexual choice may have worked. We know that our hominid ancestors did not take each other out to restaurants and films, give each other engagement rings, or wear condoms. But what can we say about how they did select mates? We'll start with a look at sexual choice in other primates, and then consider what was distinctive about sexual choice among our hominid ancestors.

Sexual Selection in Primates

In most primate species, the distribution of food in the environment determines the distribution of females, and the distribution of females determines the distribution of males. When food is so dispersed that females do best by foraging on their own, males disperse to pair up with the lone females. This gives rise to monogamous couples. It is a fairly rare pattern among primates, limited to gibbons, some lemurs, and some African and South American monkeys.

When food comes in patches large enough for several females to share, they tend to band together in small groups to find the food, and to protect each other against predators, unwanted males, and competing female groups. As long as the female band is not too large, a single male can exclude other males from sexual access to the band, which thus becomes "his." This "harem system" of single-male polygyny is fairly common in primates, being found in hamadryas baboons, colobus monkeys, some langurs, and gorillas. The competition between males to guard the female groups creates very strong sexual selection pressures for male size, strength, aggressiveness, and large canine teeth.

When food comes in still larger patches, female groups can grow too large for any single male to defend them. The males must then form coalitions, resulting in a complex multi-male, multi-female group, as in some baboons, macaques, ring-tailed lemurs, howler monkeys, and chimpanzees. Our hominid ancestors probably lived in such groups, in which sexual selection gets more complicated. Sometimes, females in multi-male groups appear to use sperm-production ability as the main fitness indicator. A chimpanzee female might mate with every male in the group every time she becomes fertile. She lets their sperm fight it out in her reproductive tract, and the strongest swimmers with the best endurance will probably fertilize her egg.

In response to this sexual selection for good sperm, male chimpanzees have evolved large testicles, copious ejaculates, and high sperm counts. Female primates face a trade-off. They can select for the best-swimming sperm by mating very promiscuously, or they can select for the best courtship behavior by mating very selectively. Or they can do a little of both, selecting a small group of male lovers for their charm and then letting their sperm fight it out.

In species that do not get completely caught up in runaway sperm competition, females can favor various male behavioral traits. Multi-male groups obviously allow greater scope for females to choose between males. If they favor dominant males, males evolve through sexual selection to compete intensely for social status by individual force or by forming coalitions. If females favor kind males, males evolve through sexual selection to groom females, protect their offspring, and guard them from other males.

Given multi-male, multi-female primate groups, how does mate choice work? Female primates can exercise choice by joining groups that contain favored males, initiating sex with them during estrus, supporting them during conflicts, and developing long-term social relationships with them. Females can reject unfavored males by refusing to cooperate during copulation attempts, driving males away from the group, or leaving the group. But female mate choice criteria remain obscure for most primate species. In contrast to modern humans, female primates rarely favor males who can provide resources or paternal care of offspring. The sporadic male care that is observed, such as watching, carrying, and protecting infants, is better described as courtship effort than as paternal care. The male is unlikely to be the infant's father, but is simply trying to mate with the infant's mother by doing her a favor.

Primate researchers still know little about what traits are preferred by male and female primates. For example, we know less about female choice in other apes than we do about female choice in the Tungara frog, the guppy fish, or the African long-tailed widowbird. Nevertheless, three kinds of female preference have been reported in primates: preferences for high-ranking males capable of protecting females and offspring from other males; preferences for male "friends" that have groomed the female a lot and have been kind to her offspring; and preferences for new males from outside the group, perhaps to avoid genetic inbreeding. Each sort of preference could be explained in terms of female choice for good genes, or female choice for material and social benefits. Although male primates have evolved an astounding diversity of beards, tufts, and colorful hair styles, there has been very little research on female choice for male appearance.

Also, there has been virtually no research on primate sexual choice for personality or intelligence. Female primates are sometimes reported to show "irrational" or "capricious" preferences that cannot be explained on the basis of male dominance, age, or group membership. Sometimes two primates just seem to like each other based on unknown features of appearance, behavior, or personality. Female primates might well be choosing males for their personalities and not just their status, but we do not know.

Most primates follow the general animal pattern of male sexual competition and female choosiness. But when the costs of male sexual competition and courtship are high, males also have incentives to be choosy. When male mate choice becomes important, sexual selection affects females as well as males. In monogamous marmosets and tamarins, females compete to form pairs with quality males and drive off competing females. In single-male harem systems, the dominant male's sperm can become a limiting resource for female reproduction, and high-ranking females prevent low-ranking females from mating through aggression and harassment. In multi-male groups, females sometimes compete to form consortships and friendships with favored males. Such patterns of female competition suggest some degree of male mate choice. When the costs of sexual competition and courtship are high, males have an incentive to be choosy about how they spread their sexual effort among the available females. Males compete much more intensely for females who show signs of fertility such as sexual maturity, estrus swellings, and presence of offspring. Like females, some male primates also develop special friendships with particular sexual partners. It may not be romantic love, but, at least among some baboon pairs, it looks pretty similar.

Our closest ape relatives, the chimpanzees and the bonobos, live in multi-male, multi-female groups in which sexual choice is dynamic, intense, and complicated. Under these relentlessly social conditions, reproductive success came to depend on social intelligence rather than brute strength. Both sexes compete, both sexes have dominance hierarchies, and both sexes form alliances. Sexual relationships develop over weeks and years rather than minutes. Many primatologists and anthropologists believe that our earliest hominid ancestors probably lived under similar social and sexual conditions. Constant sociosexual strategizing in mixed-sex groups was the legacy of our ape-like ancestors. It was the starting point, not the outcome, of sexual choice in human evolution.

Pleistocene Mating

If we could look at the Earth through an extremely powerful telescope a million light-years away, we could see how our ancestors actually formed sexual relationships a million years ago. Until NASA approves that mission, we have to combine evidence from several less direct sources: the sexual behavior of other primates, the sexual behavior of modern humans who live as hunter-gatherers, the evidence for sexual selection in the human body and human behavior, and psychological findings on sexual behavior, sexual attraction, sexual jealousy, and sexual conflict. A number of good evolutionary psychology books already review this evidence, including David Buss's *The Evolution of Desire.* A consensus is emerging about the key aspects of ancestral life, though there is still vigorous debate about many details.

Our ancestors probably had their first sexual experiences soon after reaching sexual maturity. They would pass through a sequence of relationships of varying durations over the course of a lifetime. Some relationships might have lasted no more than a few days. Given that it takes an average of three months of regular copulation before conception, very short-term partnerships would probably not produce a child. Longer-term relationships would have been much more evolutionarily important because they were much more likely to produce offspring. Indeed, in the absence of contraception the longer partnerships would almost inevitably produce a child every two or three years.

Most children were probably born to couples who stayed together only a few years. Exclusive lifelong monogamy was practically unknown. The more standard pattern would have been "serial monogamy": a sequence of nearly exclusive sexual partnerships that were socially recognized and jealously defended. Relationships may have sometimes ended amicably, but perhaps more often one partner would reject or abandon the other, or one would happen to die. This is the pattern characteristic of most human hunter-gatherers, because they do not have the religious, legal, and property ties that reinforce ultra-long-term monogamous marriages in civilized societies.

Some desirable males were probably able to attract more than one regular sexual partner. Their polygyny opened the possibility of runaway sexual selection effects. But they were probably the exception. Much more common would have been the affairs and flings that bedevil ordinary sexual partnerships. For women, there were incentives to mate with males of higher fitness than their current partner. For men, there were incentives to mate with as many females as possible (if the current partner could stand it). Yet there were probably social pressures against such dalliances from jealous partners and their families. There is plenty of evidence from evolutionary psychology that men and women have physical, emotional, and mental adaptations for short-term liaisons and adulterous affairs. The different costs and benefits of such affairs for males and females explain most of the sex differences in human psychology. In particular, the higher incentives for males to attract large numbers of sexual partners through public displays of physical and mental fitness explain why males are so much more motivated to produce such displays.

Female mate choice was powerful in prehistory. Although sexual harassment of females by males was probably common, females could retaliate by soliciting assistance from female friends, male partners, and relatives. They would not have been jailed for killing a psychopathic stalker or an abusive boyfriend. Our female ancestors lost all visible signs of ovulation, so it would not have been possible for a would-be rapist to know when a woman was fertile. Concealed ovulation reduced the male incentives for rape, and it usually protected women from conceiving the offspring of rapists. From an evolutionary point of view, it guarded their power of sexual choice. Also, rapists would have been subject to vigilante justice by the male relatives of the victim. The power of clan members to enforce good sexual behavior is often overlooked in discussions of human evolution. Once language evolved, sexual gossip would have been a deterrent against illicit affairs, sexual harassment, and reputation-destroying rape accusations. Nevertheless, the prevalence of rape in human prehistory is still subject to intense debate. The higher the actual prevalence was, the less important female mate choice would have been, and the weaker my sexual choice theory would become.

Pleistocene Flirting Versus Modern Dating

Suppose that the level of fascination, happiness, and good humor that our ancestors felt in another individual's company was a cue that they used to assess the individual's mind and character. If an individual made you laugh, sparked your interest, told good stories, and made you feel well cared for, then you might have been more disposed to mate. Your pleasure in his or her presence would have been a pretty good indicator of his or her intelligence, kindness, creativity, and humor.

Now consider what happens in modern courtship. We take our dates to restaurants where we pay professional chefs to cook them great food, or to dance clubs where professional musicians excite their auditory systems, or to films where professional actors entertain them with vicarious adventures. The chefs, musicians, and actors do not actually get to have sex with our dates. They just get paid. We get the sex if the date goes well. Of course, we still have to talk in modern courtship, and we still have to look reasonably good. But the market economy shifts much of the courtship effort from us to professionals. To pay the professionals, we have to make money, which means getting a job. The better our education, the better our job, the more money we can make, and the better the vicarious courtship we can afford. Consumerism turns the tables on ancestral patterns of human courtship. It makes courtship a commodity that can be bought and sold.

During human evolution, though, one's ability to make a good living did not automatically mean that one could buy a desired sexual partner good-quality entertainment. If you were a hominid, you would have had to do the entertaining yourself. If you did not make a desired mate laugh, nobody would do it for you. And if they did, your date would probably run off with them instead of you.

The minds of our ancestors were relatively naked compared to ours. They did not spend twenty years in formal education ornamenting their memory with dead people's ideas. They did not read daily newspapers so that they could recount human-interest stories. In courtship, they had to make up their ideas, stories, jokes, myths, songs, and philosophies as they went along. There was no masking a poor imagination with a good education, or a poor sense of rhythm with a good CD collection.

Perhaps even more importantly for long-term relationships, there was no television to keep your sexual partner amused after the first blush of romance faded. If they were bored in the relationship, there was no vicarious entertainment to be had. They either had to put up with your boring old self, or find a new lover. During the Holocene, when long-term monogamy thrived, people worked much harder and longer hours doing their planting, herding, trading, and career-climbing. There were fewer hours of leisure to fill, and more ways to fill them without talking to one another. Historically, humans did not begin to put up with lifelong marriage until they could no longer live off the land, property inheritance became the key to children's survival, and couples had economic incentives to continue cooperating long after they were no longer on speaking terms. During prehistory, there were fewer economic incentives to stay together, fewer distracting entertainments to replace lost romance, and fewer ways to insulate oneself from new sexual opportunities.

Were Fathers Important?

Single mothers may have been the norm during most of human evolution, as they were during the previous 50 million years of primate evolution. As Sarah Blaffer Hrdy has argued in her book

Mother Nature, human females have inherited a rich set of mental and physical adaptations fully sufficient to nurture their offspring with minimal assistance from males. Male help may have been a welcome luxury, but it was not a necessity.

Many Pleistocene mothers probably had boyfriends. But each woman's boyfriend may not have been the father of any of her offspring. Or he may have been the father only of the most recent baby. Even so, his typical contribution to parenting is debatable. Males may have given some food to females and their offspring, and may have defended them from other men, but as we shall see, anthropologists now view much of this behavior more as courtship effort than paternal investment.

Viewed from the broad sweep of evolution, it is unlikely that male hominids did much direct fathering. In almost all mammals and all primates, females do almost all of the child care, with very little help from males. Males could never be sure which offspring really carried their genes, whereas females could be certain. This uncertainty about paternity leads most male mammals to invest much more in pursuing new sexual opportunities than in taking care of their putative offspring.

Like all other primates, the basic social unit among our ancestors was the mother and her children. Women clustered together for mutual help and protection. Male hominids, like males of other primate species, were probably marginal, admitted to the female group only on their forbearance. Herds of young bachelor males probably roamed around living their squalid, sexually frustrated lives, hoping they would eventually grow up enough for some group of women to take them in.

The traditional view that females needed males to protect them from predators has been challenged by an increased understanding of primate and hunter-gatherer behavior. To us, our sex differences in size and strength are salient. But to a large predator looking for an easy kill, female humans would have been only marginally less dangerous than males. Adult males may be more accurate at throwing things, but females tend to go around in larger groups while foraging, with many eyes and many hands to offer mutual vigilance and protection. An ancestral female would have been much safer in a group of a dozen sisters, aunts, and female friends than with a single male in a nuclear family. Female humans were among the largest primates ever to have evolved, and among the strongest omnivores in Africa. They did not necessarily need any help from boyfriends only 10 percent taller than themselves. Female hominids seem unlikely to have displayed the exaggerated physical vulnerability expected of women under patriarchy. When you picture ancestral females facing predators, do not imagine Marilyn Monroe whimpering and cowering. Imagine Steffi Graf brandishing a torch in place of a tennis racket.

The same group-protection effect would have guarded females against sexual predators. Ancestral women could protect one another from harassment and rape, just as other female primates do. From a female's point of view, a strong male partner would be a mixed blessing. He could fend off unwanted attention from other males, but he could also beat you up if he got jealous or angry. Women consistently show preferences for tall, strong males in mate choice studies, but this may reflect a preference for good genes and high fitness, rather than a preference for a male capable of physical violence and intimidation that might get turned against her or her children.

Interviews with contemporary hunter-gatherer women by anthropologists such as Marjorie Shostak reveal that these women view many men as more trouble than they're worth. If the men are hanging around, they usually eat more food than they provide, and demand more care than they give one's children. If they have very high fitness, then their good genes, good sex, and good conversation might compensate for their messiness and lethargy. But if they are only average, their potential for sexual jealousy and violent irritability may render them a net cost rather than a benefit.

On the other hand, David Buss and other evolutionary psychologists have amassed considerable evidence that modern women generally favor tall, strong, healthy, and self-confident men, all else being equal. These traits may be favored because they would have correlated with good hunting abilities and protection abilities under ancestral conditions. However, as we shall see in the next chapter, many of these traits also reveal good genes—they are genetically heritable, and they work as effective fitness indicators. It is not yet clear whether the genetic or non-genetic benefits of such traits were more important to women. Mate choice mechanisms should evolve to capture both sorts of benefit whenever possible, so they may be difficult to disentangle.

There is still much debate about the importance of fathers in human evolution. Men show some signs of having been selected as good and helpful fathers, but our paternal instincts have not been well researched yet. Modern fathers form strong emotional attachments to their children, and this is probably an evolved propensity. A few of them even spend almost 20 percent as much time doing child care as their female partners do. Recent surveys show that Japanese fathers are starting to play with their children for almost seven minutes a day. That is a relatively high amount of paternal care compared to other male mammals. But to better understand the evolution of fathers, we need a closer look at how courtship may have overlapped with parenting.

Combining Courtship and Parenting

Before contraception, our female ancestors would have produced their first child by around age 20, within a few years of reaching sexual maturity. (Female puberty probably happened several years later in prehistory than it does now, because the modern fat-rich diet artificially hastens puberty and increases teenage fertility.) Before legally imposed monogamous marriage, individuals probably passed through several sexual relationships during their reproductive years. These two patterns imply that most courtship during most of human evolution occurred between adults who already had children by previous relationships. Without nannies, nurseries, or schools, those children would have been hanging around their mothers almost all the time. (In the wild, no primate female ever grants parental custody of her children to their father after they split up.) Where there were women, there were usually already children. In modern Western societies we forget how parenting and courtship must have overlapped because we have children later in life, have very few of them, and exclude them from adult social life.

Female hominids must have juggled their courtship efforts with their mothering. Some of their courtship displays may have originated by turning normal motherly duties into better fitness indicators and entertainments. If they must tell stories to entertain their children, and if potential male mates are within earshot, they might as well make the stories appeal on both the child and the adult levels. If they must feed their children, and they want to attract a man, they might as well forage for something unusually tasty. Male mate choice almost never had the luxury of favoring a woman who did not yet have any children, who could spend all her time frolicking and canoodling. The important variable was not whether a female already had children, but whether she was a cheerful mother or a careworn mother, a beautiful mother or an ugly mother, an intelligent mother or a boring mother. Sexual competition between females was mostly sexual competition between mothers.

Moreover, mothers probably cared about the views of their children in choosing new sexual partners, so female choice must have intertwined with children's choice. Kids who hated their mother's new boyfriend might have destroyed his chances of sustaining a successful relationship. Mothers had good reasons to listen to their children's likes and dislikes, because their children were the vehicles carrying their genes. The children were every mother's paramount concern. A healthy child in hand was worth two male lovers in the bush. This put male hominids in an unusual position: their courtship had to appeal not only to mothers but to their children. This has a surprising implication. If children's judgments influenced mate choice, then they influenced sexual selection, and children's preferences indirectly shaped the evolution of adult male humans.

So, what did those hominid kids do to us? They did not make

male humans as good at parenting as the average female mammal, but they made them better fathers than in almost any other male primate species. Men bring children food, make them toys, teach them things, and play with them. Their willingness to do this even for step-children could be viewed as a side-effect of a male adaptation for taking care of their own genetic offspring. But perhaps fatherly support and protection of step-children was the norm in the Pleistocene. If typical sexual relationships only lasted a few years, men were much more likely to be playing with some other guy's children than their own. Many evolutionary psychologists have pointed out that what looks like paternal effort may actually have evolved through sexual choice as courtship effort. Men attracted women by pleasing their kids.

This is not to say that step-fathers are all sweetness and light. Evolutionary psychologists Martin Daly and Margo Wilson have found that men in every culture are about a hundred times more likely to beat and kill their step-children than their genetic children. There are clear evolutionary reasons for that. When male lions and langur monkeys mate with a new female, they routinely try to kill all of her existing offspring. Those offspring do not carry the males' genes, so by killing them the males free the females to conceive their own offspring, who will carry their genes. The risk of infanticide by males is a big problem for many female primates. Yet is it much less of a worry for modern women. I want to highlight how kind most human step-fathers are compared with other male primate step-fathers. Not only do we consistently fail to kill our step-children like lions try to, we sometimes take reasonably good care of them. Surprisingly, human fathering instincts may have evolved through sexual selection for pleasing the existing children of potential female mates. Of course, where those existing children happen to be ours because we are still in a long-term sexual relationship, there are extra genetic incentives to be good fathers.

Where Sexual Choice Did Its Work

Mating among our ancestors was complicated, flexible, and

strategic. When we talk about their "mating pattern," this is just a generalization across a lot of individual strategic behavior. The individual sexual choices, not the aggregate mating pattern, drive sexual selection. To describe our ancestors as following mating patterns like "moderate polygamy" and "serial monogamy" is just a useful shorthand for identifying these sexual selection pressures.

For sexual choice to have any evolutionary effect, different individuals must produce different numbers of surviving offspring by virtue of their sexual attractiveness. How did the most attractive hominids leave more offspring? When we focus on the polygynous aspects of ancestral mating, it is easy to see. The most attractive males simply inseminate a larger proportion of females, and the least attractive males inseminate fewer. The next generation will inherit many genes from the most attractive males, and none from the least attractive. Polgyny raises the possibility of runaway sexual selection, which is driven mostly by differences in male reproductive success. Also, polygyny helps explain sex differences. The higher variation in reproductive success among males explains why male humans are so keen to show off, to dominate culture and politics, and to broadcast indicators of their fitness to any female who might listen. To the extent that our ancestors were polygynous, there were sexual selection pressures for males to display more intensely then females.

However, we should not assume that sexual selection requires polygyny. As Darwin appreciated, the sexual choices that lead to monogamous pairs can also be crucial. Is it possible that sexual selection can produce equal mental capacities for courtship in both sexes? How can the sexual choices that create monogamous couples possibly have any evolutionary effects? Sexual selection depends on differences in reproductive success, and at first glance monogamy looks as if it produces no such differences.

Suppose that sexual choice among our hominid ancestors worked as follows. Male and female hominids both tried to attract the best sexual partner they could. If they liked that partner's company, they hung out a lot together, had a lot of sex, and produced a child. If they still liked each other after the baby arrived, they stayed together and produced another one. If they did not, they separated and looked for the best new partner they could find. Most hominids spent most of their lives in some kind of sexual relationship with somebody. Most sexual relationships longer than a few months produced at least one child.

Sexual Selection When Everyone Finds a Partner

To see how sexual selection can work even when everyone pairs up into couples, we need a thought experiment. Like all good thought experiments, it will be simplistic, unrealistic, and cartoon-like. But it will give us a surprising result. In this imaginary scenario, every hominid individual finds a sexual mate, every relationship is totally monogamous and permanent, and every relationship produces an identical number of babies. And yet, as long as sexual choice favors fitness indicators, sexual choice can still drive sexual selection by producing unequal numbers of grandchildren. Here's how it works.

Imagine a tribe of hominids, half of them male and half female, all single, all just reaching sexual maturity at the same time. Some males have higher fitness than other males, and they advertise their higher fitness using fitness indicators such as vigorous dancing, intelligent conversing, or realistic cave-painting. Some females have higher fitness than other females, which they advertise through the same sorts of fitness indicator. Fitness is genetically heritable, so higher-fitness parents generally have higher-fitness offspring. The tribe has a tradition of strict monogamy and no infidelity. Every individual has to pick a partner once and stick with them until they die. Both sexes exercise mate choice, accepting and rejecting whomever they want.

What will happen? Each individual wants to attract the highest-fitness mate they can, because they want the best genes for their offspring. There will be a sorting process. Probably, the highest-fitness male will court the highest-fitness female first. If she is sensible, she will accept him, and they will pair off, leaving the rest of the tribe to sort themselves out. The second-highest-fitness male is disappointed. He wanted the highest-fitness female, but could not attract her. He must settle for the second-highest-fitness female. She is also disappointed, because she wanted the best male. But she settles for male number two, because she cannot do any better. Perhaps they fall in love, thanking their lucky stars that they did not end up with the cold and snooty number ones, or the repulsively inferior number threes. Now the third-highest-fitness male is doubly heart-broken. Golden female number one and silver female number two have both ignored him, leaving him to court bronze female number three. He can't do any better, and neither can she, so they pair off. And so on. Eventually, the whole tribe sorts itself into mated pairs of roughly equal fitness.

The fitness matching does not result from any individual's preference for a similarly ranked mate. Instead, it results from the interaction of everyone's preferences during the sorting process. Everybody would prefer a higher-fitness mate rather than a same-fitness mate. But the opposite sex feels the same way too. For a male to mate above his fitness, a female would have to mate below her fitness. Her response to his offer will be "Dream on, loser." Likewise for females trying to mate above their fitness. Individuals have no realistic hope of mating far above their own fitness level, or any willingness to mate below their fitness.

The result will be that mated pairs will correlate highly for fitness. If height correlates with fitness, they will be of similar height. If intelligence correlates with fitness, they will be similarly bright. If facial attractiveness correlates with fitness, they will be similarly beautiful. This is basically what we see in modern human couples: a fairly high degree of "assortative mating" for fitness indicators.

After the mated pairs start having sex, babies start arriving. To make this thought experiment challenging, let's look at the situation where sexual selection seems weakest, and assume that every pair has exactly the same number of babies, say four babies per pair. During most of human evolution, probably only 50 percent of infants survived to sexual maturity, so two babies surviving out of four for every two parents will keep the population size stable. The question is, which mated pairs will contribute the most genes to future generations?

At first glance, it looks as if each pair should contribute the same number of genes, since they have the same number of babies. But we already know that mated pairs differ in their heritable fitness. That is what they were being choosy about when they were sorting themselves into pairs. So, the babies of higher-fitness couples will inherit higher-fitness genes. By definition, higher fitness leads to a better chance of surviving to sexual maturity. The offspring of male number one and female number one may have a very high chance of surviving. The offspring of the lowest-fitness male and the lowest-fitness female may only have a very low chance of surviving. By the time the babies' generation grows up, there will be more surviving offspring of high-fitness parents than of low-fitness parents. In fact, the babies' generation will have a higher average fitness than their parents' generation did.

Evolution just happened. But did sexual selection happen? Things get a little complicated here, because there are two effects at work.

Fitness Spreading

One effect of fitness matching is to increase the variation in fitness in the next generation. In fact, it creates the widest possible fitness differences between babies. Fitness matching by parents leads to fitness spreading among offspring. Consider the extremes of the fitness spread. The only way to produce a baby of the highest possible fitness given the parents available, would have been for the highest-fitness male to mate with the highest-fitness female. That is exactly what happened, through the mating market. And the only way to produce a baby of the lowest possible fitness would have been for the lowest-fitness male to mate with the lowest-fitness female. Again, that is exactly what happened. Fitness matching does not just increase the variation in fitness a little bit. It increases that variation as much as any mate choice process could, with or without monogamy.

The fitness-spreading effect is important because it creates a very tight link between sexual selection and natural selection. The power of natural selection is proportional to the fitness spread that is available in a population. Bigger fitness differences between babies lead to faster evolution. By creating the largest possible fitness spread, fitness matching gives natural selection the greatest diversity of raw material to work on. Psychologists Aaron and Steven Sloman emphasized the importance of this effect in an important paper they published in 1988.

From a genetic point of view, fitness matching concentrates harmful mutations from low-fitness parents in their low-fitness babies. When those babies die, they take a lot of harmful mutations with them. Fitness matching also concentrates helpful mutations (which are much rarer) in high-fitness babies. When those babies thrive at the expense of lower-fitness competitors, the helpful mutations increase their share of the gene pool. This is a heartlessly unromantic view of sexual selection's effects, but evolution is heartless.

From Fitness Matching to Fitness Indicators

The fitness-spreading effect is interesting, but it doesn't take us very far in understanding the evolution of the human mind. To do that, we have to ask how fitness matching affects the fitness indicators themselves. What follows is admittedly a subtle and speculative argument, but one I think is critical to understanding how sexual selection shaped the human mind.

In the above description of fitness matching, it was assumed that individuals could perceive each other's fitness with perfect accuracy. But it is not that simple. Our hominid ancestors did not have portable DNA sequencing laboratories to measure the mutation load of every potential mate. They had to make do with fitness indicators such as sexual ornaments and courtship displays. By definition, fitness indicators have some correlation with fitness, but it is never a perfect correlation. The handicap principle keeps indicators relatively honest, but it cannot keep them perfectly honest, so there will always be a discrepancy between true fitness and apparent fitness. The evolution of fitness indicators is driven by this discrepancy.

Consider the mating market from female number two's perspective. She is the second-highest-fitness female hominid in the tribe. She would love to get together with male number one and have his higher-fitness babies, who will survive better and attract better mates. But female number one stands in the way, seducing male number one with her high-fitness charms. (For the moment, we are still assuming strict monogamy and no adultery, so female number two cannot just have an affair with male number one.)

What can female number two do? She cannot raise her true heritable fitness, because on the African savanna she has no access to retroviral germ-line genetic engineering. But she could produce an appearance of higher fitness by allocating more energy to her fitness indicators. If she had a mutation that increased the quality of one of her fitness indicators, even at the expense of her other adaptations, she might look better than female number one. In fact, she would become female number one, in terms of apparent fitness. She could attract male number one, and produce high-fitness babies. She might produce the same number of babies she would have had with male number two, but now her babies have higher fitness, and are more likely to survive. Even though, according to our assumption, she has produced no more children than any other woman, she will produce more grandchildren who will carry her mutation. Her granddaughters and grandsons would inherit her propensity to allocate more energy to their sexual ornaments and courtship displays. If those displays included evolutionary novelties such as art, music, and language, sexual selection would improve their performance. This is how fitness matching can push fitness indicators to evolve. This is how sexual choice can drive sexual selection, even under strict monogamy.

Now, step back from female number two's predicament and consider the general point. Here we have a hominid tribe that would make Puritans look sinful. They are perfectly monogamous, they have no adultery, and they all have exactly the same number of children. Yet even here, under the most impossible-looking conditions, sexual selection still works. It still favors more extreme, costlier, more impressive fitness indicators such as sexual ornaments and courtship displays. Sexual selection still works on fitness indicators because fitness still means something: some babies still survive better than others because they have higher fitness. Since fitness matching pays evolutionary dividends to those who have high apparent fitness, there are incentives for displaying the most extreme fitness indicators you can afford. The handicap principle will keep the fitness indicators within reasonably honest limits. It can keep low-fitness pretenders from displaying very high apparent fitness, but it cannot keep high-fitness competitors from escalating their sexual arms race. As long as there is some natural selection going on, fitness matching alone should suffice to drive sexual selection for indicators.

This fitness matching theory may sound speculative, but it is just a variation of Darwin's theory of sexual selection in monogamous birds. Darwin faced the same problem: how to explain sexual ornaments that are equally extreme in both sexes in species that form monogamous pairs. He proposed a fitness matching process that relied on the fittest female birds arriving first at the best nesting sites in each breeding season, mating with the fittest male birds, and producing higher-fitness offspring who are more likely to survive. Sexual selection theorists such as Mark Kirkpatrick have shown that Darwin's model can work as long as fitness remains heritable and sexual choice favors reliable fitness indicators. If fitness matching can explain ornamentation in monogamous birds, perhaps it can explain courtship abilities in relatively monogamous apes like us.

Sexual Selection Without Sex Differences

The pure fitness matching process would not produce any sex differences. All else being equal, males and females would evolve fitness indicators to precisely the same degree. This is because under strict monogamy they would have equal incentives for displaying their fitness and for selecting mates based on fitness. Fitness matching tends to promote sexual equality in the indicators it favors. This is one reason why it has the potential to be so important for human evolution. The sexual egalitarianism makes it an attractive model for explaining traits that are ornamental, costly, and sexually attractive, yet do not show the sex differences predicted by traditional models of sexual selection.

How many traits have these features predicted by the fitness matching model? Many traits in many species look ornamental and costly, show minimal sex differences, and probably influence mate choice. However, biologists since the 1930s have usually called such traits "species recognition markers." They assumed, following the tradition of equating sexual selection with a mechanism for producing sex differences, that such traits simply advertise one's species rather than one's fitness. For the last fifty years, whenever a biologist noticed something that exists in both sexes, which would have been called a sexual ornament if it existed only in males, it was called a species recognition marker. If the marker was displayed vigorously by both sexes during mutual courtship, biologists would say that the animals are performing a "pair-bonding ritual." This terminology obscured the fact that one individual would often walk away from the ritual, unimpressed by his or her would-be partner. The evidence for mutual choice was there, but most biologists neglected Darwin's theory of sexual selection in monogamous species.

Birds offer many examples. If, among emus, only males had bright blue bare patches on their cheeks and necks, biologists would probably have called the patches sexual ornaments. But since females have them too, they are usually relegated to the status of species recognition markers. Likewise for the dramatic yellow eyebrow-tufts sprouting from both male and female rockhopper penguins. And the 11-foot wingspans of both male and female wandering albatrosses, which are displayed during mutual courtship by stretching the black tips of the white wings as far apart as possible for the inspection of the opposite sex. All, we are told, for mere species recognition. This viewpoint implies that the hours of mutual conversation during human courtship are likewise nothing more than a way for us to tell that the other individual is a human rather than a chimpanzee. Amotz Zahavi has mocked the species recognition idea as attributing a very high degree of stupidity and very poor mate choice to animals. I agree with his view. These same animals show good discrimination ability when it comes to food and predators, so why should they need such dramatic markers to tell whether a potential mate is of their own species? Fitness matching, a form of mutual mate choice based on fitness indicators, may be a more sensible explanation for most sexual ornaments that show very small sex differences.

In Search of a Few Good Hominids

The question remains of how our ancestors actually made their sexual choices. Perhaps during large tribal gatherings, they formed huge mixed-sex aggregations like sage grouse, where individuals could weigh up hundreds of prospects. This would have made mutual choice extremely easy. However, such Pleistocene singles bars were probably rare.

Much more likely, individuals would encounter a slow trickle of new sexual possibilities, one at a time. The search for a good sexual partner was sequential and opportunistic. Success would depend on one's ability to manipulate which band one joins, and who joins one's band. (A band is the small group of individuals with whom a hominid would forage and spend most nights; clans and tribes are larger sets.) New individuals might join an existing band. The band may encounter other bands at water sources. Individuals might leave their band, looking for new groups that offer more sexual opportunities.

Contact between bands may have been tense and brief, with the threat of violent confrontation balanced against the possible benefits of trade, gossip, and the exchange of sexual partners. Selection would have favored a capacity for very fast decisions about which individuals were attractive enough to pursue. These snap judgments could have been based on information like physical appearance, bodily ornamentation, apparent social status, and public display behavior (such as sports, music, and story-telling). Our ability to judge the physical attractiveness of a human face in a seventh of a second is a legacy of selection for such fast decision-making. Since males would usually have been more motivated to pursue sexual prospects, they would have been more active in this initial phase of searching through bands, looking for attractive potential mates, and trying to switch bands to court good possibilities.

Once mutually attracted individuals arranged to be in the same band, they could split off into temporary courting pairs. Their interaction would resemble the consortships formed by chimpanzee pairs who go off into the bush together for several days. During this most intense phase of courtship, hominids could get to know each other much better, bringing into play all of the psychological levels of courtship discussed in this book. Before language evolved, they would have groomed each other, played, canoodled, shared food, and done all the usual primate things to form social relationships. After language, they would have talked endlessly. During these consortships, the male would usually have been trying to copulate because he would have little to lose from a short-term sexual relationship. If he succeeded, he might grow bored and go away, or he might stay around.

Male and female mate choice waxed and waned in importance at different stages of courtship. Basically, males would scan for physically attractive females and pursue them, trying to establish consortships. This would be a major stage of male mate choice, subjecting females to intense sexual selection for immediate physical appeal. Once a male tried to approach a female to form a consortship, the first stage of female mate choice would be triggered. On the basis of his appearance and behavior, she would reject him (usually) or provisionally agree to continue interacting. This would impose sexual selection on males to create a positive impression during the first few minutes of interaction. After several hours or days of consorting, the female would decide whether to have sex. If she agreed, they would probably copulate frequently for several days or weeks. At that stage, male mate choice would once again reassert itself: will he stay with this female, or grow bored and abandon her in search of someone who would make a more interesting long-term partner? The female would be deciding the same thing: does he offer anything beyond a few orgasms and some good times?

Very Simple Rules Can Lead to Very Good Sexual Choices

How smart did our ancestors have to be to make all these complicated mate choices? A cognitive psychologist might try to construct mathematical models of how all the information about sexual cues gets integrated, and how all the individuals get compared. This makes the mate choice task look daunting. However, my research on simple rules for mate choice suggests that very good sexual choices can result from very fast, very simple decision rules.

Fitness indicators themselves make sexual choice simple. When a female long-tailed widowbird chooses a mate, she can get a pretty good estimate of his fitness simply by looking at the length and symmetry of his tail feathers. She does not need a complete DNA profile highlighting all his mutations—the tail is all she needs to see. The fitness indicators that our ancestors evolved also made sexual choice much easier. They could just pay attention to a few cues like height and facial appearance, and get a pretty good estimate of an individual's fitness. Each trait that we consider sexually attractive already summarizes a huge amount of information about an individual's genes, body, and mind.

We do not need to combine the information about these sexual traits in very complicated ways, either. It might seem difficult to compare two possible mates who differ in dozens of ways. It seems that the mathematically correct procedure would be to take each of their features, multiply it by its importance, add up all the results, and then compare the total score for each individual. But this is not necessary. Psychologist Gerd Gigerenzer and his colleagues have found that if you have to pick between two prospects based on a number of features, you can make extremely good decisions by doing something much, much simpler. You can rank the features you find most important, then compare the prospects on each feature until you find a feature where one prospect is clearly superior. For example, if you think intelligence and beauty are the most important two features in a sexual partner, you can just go down your list and compare each prospect. Is one significantly more intelligent than the other? If so, pick the bright one. If not, then is one significantly more physically attractive than the other? If so, pick the beautiful one. If not, choose randomly, because it doesn't matter. Gigerenzer's team has a lot of evidence that this very simple rule, which they call "Take the Best," makes decisions almost as good as the most sophisticated mathematical decision rules in almost every situation. It has astonishing power as a decision rule, yet it is very simple. If our ancestors used a rule of thumb like Take the Best to choose mates, they could have made very good decisions without needing to process a great deal of information using very complicated rules.

Although sexual decision-making can itself be fast and efficient, it sometimes takes time to acquire the relevant information about a potential mate. If a woman is interested in assessing a man's personality, intelligence, and experiences, it may take weeks of conversation before she has (unconsciously) gathered all the information she needs to fall in love. As we shall see in Chapter 10, conversations during courtship are how we learn the most about potential mates, and these conversations take time. Insofar as men may be satisfied with certain minimal standards of physical appearance before their sexual interest is aroused, their sexual decision-making may appear faster—but only because physical appearance can be judged much faster than character. When it comes to making long-term sexual commitments based on traits that are more than skin deep, men may take even longer than women.

Another challenge is to decide when to form a serious relationship while one is searching through a sequence of encounters and consortships. Economists and statisticians have developed mathematical models of optimal search that look appropriate. But here again a simple rule can do much better. The standard optimal search strategy is called the 37 percent rule. It is useful when you are looking for the best candidate for a position, and you encounter the candidates one at a time, and you have to offer the position on the spot to the first candidate you like, without going back to previously interviewed candidates. This is somewhat like looking for a long-term mate. The 37 percent rule says that you should estimate how many total candidates are likely to apply for the position, interview the first 37 percent of them, and remember the best out of that initial sample. Then, keep interviewing until you find a candidate who seems even better than that. Once you find that better candidate, stop searching and stick with that one. The trouble with this rule is that the time and energy costs of searching can grow very large if you have a large number of possible candidates. For single New Yorkers, it is infeasible to date 37 percent of Manhattan's population before finding a spouse.

In our research on mate search strategies, colleague Peter Todd and I found that a rule we call "Try a Dozen" performs as well as the 37 percent rule under a wide range of conditions. Try a Dozen is simple: interview a dozen possible mates, remember the best of them, and then pick the very next prospect who is even more attractive. You do not have to estimate the total number of potential mates you will encounter in your reproductive lifetime; you only have to bet that you will meet at least fifty or so. Humans seem to follow something like the Try a Dozen rule: we get to know a number of opposite-sex friends during adolescence, fall in love at least once, remember that loved one very clearly, and tend to marry the next person who seems even more attractive. Each individual is "satisficing"–looking for someone who is pretty good and good enough, rather than the absolute best they could possibly find. But at the evolutionary level, these satisficing rules impose sexual selection that is almost as strong as the most complicated, perfectionist decision strategy.

In general, very simple rules of thumb can result in sexual choices that are almost as good as the best strategies developed through mathematical analysis. Our ancestors did not have to have sexual supercomputers in their heads in order to make very good sexual choices under Pleistocene conditions of great uncertainty, limited information, and potential deception. Sexual selection does not require a sophisticated set of sexual choice rules. What matters is how efficient the rules are at distinguishing between mates. If very simple rules can make fairly good sexual decisions, then, across many matings and many generations, those rules can impose very strong sexual selection.

Indicators for Qualities Other than Fitness

When trying to attract a sexual partner, heritable fitness is not the only thing worth advertising. When males and females cooperate to rear offspring, they should care about more than each other's good genes. They should seek mates in good health because they are more likely to survive as partners and parents. They should seek mates capable of efficient cooperation and coordination, so they make an effective team. Since health and future cooperation cannot be assessed directly, they must be estimated using indicators such as energy level and kindness. Those indicators can evolve according to the same principles as fitness indicators.

Usually, there is a lot of overlap between basic fitness and these other qualities. Condition-dependent indicators can advertise both heritable fitness and the aspects of bodily and mental condition that are important for shared parenting. An individual who is grossly incompetent at finding food may have bad genes, bad condition, and bad parenting potential.

In principle, sexual choice could sometimes put non-heritable qualities ahead of heritable fitness. If the environment is so demanding that a female simply cannot raise a child by herself, then she might favor an attentive, experienced father, even if he has a lower general fitness than a charming athletic genius who is hopelessly incompetent with babies. However, she might still prefer to have an affair with the genius and let the experienced father raise the resulting child. New DNA methods for establishing paternity have shown that this sort of eugenic cuckoldry is surprisingly common in birds previously thought to be monogamous, and in humans.

Until recently, evolutionary psychology emphasized the non-genetic benefits of mate choice. This emphasis may have come in part from sexual selection terminology favored by biologists in the 1980s. Food gifts, nests, territories, and fertility were termed the "direct" benefits of mate choice, and good genes were termed the "indirect" benefits; it sounds more secure to receive a direct than an indirect benefit. In particular, leading evolutionary psychologists such as Don Symons, David Buss, and Randy Thornhill focused on the material benefits that high-status men could offer women, and the fertility benefits that healthy young women could offer men. This has been a powerful research strategy for explaining many sex differences in human mating behavior.

However, many male human courtship behaviors that appear to give purely material benefits to females may have evolved mainly as fitness indicators. Males of many species give females food during courtship. Male scorpionflies give females the prey they have caught. Our male ancestors probably gave females a share of the meat from the hunt. Until recently, men in modern societies brought home almost all of the money necessary to sustain their families. Don't females in all cases simply want a good meal instead of good genes? I think the analogy is deceptive. Male scorpionflies give females a significant proportion of all the calories the female will need to produce her next batch of eggs. Modern men used to give women all the money they needed to live in a market economy. But the meat provided by our male ancestors may have been only a minor contribution to the energy needs of a mother and her children. A pregnant hominid would have needed about four pounds of food a day for 280 days, about a thousand pounds in total. If a male hominid gives her ten pounds of meat during a month-long courtship, that's fairly generous by modern hunter-gatherer standards, but it is less than 1 percent of the food she will need just during the pregnancy.

Of course, given a choice between a fitness indicator that offers zero material benefits (such as an impressive courtship dance) and one that happens to produce a material benefit (such as an impressive hunting success), evolution may favor females who appreciate the material benefit. From a fitness indicator viewpoint, the material benefits simply bias evolution to favor fitness indicators that happen to deliver practical benefits in addition to information about mutation load.

Likewise, male defense of good territories may have evolved as a fitness indicator as well as a material benefit. Generally, female animals forage where they want, exploiting the available food resources. Males follow the females around and try to mate with them. The strongest males often succeed in driving the weakest away from the prime food-patches where the females have already decided to forage. Since the females might as well prefer a stronger to a weaker male, they might as well mate with the male who happens to be defending their food-patch. To a human observer used to the idea of land ownership, it might look as if the strong male has "acquired ownership" of the territory, which he generously allows the females to use. Perhaps even in the male animal's mind, he "owns" the territory. But to the females, they are just foraging wherever they want. The males may be running around and fighting each other, and large, muscular males may happen to last longer and stay closer to the females. The females have little incentive to go chasing after the smaller, weaker males that were driven away, so they may tend to mate with the stronger males. The females thus use the male's ability to defend the territory from other males as a fitness indicator. Sometimes the strategies of sexual choice are so efficient that they hardly look like active sexual choice at all. As long as the females do not stumble across any male trait that is a better fitness indicator than resource-defense ability, it may look as if the male automatically wins "the right to mate" by "owning the territory." But that would be missing the point. The females may be using the cue of resource-defense ability mainly to get good genes, not to get food.

In modern market economies people put a high value on wealth indicators during courtship. This can be rational, given the range of goods and services that money can buy, and the difference it can make to one's quality of life. As Thorstein Veblen argued a century ago, modern culture is basically a system of conspicuous consumption in which people demonstrate their wealth by wasting it on luxuries. Wealth indicators follow the handicap principle just as fitness indicators do, but this makes it easy to mistake one for the other. David Buss has amassed a lot of evidence that human females across many cultures tend to prefer males who have high social status, good income, ambition, intelligence, and energy—contrary to the views of some cultural anthropologists, who assume that people vary capriciously in their sexual preferences across different cultures. He interpreted this as evidence that women evolved to prefer good providers who could support their families by acquiring and defending resources. I respect his data enormously, but disagree with his interpretation.

The traits women prefer are certainly correlated with male abilities to provide material benefits, but they are also correlated with heritable fitness. If the same traits can work both as fitness indicators and as wealth indicators, so much the better. The problem comes when we try to project wealth indicators back into a Pleistocene past when money did not exist, when status did not imply wealth, and when bands did not stay in one place long enough to defend piles of resources. Ancestral women may have preferred intelligent, energetic men for their ability to hunt more effectively and provide their children with more meat. But I would suggest it was much more important that intelligent men tended to produce intelligent, energetic children more likely to survive and reproduce, whether or not their father stayed around. In other words, I think that evolutionary psychology has put too much emphasis on male resources instead of male fitness in explaining women's sexual preferences.

Age and Fertility

The most important quality that indicators advertise other than heritable fitness is age. Obviously, age is not directly heritable. A 40-year-old woman will give birth to a nine-month-old, just as a 20-year-old woman will. However, age has a dramatic effect on fertility, especially in women. Individuals before puberty are infertile. Female adolescents are significantly less fertile than 20-

year-olds. Female fertility declines gradually during the thirties, and declines steeply after age 40. Women after menopause are infertile. This female fertility profile is a basic fact of life to which male mate choice systems have adapted. Youth is an important cue of fertility.

There may have been male hominids who preferred to start exciting relationships with wise, fulfilled, 60-year-old females. But if they did so exclusively, they would have left no offspring to inherit that preference. Any sexual choice mechanism that preferred infertile individuals to fertile individuals would have died out in one generation. Since male sperm production ability declines more slowly with age, female mate preferences need not have paid so much attention to a man's age as a cue of his reproductive ability. This reasoning, as developed by Don Symons, David Buss, and other evolutionary psychologists, explains the universal, cross-cultural pattern that men care more about a partner's age than women do, men generally preferring partners younger than themselves, and women generally preferring partners older than themselves.

However, male hominids may not have been quite so youth-obsessed as men from agricultural, pastoral, and modern civilizations. In most cultures with recorded history, men were under social, legal, economic, and religious pressures to stay monogamously married for life. The younger their bride, the more offspring they could produce. This put a huge premium on youth, and men competed to claim young women before another man could.

A woman's youth may not have been quite so crucial in the Pleistocene, as long as the woman was still reasonably fertile. If our hominid ancestors had several medium-term relationships in sequence, males need not have been so picky about female age. If the relationship was likely to end after five years—as anthropologist Helen Fisher has argued that they usually did in prehistory—it would have mattered little whether she was 10 years or 30 years away from menopause.

During her reproductive years, a woman's age does have a

negative correlation with her fertility. But under challenging Pleistocene conditions, age would have had a positive correlation with heritable fitness because low-fitness individuals would have died younger. Any woman who managed to reach her mid-thirties and raise several children successfully, while staying physically and psychologically attractive, might have made a better genetic bet for a choosy male than an untested teenager of unproven fertility. Other male primates tend to shun adolescent females without offspring, and prefer older, high-ranking females with offspring who have already demonstrated their fertility, survival ability, social intelligence, and mothering skills.

There is strong evidence from evolutionary psychology that men in modern societies generally prefer the physical appearance of women around 20 years old to those who are older (or younger). But I have argued that this preference may have been amplified somewhat by the economic and religious pressures for monogamy since civilization arose, which makes finding a young bride crucial to a man's reproductive success.

More importantly, there has been much less research on the age at which women's minds are most attractive. Perhaps mature men tend to find young women beautiful but boring, and older women slightly less physically attractive but much more interesting. If so, we should not view the preference for youthful appearance as any less of a legitimate adaptation than the preference for a worldly mind. Data gathered by Doug Kenrick shows that older men generally prefer women closer to their own age—in their mid-thirties rather than their early twenties, for example—as long-term sexual partners. Presumably this is because women in their mid-thirties are typically more intriguing, multifaceted people who display the mental aspects of their fitness in richer ways that can be more reliably assessed. Evolutionary psychology has rightfully emphasized the strong male human interest in young female bodies, but I think its scope should be broadened to include the romantic interest aroused in both sexes by mature, worldly minds.

In any case, chronological age, like heritable fitness, could not be perceived directly during human evolution. To distinguish children from adults, our ancestors had to rely on cues of sexual maturity such as male musculature, beard growth, and voice pitch, and female breast and hip development. To distinguish young adults of peak fertility from other adults of declining fertility, they had to rely on age cues such as wrinkles, gray hair, sagging skin, slow gait, and memory loss.

Like fitness indicators, age indicators leave some room for deception. This may have some relation to our apparent "neoteny," which means that we have, it has been argued, retained some of the physical and mental traits of juvenile apes into our adulthood. Our faces look more like the faces of very young chimpanzees than they do like those of adult chimpanzees. Our playful creativity resembles the behavior of young primates more than it does the stern, lazy brutality of adult apes. Stephen Jay Gould has argued that our neotenization was a key trend in human evolution, and he sees our behavioral flexibility as a side-effect of our general neoteny.

But neoteny can be viewed very differently. Our neotenous features may have evolved through sexual choice as somewhat deceptive cues of youth. If male hominids preferred younger, more fertile females to older, less fertile females, then there would have been sexual selection pressures on females to appear physically and behaviorally younger than they really were. They could do this by evolving younger-looking faces, and by being more playful, creative, spontaneous, and uninhibited throughout their adult life. The result would be neotenized female hominids. The same argument could apply to males, insofar as female choice favored signs of youthful energy. (It is not clear why our lineage evolved these neotenous youth-cues while other primates did not—one could invoke sexual selection's unpredictability, though that is not a very satisfying explanation.) In my view, Gould's neoteny theory identified a set of somewhat deceptive youthfulness indicators that must have evolved through some form of sexual or social selection. It is not a competing theory of human evolution, but a description of some physical and psychological trends that still require an evolutionary explanation.

Apparent preferences for youth are not as simple as they seem. It is often hard to distinguish indicators of youth from indicators of fitness. This is because fitness indicators usually work by being very dependent on condition, and condition is highest during the flower of youth. All things being equal, any mate choice mechanism that evolved to favor a condition-dependent indicator will tend to favor youth over age simply because youths will display the indicator in a healthier condition. However, the fact that women often prefer older men suggests that mate choice mechanisms can easily evolve to compensate for this youth-bias whenever it proves maladaptive.

Fitness Indicators for People Other than Mates

Sexual selection was not the only kind of social selection during human evolution. For humans, as for most primates, all kinds of social relationships affect survival and reproduction. In forming and maintaining many of these relationships there are good reasons to advertise one's fitness, just as one does to potential sexual partners. Friends of higher fitness may survive longer, offer more competencies, and give better advice. Allies of higher fitness may help one to win fights and wars. Trading partners of higher fitness may live longer, travel longer distances to acquire more valuable commodities, and have the social intelligence to keep their promises. None of these social relationships entails any merging of genes, so they are not subject to positive-feedback processes as powerful as runaway sexual selection. But they still offer plenty of scope for all kinds of socially selected indicators to evolve.

We can often use the same fitness indicators in non-sexual relationships as we do in sexual relationships. If vigorous dancing all night displays our physical fitness to potential mates, it equally displays our fitness to potential friends and allies. Whenever a fitness indicator evolved in our ancestors through sexual selection, it was probably generalized to other social relationships rather quickly. Conversely, any indicator that evolved in the context of friendships or tribal alliances could easily have been modified for courtship.

The overlapping use of fitness indicators in sexual and non-sexual relationships is why making friends so often feels like a variant of sexual courtship. There is the same desire to present oneself to best advantage, emphasizing skills, downplaying weaknesses, revealing past adventures, investing extra energy in the interaction. This does not mean that friendships always have a sexual undercurrent, or that friendship is maintained through some kind of sexual sublimation. It simply means that the same principles of self-advertisement work in both kinds of relationship. If friendships gave important survival and social advantages during human evolution, and if our ancestors were choosy about their friends, then many of our fitness indicators may have evolved for friendship as well as for sexual relationships.

An especially important non-sexual relationship is that between parents and offspring. Children often compete to display their fitness to their parents, older siblings, and older relatives. They may shout "Hey dad, look at this!," and then try to do something that is challenging for a child of their age and abilities. At first glance it seems odd that they should bother. According to modern social norms, parents are supposed to love their children unconditionally, regardless of their fitness or abilities. But Pleistocene Africa did not always permit such unconditional support. Times were sometimes tough. Just as birds often have to choose which chick gets the worm and which starves, human parents may have had to choose how much support to invest in a particular child. Evolutionary psychologists Martin Daly and Margo Wilson have called this the problem of "discriminative parental solicitude." Parents must sometimes discriminate about which child deserves their solicitude. Older children are often favored because they have already survived the risky phase of infancy. But parents may also be sensitive to a child's fitness, which mean its prospects of successful survival and reproduction. Investment in a very low-fitness child means investment in an individual very unlikely to pass one's genes on to grandchildren. For better or worse, evolution considers that an unwise investment, and favors a more discriminating attitude. In every culture, children with physical deformities and serious psychological disorders are at enormously greater risk of neglect, abuse, beating, and infanticide by parents.

Given parents who discriminate between children based on their apparent fitness, children have incentives to evolve fitness indicators. As when people initiate friendships, children can use many of the same strategies that work in courtship, without there being any hidden sexual motive to the display. This is where I believe Freud went wrong with his hypotheses about Oedipus and Electra complexes. He observed a set of fitness indicators that children directed at parents—energetic play, humorous story-telling, flirtatious conversation—and inferred a secret children's desire to have sex with their parents. That inference seems evolutionarily incredible. Presumably our hominid ancestors evolved a set of sexual choice mechanisms for judging the fitness of potential mates. Perhaps children found it convenient to play upon some of the same mechanisms to advertise their fitness to their parents, to solicit more attention and care. This does not mean that children want incest—it means that they want parental support.

Gay Hominids?

Homosexuality has not been mentioned so far in this book. My heterosexual emphasis comes not from homophobia, religious conviction, or moral conservatism. My subject is human evolution, and homosexual behavior is just not very important in evolution. Not a single ancestor of any living human was exclusively homosexual. Any hominid that was would not have produced any offspring, and would not have become anyone's ancestor. There may have been many gay and lesbian hominids, but if they were exclusively homosexual, they are not our ancestors, and we are not their descendants. In any case, it is unlikely that there were many exclusively homosexual hominids. Any genetic propensity towards exclusive homosexuality would have been eliminated in just one generation of selection. No biologist has ever offered a credible theory explaining how exclusive homosexuality could evolve in a sexually reproducing species. Its existence in 1 or 2 percent of modern humans is a genuine evolutionary enigma that I cannot explain.

There is no such evolutionary problem with bisexuality, in which individuals enjoy sex with both sexes. Certainly bisexual behavior occurs in other species. Bonobos (previously known as "pygmy chimpanzees") engage in a lot of sexual activity with same-sex individuals, including kissing, genital rubbing, and genital licking. This does not impair their heterosexual reproduction in the slightest. Evolution does not respect our hunger for simplistic political categories of sexual behavior, in which every individual can be put on a continuum of "sexual orientation." Ordinary bonobos enjoy heterosexual behavior, and homosexual behavior, and they have lasted a million years as a species, about ten times longer than we have so far. There is nothing "unnatural" about homosexual behavior.

Moreover, many male humans with strong homosexual desires get married and produce offspring, as Oscar Wilde did. Many female humans with strong lesbian desires produce children too. Evolution has no moralistic motive to punish homosexual behavior. As long as homosexual behavior does not displace heterosexual behavior, it has little impact on evolution. Homosexual behavior—as an adjunct to heterosexual behavior—would be expected to evolve whenever its fitness benefits (making friends, appeasing threats, making peace after arguments) exceed its costs (energy, time, and the increased risk of sexually transmitted disease).

Our hominid ancestors might have been almost exclusively heterosexual, like chimpanzees, or very homoerotic like bonobos. We do not know. Even male chimpanzees hold each other's penises for comfort when they are frightened. Perhaps, like bonobos, humans evolved some adaptations for homoerotic flirtation and same-sex sexual friendships. If the social benefits of homosexual relationships were strong enough, homosexual preferences could, in principle, have shaped human physical appearance and mental capacities. However, these preferences had no direct reproductive consequences, so they would have had much weaker evolutionary effects than heterosexual preferences. As a result, we have to focus on heterosexual behavior when considering the role of sexual choice in the mind's evolution.

Mate Choice and Courtship as Social Events

Sexual choice and courtship in human evolution was not just a matter of boy meets girl. We have seen that our ancestors were highly social primates living in groups with children, relatives, and friends. Sexual relationships began and ended within family and tribal contexts.

If mate choice favors good genes, it can be useful to meet a potential mate's blood relatives, because they share some of the same genes. An individual's kin give additional information about their heritable fitness. If an intelligent man has foolish brothers or a beautiful woman has ugly sisters, this may lower their attractiveness as potential parents of one's children. Siblings share half of their genes, as do parents and offspring. The apparent fitness of a woman's mother or daughter carries half as much information about the woman's own genetic quality as her own fitness indicators. Given two sexual prospects who appear to display equal fitness, the one whose relatives appear healthier, brighter, more attractive, more fertile, and more successful probably has higher actual fitness. Since our ancestors tended to live in kin groups, there were plentiful opportunities for mate choice to take into account this sort of kin quality. Our mate choice systems would have evolved to exploit this gold mine of genetic information.

If sexual choice paid attention to the fitness of a potential mate's relatives, then those relatives would have been under sexual selection to display high fitness. This would have been a much weaker pressure than ordinary sexual selection, but it could still have been significant in shaping our instincts for display. If parents could help their offspring attract better mates by appearing intelligent, healthy, and successful, then the copies of their own genes that are carried in their offspring would benefit. Likewise, if children could help their mothers appear more attractive by demonstrating that they carry good genes, then the copies of their genes in their mothers would be passed on to larger numbers of half-siblings. Any courtship effort that helps your relatives to find good mates helps your own genes to spread. (Of course, there may be conflicts of interest between relatives over these courtship displays, as when adolescents wish that their parents made more effort to act reasonably cool when their friends visit, or divorced parents wish that their adolescents would behave better towards potential step-parents.)

Mate choice that takes into account the qualities of a potential mate's relatives would have favored hominids who spread their courtship effort out across their lifetimes. In childhood and old age their courtship would be vicarious, carried out on behalf of their relatives. In the prime of life it would be mostly for themselves, but also for their sexually active relatives. We should not expect to see fitness indicators used exclusively after puberty and before menopause, only that they are then directed at different targets.

Vicarious, collective courtship by relatives might explain why humans are so good at producing certain kinds of cooperative display. Evolutionary psychologists have usually assumed that human cooperation evolved for survival benefits. Cooperation can certainly help the cooperators survive better—if they are doing something that is actually useful together. But what about religious rituals, dances, and feasts that have high time and energy costs and no credible survival payoffs? Consider the huge Thanksgiving feasts that American families organize when a daughter first brings home a potential husband. The family members are not improving their collective survival chances; they are improving the daughter's mating prospects by demonstrating their wealth, health, family size, and other aspects of familial fitness. The prodigious waste of uneaten turkey even follows the predictions of the handicap principle. Across many cultures, marriage rituals serve similar functions, wasting vast resources so that a kin group can display its fitness to a group of possible in-laws. American families also advertise their wealth and status by producing costly rituals when one of them reaches sexual maturity—as in bar mitzvahs, debutante balls, and "sweet sixteen" parties. Rich parents even advertise familial fitness by paying over a hundred thousand dollars for each child to attend a private university, whereas in Britain, they pay even more for pre-university private schooling.

Modern human families compete to attract good mates for their young people. Perhaps Pleistocene kin groups and tribes did so too, inventing various rituals, myths, legends, totems, and dances to display their superiority over other groups competing in the same sexual market. To the extent that mating occurred across group boundaries, cooperative group activities may have evolved as collective courtship displays through sexual selection. This may explain the observation by anthropologists Chris Knight and Camilla Power that a great deal of human ritual behavior consists of collective displays by female relatives on behalf of their youngest female kin when they reach sexual maturity. Of course, once the mental capacities for collective fitness displays evolved through sexual selection, those capacities might prove useful for other functions as well, such as intimidating rival groups competing for the same territories and resources.

So much for collective courtship. As for collective sexual choice, each individual's mate choice decisions probably took into account the views of their parents, siblings, offspring, and companions. Sometimes they may have immediately discounted this advice, realizing that their relatives' interests did not coincide with theirs. But sometimes other individuals would have offered useful information about a potential mate. They may have interacted with the prospect in other contexts, or heard useful gossip. Older relatives may have offered words of wisdom from their past experience of sexual choices and sexual relationships. During the Pleistocene, when social conditions were less volatile than today, one generation's experiences of courtship and parenting would have been much more relevant to the next generation. Before the evolution of language, relatives could have revealed their attitudes about a sexual prospect through the usual primate signals: threat displays and attacks, or friendly grooming and food sharing. After language evolved, the relative merits of sexual rivals must have become subjects of impassioned discussion. Parents may have been especially vocal about their views, because the sexual choices of their children were so important to the number and quality of grandchildren who would carry their genes. However, parental influence on sexual choice does not imply some sort of arranged marriage system in which sexual selection no longer operates. On the contrary, by integrating information from several individuals our ancestors could have made much more accurate estimates of each prospect's strengths and weaknesses, driving sexual selection more strongly in particularly evolutionary directions.

Biologists have not developed models of how sexual selection works when mate choice and courtship are socially distributed. I would guess that the runaway process would not work so strongly when sexual preferences and sexually selected traits are spread across different bodies. It would have a harder time establishing the genetic correlations between preferences and traits that drive runaway. However, there may be fewer such problems with sensory bias effects and preferences for fitness indicators. For example, the sensory and cognitive biases of friends and relatives could influence an individual's sexual choices just as their own biases would. Ornaments and indicators could still evolve even if parents were choosing sexual partners for their children, and even if aunts were producing courtship displays on behalf of their nieces.

Afrocentrism

It should go without saying, but I'll say it anyway: all of the significant evolution in our species occurred in populations with brown and black skins living in Africa. At the beginning of hominid evolution five million years ago, our ape-like ancestors had dark skin just like chimpanzees and gorillas. When modern *Homo sapiens* evolved a hundred thousand years ago, we still had dark skins. When brain sizes tripled, they tripled in Africans. When sexual choice shaped human nature, it shaped Africans. When language, music, and art evolved, they evolved in Africans. Lighter skins evolved in some European and Asian populations long after the human mind evolved its present capacities.

The skin color of our ancestors does not have much scientific importance. But it does have a political importance given the persistence of anti-black racism. I think that a powerful antidote to such racism is the realization that the human mind is a product of black African females favoring intelligence, kindness, creativity, and articulate language in black African males, and vice versa. Afrocentrism is an appropriate attitude to take when we are thinking about human evolution.

7

Bodies of Evidence

By primate standards, humans look strange, even after we step out of our sport utility vehicles. Compared with other apes, we have less hair on our bodies, more on our heads, whiter eyes, longer noses, fuller lips, more expressive faces, and more dextrous hands. In most species, sexual ornaments like long head hair, hairless skin, and full lips would have evolved only in males, because females would have been the choosy sex. Males have few incentives to reject any female mates. The fact that both human sexes evolved distinctive sexual ornaments shows that both female choice and male choice was important in human evolution. If both sexes were choosy about bodies, they might also have been choosy about minds.

Not only do we look different from other apes, but each human sex also has distinctive body traits shaped by sexual selection. Men are taller and heavier on average than women, with more upper body strength, higher metabolic rates, more hair, deeper voices, and slightly larger brains. Some of these traits may have evolved for sexual competition against other males. But male bodies are also living evidence of the sexual choices made by ancestral females. Men grow beards, and possess penises that are much longer, thicker, and more flexible than those of other primates. These are more likely to reflect female choice than male competition. Women also evolved to incarnate male sexual preferences. Women have enlarged breasts and buttocks, narrower waists, and a greater orgasmic capacity than other apes.

Sexual selection has also made male bodies grow according to a higher-risk, higher-stakes strategy. For males there is a higher incidence of birth defects, more death in infancy, higher mortality at every age, earlier senescence, and greater variation in health, strength, body size, brain size, and intelligence. This risky, go-for-broke strategy suggests that sexual competition among males was often a winner-takes-all contest. It was better to take a big gamble on producing the most attractive image during a short peak, rather than aiming to create a mediocre impression over a long period of time.

Our bodies are rich sources of evidence about sexual selection pressures because they are visible, measurable, easily comparable with those of other species, and relatively undistorted by human culture. In recent years much nonsense has been written by post-modern theorists such as Michel Foucault about the "social construction of the body," as if human bodies were the incarnation of cultural norms rather than ancestral sexual preferences. These theorists should go to the zoo more often. What they consider a "radical reshaping" of the human body through social pressure is trivial compared to evolution's power. Evolution can transform a dinosaur into an albatross, a four-legged mammal into a sperm whale, and a tiny, bulgy-eyed, tree-hugging, insect-crunching proto-primate into Julia Roberts—or Arnold Schwarzenegger. Selection is vastly more powerful than any cosmetic surgeon or cultural norm. Minds may be sponges for soaking up culture, but bodies are not.

The most sexually selected parts of our bodies have been neglected in theories of human evolution because they don't fossilize. Sexual choice sculpts body ornaments out of muscle, fat, skin, and nerves, often without leaving many clues in the bones. This makes it hard to know when and where these traits evolved. We don't know how hairy our ancestors were a million years ago, whether *Homo erectus* males had huge penises, or whether Neanderthal females had large breasts. But we do know that our body's sexual ornaments are universal across human groups, so they must have evolved at least 60,000 years ago or so, when human groups colonized different areas of the world. In these respects our bodily ornaments are like many of our mental adaptations for courtship: we don't have much fossil evidence about their antiquity, but we can infer a lot from their modern human form and their absence in closely related ape species.

If sexual selection drove our bodily divergence from other apes, it may have driven our mental divergence as well. René Descartes saw a dichotomy between body and mind, but sexual choice judges them as a package. As Walt Whitman put it in his 1855 poem "One's-Self I Sing:"

> Of physiology from top to toe I sing,
> Not physiognomy alone nor brain alone is worthy for the Muse,
> I say the Form complete is worthier far,
> The Female equally with the Male I sing.

Penises, clitorises, breasts, and beards are fascinating not only in their own right, but also for what they reveal about sexual selection among our ancestors.

Which Body Traits Evolved as Sexual Ornaments?

Many of our body traits such as penises, breasts, buttocks, beards, head hair, and full lips show the hallmarks of sexual selection through mate choice. They are uniquely amplified in our species. Many of them show large sex differences. Mostly, they appear or enlarge only after puberty, and become more engorged with blood during sexual arousal. All around the world they are clearly valued as sexual signals, and are made more conspicuous through embellishment and make-up. They probably evolved partly as fitness indicators and partly as ornaments through runaway or sensory preferences. A body trait does not have to fulfill all of these criteria to qualify as a sexually selected ornament or indicator, but the more the better. Many of these criteria work for mental traits as well as body traits, so we'll be using them often throughout the rest of this book.

As we have seen, sex differences are highly diagnostic of sexual selection. Traits found in one sex but not the other usually result from sexual selection. Yet sexual selection does not always produce sex differences. Where we find sex differences it is likely that sexual selection has affected at least one sex, but even if we do not find sex differences in an ornament, sexual selection may still have affected both sexes.

A common fallacy is to argue that sex hormones are sufficient to explain sex differences. This is not an alternative to sexual selection, it just identifies a mechanism that sexually selected genes use to produce sex differences. For example, the hormone testosterone is a simple molecule that cannot by itself carry instructions for growing a complex trait such as a penis or a beard. Rather, the genes for growing penises and beards have evolved the ability to be switched on in response to testosterone, because testosterone tells these genes that they happen to be in a male body in this generation. (Most genes underlying distinctive male and female traits are present in both sexes, but are activated only by the cascade of sex hormones during fetal development.) A trait's sensitivity to sex hormones is itself a product of sexual selection.

Active display in courtship is a good sign that sexual selection has shaped a trait. Since courtship is restricted to sexual maturity, any trait that grows only after puberty is likely to be a result of sexual selection. Prepubescent girls don't grow breasts because they would be physiologically expensive and encumbering. Only when attracting a mate becomes a potentially adaptive thing to do, do the breasts sprout. Likewise for male beards and other body hair, male penises, male upper-body musculature, and many other traits. On a shorter time-scale, some bodily ornaments change their state during sexual arousal, the most intense phase of courtship. The penis grows erect and larger. A sexual flush spreads over a woman's neck, chest, and breasts. The breasts, lips, and labia engorge with blood. Traits that attain their full form only during sexual maturity and sexual arousal probably evolved through sexual choice.

Is the trait still viewed as sexually attractive today, across human cultures? Traits shaped by prehistoric sexual choice should still be considered sexually attractive today, insofar as our sexual choice mechanisms remain similar. If a bodily trait is considered sexually attractive across a wide range of cultures and historical epochs, the trait was probably viewed that way during human evolution. The manifest sexual appeal of female breasts and buttocks, for example, seems subjectively obvious to all heterosexual male humans, and that obviousness is good evidence for these traits having arisen through male mate choice. Around the world, the same bodily traits tend to be emphasized with special clothing and ornamentation when individuals wish to appear attractive, the same traits are covered when they wish to avoid sexual harassment, and the same traits are mutilated as punishment for sexual offenses.

When anthropologists claim that standards of beauty vary capriciously from one culture to another, they are usually studying the wrong traits in the wrong ways. Individuals of different cultures may like skin of different shades, but they all prefer clean, smooth, unwrinkled skin. Women differ in the exact male height they prefer, but almost always prefer a man taller than themselves. Different ethnic groups may prefer different facial features, but all prefer faces that are symmetrical and averagely shaped for their population. If you don't look for the universals of human beauty at the right level of description, you will not find them.

There is another test we have seen throughout this book: traits that are unique to one species are often the outcome of sexual selection. This is because traits shaped by natural selection that prove useful for survival tend to make a species successful, and successful species tend to split apart into daughter species. The species turns into a genus (a group of closely related species), and the useful trait is shared by all members of the genus. Sexual ornaments do not usually increase survival success, however, so each particular ornament tends to stay restricted to one species.

Even within a species, sexual selection produces diversity between populations. In humans, the runaway effect can take different populations ("ethnicities," "races") off in different evolutionary directions, ornamenting them with different face shapes and body traits. Where the divergence has no apparent relationship to different climates or ecological challenges, it probably arose through sexual selection. Human populations differ markedly in skin color, eye color, hair length, facial features, breast size, and penis size. Darwin took such differences as evidence for such traits having diverged rapidly and recently through sexual selection, but he may have overstated his case. Natural selection can account for some latitude trends, explaining why skins got lighter, noses got larger, and bodies got shorter and thicker as human populations migrated from equatorial zones to colder climates. However, latitude and climate cannot account for most of the subtler differences between populations. Most differences in eyes, hair, facial features, and the sizes of breasts, buttocks, and penis are more likely to be consequences of sexual choice focusing on different traits in different populations.

Because sexual choice often shapes traits to work as fitness indicators, it can also produce traits that show large differences between individuals within the same population. If male choice selected female buttocks as reliable indicators of fertility, health, and youth, we should not expect all females to have identical buttocks, for that would make the trait useless as an indicator. Evolutionary psychologists are discovering that many human body traits advertise a particular aspect of fitness called "developmental stability." This refers to an individual's ability to grow a trait in a normal form despite the mutations they may be carrying, and despite the environmental challenges (poor nutrition, parasites, injuries) that they may encounter during development inside and outside the womb. For traits that normally grow symmetrically, like faces and breasts, the exact degree of symmetry can be a powerful indicator of developmental stability, which in turn is a major component of fitness. (Symmetry is just one way to measure developmental stability—it could also be measured by comparing the similarity of identical twins who have grown from the same genes, for example.) Bodily symmetry is biologically important because it is one of the easiest components of fitness for biologists to measure, and for animals to assess when choosing mates. The symmetry of sexual ornaments is an important determinant of sexual attractiveness in many species, including our own. Many of our bodily ornaments, not least faces and breasts, probably evolved in part as symmetry indicators.

We can use these criteria to identify parts of the human body that probably evolved through female choice, male choice, or both. The more evidence we find for mutual choice having shaped the body, the more reasonable it becomes to suggest that mutual choice shaped our minds as well, without creating large sex differences in mental abilities.

The Evolution of the Penis

Sexual reproduction does not really require many sex differences. Males must make sperm, and females must make eggs. But males do not have to grow penises, and females do not have to grow clitorises. Male frogs and birds do not have penises. Genitalia are products of sexual choice, not requirements for sexual reproduction. The traditional distinction between "primary" sexual traits (such as penises) and "secondary" sexual traits (such as beards) is misleading. Perhaps for reasons of Victorian propriety, Darwin wrote as if female choice applied only to the secondary sexual traits. But modern biologists view penises themselves as targets of sexual choice. Biologist William Eberhard has argued convincingly that male genitals in a wide range of species are shaped as much by female choice as by the demands of sperm delivery.

Adult male humans have the longest, thickest, and most flexible penises of any living primate. The penises of gorillas and orangutans average less than two inches when fully erect, and those of chimpanzees average only 3 inches. By contrast, the average human penis is over 5 inches when erect. The longest medically verified human penis was about 13 inches when erect, more than twice the average length.

Even more unusual than the length of the human penis is its thickness. Other primate penises are pencil-thin, whereas the erect human penis averages over one inch in diameter. Also, most other primates have a penis bone called the "baculum," and achieve erections mostly through muscular control, like a winch raising a rigid strut. The penis bone is typical of most mammals. By contrast, the male human relies on an unusual system of vasocongestion. The penis fills with blood before copulation, like a blimp inflating before flight.

Although it is larger than any other primate's, the human penis has plenty of rivals in more distantly related animals. Blue whales and humpback whales have penises eight feet long and one foot in diameter. Bull elephants have penises around five feet long. Boars have 18-inch penises that ejaculate a pint of semen. Hermaphroditic snails have penises about as long as their entire bodies. Stallions, like men, use blood rather than muscular contraction to fill their much larger penises. Dolphins have voluntary control over the tips of their man-sized penises, which can swivel independently of the shaft. Male genitals are even stranger among the invertebrates, sporting a dizzying variety of sizes, flagella, lobes, bifurcations, and other ornaments, apparently designed to stimulate invertebrate female genitalia in as many different ways as there are species.

Didn't penises evolve just to deliver sperm? Sperm competition is certainly one of the most important forms of reproductive competition. If two males copulate with a female when she is fertile, their sperm are in competition. Only one, at best, will fertilize her egg. The male with the fastest, longest-lasting, most numerous sperm is more likely to pass on his good-sperm genes to his sons. Heritable differences in sperm quality and sperm delivery equipment will be under intense selection. Male humans show many adaptations for sperm competition, both physical and mental. For example, some studies have shown that when a woman returns home from a long trip, her partner tends to produce a much larger ejaculate than normal, as if to overwhelm any competitor's sperm that may have found its way into his unwatched partner's vagina.

However, comparisons of male testicles across species reveal that penises did not evolve purely for spermatic firepower. Among primates, the intensity of sperm competition correlates much more strongly with testicle size than with penis size. For example, male chimpanzees face much greater sperm competition than humans. When female chimps ovulate, they copulate up to fifty times a day with a dozen different males. In response, male chimps have evolved huge, 4-ounce testicles to produce sperm, but only small, thin penises to deliver it. At the other extreme, male silverback gorillas guard their harems vigilantly and violently, and tolerate no sperm competition, so they have evolved very small testicles. Humans have moderately sized testicles by primate standards, indicating that ancestral females copulated with more than one male in a month fairly often. Sequential fidelity to different men in different months would not produce any sperm competition, because each egg would be exposed only to one man's sperm. The fact that male human testicles are larger than those of gorillas is one of the strongest pieces of evidence that ancestral females were not strictly monogamous.

For sperm competition, sperm count and ejaculate volume are more important than penis length or thickness. A thick penis might tend to keep a competitor's sperm inside a female rather than allowing it to wash out. A long penis tends to overshoot the cervical opening rather than meet it accurately. Many species adapted for heavy sperm competition evolve penises with scoopers, scrapers, suckers, and flagella for removing rival sperm. If sperm competition were the driving force behind penis evolution, males might have evolved scary-looking flagellated genitals. Men would copulate by inserting their equipment, instantly flooding the cervix with half a pint of semen, and then lying on top of the woman for the next three days to make sure no rivals have the chance to introduce competing sperm. I understand that such behavior is quite rare.

Size Mattered

Male scientists have traditionally viewed the penis as a sperm-delivery device or a symbol of dominance in male competition. They neglected to consider the possibility that the penis evolved through female choice as a tactile stimulator. One popular theory, developed in the 1960s, was that human penile displays evolved to intimidate rival males rather than to attract females. This is an odd idea, given that in most ritualized threat displays males advertise features related to fighting ability. Dominant gorillas intimidate subordinates with their awesome muscles and sharp teeth, not their one-inch penises. I suspect that heterosexual male scientists find it difficult to think of the penis as something that evolved through sexual choice because it felt good inside one's body.

Most female scientists have been equally reluctant to suggest that penis size or shape was important to the sexual satisfaction of ancestral females. In her book *Mystery Dance*, biologist Lynn Margulis argued that "penis dimension is neither the major determinant of female sexual pleasure nor is a big penis a guarantee of female pleasure." Other women who wish to avoid perpetuating the myth that penis size is all-important go to the opposite extreme and claim that modern women do not use penis size at all as a mate selection criterion, so neither did our ancestors. Nonetheless, I suspect that few modern women would be happy with a sexual partner who had a penis of chimpanzee design—less than three inches long, half an inch thick, and rigid with bone. Of course, no single sexually selected trait is a guarantee of satisfaction. Sexual selection works on the principle of all else being equal. Given two otherwise identical hominid males, if female hominids consistently preferred the one with the longer, thicker, more flexible penis to the one with the shorter, thinner, less flexible one, then the genes for large penises would have spread. Given the relatively large size of the modern human penis, it is clear that size mattered. If it had not, modern males would have chimp-sized sexual organs.

So, why did picky female hominids start selecting for larger penises? Perhaps upright walking gave females a better view of male genitals. Anthropologist Maxine Sheets-Johnstone has argued that bipedalism may have evolved in part because it makes penile display more effective. She observed that in other primates, bipedal standing and walking are most often done by males displaying their penises to potential mates. Bipedal genital displays to strangers are now considered a criminal offense rather than a legacy of primate courtship. Likewise, the male open-legged sitting position, still universal across cultures, resembles open-legged penile displays by chimpanzees. If Sheets-Johnstone is right that bipedalism originated as a form of male sexual display, then here is another example of an evolutionary innovation originating through sexual selection and later proving useful for survival.

Against the visual display idea, however, is the fact that human penises are a rather sorry spectacle. We have not evolved a bright purplish-pink scrotum and a bright red penis with a yellow tip, as one species of mandrill has. Male vervet monkeys have a blue scrotum and a red penis set off against white hair. When primate penises are selected for visual appearance, they evolve much more color, and females seem to consider them much more attractive. The male human penis does not appear to be especially well adapted for producing auditory, olfactory, or gustatory stimulation. That leaves the sense of touch as the medium for female choice.

Female Choice Continued After Copulation Began

The role of female choice in penis evolution is revealed in the way the penis is used during copulation. Biologist William Eberhard has argued that copulation is not the end of courtship, but rather its most intense phase. In most species, female choice does not end when a male penis first enters, but can continue until sperm actually reach a fertile egg. Eberhard calls this "copulatory courtship." Some female insects can store the sperm of several males for weeks and use it when they want to fertilize their eggs. Many female mammals (unconsciously) squeeze the ejaculate of some males back out after copulation—a process called "flowback"—as if rejecting sperm from males whose copulation is not up to their standard. In a human female with concealed ovulation, a male's sexual ability may influence whether she keeps copulating with him, and that will determine his likelihood of producing offspring with her. If she rejected him after one or two unexciting encounters, he is very unlikely to father her children.

The duration and intensity of copulatory courtship in a species is a clue to the power of female choice. If efficient sperm delivery were the only point of copulation, a single thrust would be sufficient. Tomcats use this hit-and-run strategy. Copulation in most birds is very brief, and this absence of copulatory courtship is probably why birds have not evolved penises. Most primates make several separate "mounts" and several thrusts per mount before ejaculating. Copulatory thrusting seems designed to maximize the intensity, duration, and rhythmicity of tactile stimulation delivered to the female genitals. Delivering stimulation in addition to delivering sperm suggests that female choice has been important.

Copulatory courtship was probably especially important among hominids. Continuous sexual receptivity and concealed ovulation gave our female ancestors an unprecedented opportunity for testing males as sexual partners, while running a lower risk per copulation of unwanted pregnancy than any other primate did. Sex during menstruation, pregnancy, and breast-feeding would also have given ample opportunity for judging potential long-term lovers by their copulatory skills.

In species that do not use copulatory thrusting, especially insects, penises evolve more obvious tactile stimulators: nubs, spikes, ridges, curls, barbs, hooks, and flagella. Male insects often try to push each other off during copulation, so copulatory thrusting would risk disengagement. Better to lock the genitals together and have internal flagella to excite the female. With primates, it is not so common for male rivals to swarm over females knocking each other off. This allows couples a bit more copulatory leisure, with more complex movements favoring simpler penis designs. The human penis is especially streamlined because ancestral females apparently favored whole-body copulatory movement over the flagellar vibrations favored by female insects. Perhaps whole-body copulatory movements, requiring much more energy than waving a couple of vibrators on the end of the glans, were better indicators of physical fitness. It is not clear whether many middle-aged men do actually have heart attacks during vigorous sex with mistresses, but this plausible risk reveals the energetic costs of human copulation, and one way that female demands for tactile stimulation separate the healthy from the unhealthy. The loss of the baculum (penis bone) also reveals female choice for tactile stimulation. Since male human penises become erect with blood rather than muscle and bone, this gives them more flexibility, and permits a greater range of copulatory positions. Although bonobos also enjoy face-to-face copulation, their positional variety pales in comparison to the *Kama sutra.* Human penises evolved as tactile stimulators for use in copulatory courtship. Further research may clarify whether penises and copulatory courtship evolved mostly as fitness indicators or just as sexually selected entertainment.

Female hominids may not have preferred thicker, longer, more flexible penises per se. They may simply have liked orgasms, and larger penises led to better orgasms by permitting more varied, exciting, and intimate copulatory positions. This rather contradicts the view of the penis as a symbol of male domination. If we were a species in which males dominated the sexual system, we would have one-inch penises like dominant gorillas. The large male penis is a product of female choice in evolution. If it were not, males would never have bothered to evolve such a large, floppy, blood-hungry organ. Ancestral females made males evolve such penises because they liked them.

The Penis and the Brain

Why have I paid so much attention to the evolution of the penis? One reason is its importance as a genetic conduit. Every gene in every human body has passed through thousands of penises over thousands of generations of human evolution. Equally, every gene has passed down through thousands of eggs inside female ancestors who chose to copulate with particular males. In sexually reproducing species, copulation is the genetic gateway from one generation to the next, which is what makes it so important evolutionarily, physically, and psychologically.

The penis is an easy trait to study because it is visible, measurable, and directly comparable to the analogous organs of other species. Yet even for such a simple trait we have seen how the biases of male and female scientists may have influenced their views on penis evolution. We have considered both the sperm competition model and the "symbol of dominance" model for penis evolution. I could have mechanically run through the checklist of criteria for identifying sexually selected traits, but that would get rather tedious for every adaptation I shall be assessing in the rest of the book. The penis's fit to the criteria is rather obvious anyway: the penis shows distinct sex differences (it is much larger than the homologous female organ, the clitoris), grows mainly after puberty, is used during copulatory courtship, is considered sexually attractive by internal touch if not by sight, and differs markedly between species.

Physical organs shaped by sexual choice can also be seen as metaphors for mental organs shaped by sexual choice. Just as the human penis has been misunderstood as nothing more than plumbing for delivering sperm, the human mind has been misunderstood as wiring for processing information. In both cases, I argue that the organ evolved for the stimulation it can deliver, not to solve some straightforward physical problem of insemination or toolmaking. The sexual choice that mattered did not focus directly on the physical form of the organ, but on the shared experiences it could generate. Ancestral females did not apparently favor penises directly as visual ornaments, but favored them indirectly for the copulatory pleasure that they afforded, so they came back for more. Perhaps our ancestors did not favor intelligence and creativity directly, but indirectly: for how they contributed to having a great time with someone. If the penis really did evolve through female choice as a copulatory stimulator, then it should be considered not just a physical organ that reaches inside the body, but a psychological organ designed to reach inside the pleasure systems of another individual. It happens to have a physical form only because the other individual's pleasure systems happen to be connected to tactile sensors.

The Clitoris and the Orgasm

In most species in which males have a penis, females have a homologous organ called the clitoris. "Homologous" means that both organs grow from the same kinds of cells in the fetus. Anatomically, the human clitoris has the same three-part columnar structure as the penis: a glans, a shaft, and bifurcating roots. The main differences are that the penis is much larger overall, its shaft protrudes much more from the pelvis, it keeps blood from flowing back out when aroused, and it has a tube down the axis for urine and semen.

The human clitoris shows no apparent signs of having evolved directly through male mate choice. It is not especially large, brightly colored, specially shaped, or selectively displayed during courtship. By contrast, in spider monkeys the clitoris is almost as large as the penis, protruding nearly an inch. In hyenas, the female clitoris is larger than the male penis, and seems to play a role in female competition. The human clitoris could easily have evolved to be much more conspicuous if males had preferred sexual partners with larger, brighter clitorises. Its inconspicuous design combined with its exquisite sensitivity suggests that the clitoris is important not as an object of male mate choice, but as a mechanism of female choice. It helps to select for males who provide pleasurable foreplay, copulation, and orgasms, and such discriminative power is just what we should expect from an organ of female choice. Yet this has led to all sorts of confusion among evolutionists.

Some male scientists, such as Stephen Jay Gould and Donald Symons, have viewed the female clitoral orgasm as an evolutionary side-effect of the male capacity for penile orgasm. They suggested that clitoral orgasm cannot be an adaptation because it is too hard to achieve. Sigmund Freud suggested that clitoral orgasm was a sign of mental disorder, and counseled his female clients to learn how to have purely vaginal orgasms. Other male scientists such as Irenaus Eibl-Eibesfelt and Desmond Morris have viewed female orgasm as a reinforcement mechanism for promoting long-term pair-bonding that keeps a female faithful to her mate. They also wondered why clitorises have such trouble provoking orgasm. They assumed that if clitorises worked properly like penises, they should just do their job of promoting marital satisfaction without so much copulatory effort.

These men seem to have overlooked the possibility that clitoral orgasm is a mechanism for female choice rather than pair-bonding. Mechanisms for choice have to be discriminating: they must fire off excitedly when given the right stimulation, and emphatically must not fire off when given inferior input. As a mechanism for female choice, we would not expect female clitoral orgasm to respond to every male copulation attempt, however inept, lazy, inattentive, brief, and selfish. It is possible for a woman's vagina to become lubricated during unwanted sex to avoid injury, but women under such conditions practically never have orgasms. This is strong evidence of clitoral orgasm's role in female choice.

From a sexual selection viewpoint, clitorises should respond only to men who demonstrate high fitness, including the physical fitness necessary for long, energetic sex, and the mental fitness necessary to understand what women want and how to deliver it. The choosy clitoris should produce orgasm only when the woman feels genuinely attracted to a man's body, mind, and personality, and when the man proves his attentiveness and fitness through the right stimulation.

Not surprisingly, female scientists have held the clitoris in higher regard than have male scientists. Helen Fisher, Meredith Small, and Sarah Blaffer Hrdy have viewed the clitoral orgasm as a legitimate adaptation in its own right, with major implications for female sexual behavior and sexual evolution. Lynn Margulis has pointed out that female orgasm leads to female choice, and female choice is how females influence the evolutionary trajectory of their species. Natalie Angier's recent book *Woman: An Intimate Geography* stressed the clitoral orgasm's role in sexual choice: "She is likely to have sex with men she finds attractive, men with whom she feels comfortable for any number of reasons, and thus to further her personal, political, and genetic designs." I agree that the clitoris is an adaptation for sexual choice, and want to go one step further in considering its design within a sexual selection framework.

The sex difference between penis and clitoris can be viewed as a physical manifestation of Fisher's runaway process: a highly developed male trait (the penis) designed to stimulate, and a highly discerning female preference (the clitoral orgasm) designed to respond selectively to skillful stimulation. If this runaway model is right, then there was a sort of stimulatory arms race between the human penis and the human clitoris. The penis evolved to deliver more and more stimulation, while the clitoris evolved to demand more and more.

This tension explains why women and men are not well adapted to giving each other easy, simultaneous, repeated orgasms. If the function of orgasm were simply to reinforce monogamous pair-bonds, why should evolution make female orgasm so difficult and male orgasm so easy during vaginal intercourse? If female orgasm is a side-effect of male orgasm, why does it just happen to work when an attractive man provides a lot of foreplay and deep, slow copulatory thrusting, but not so well when sex is hurried or the partner is undesirable? Surely, sexual selection theory offers insight into this ancient human mystery. Female orgasm seems poorly designed as a pair-bonding mechanism, but it is perfectly designed as a discriminatory system that separates the men from the boys.

Yet the image of an evolutionary arms race between penis and clitoris is not quite accurate. The female mechanism for assessing penis size is not the clitoris itself, but the ring of nerves around the entrance to the vagina, which sense circumference. The clitoris does something more sophisticated, assessing the male's ability to move in pleasurable, rhythmic ways during copulation. Also, clitoral stimulation usually leads to orgasm only when the female mind is feeling erotic about the man and the situation. Human female orgasm depends on an interaction between the clitoris, the hypothalamus (the brain's emotional center), and the cerebral cortex (the brain's cognitive center). The clitoris is only the tip of the psychological iceberg in female choice. Having a mate with a large penis is not enough. To be fair, the penis is not just an insensate stimulator either. It is also a mechanism for male mate choice. If it is happy, its owner may be more likely to stay in a long-term relationship with a woman.

Tragically, while scientists in developed countries spent decades debating whether clitorises are legitimate adaptations, over a hundred million clitorises were cut out of African girls by village women precisely so that the girls would not be tempted to exercise their powers of sexual choice. Currently, another two million girls a year are genitally mutilated in countries such as Egypt, Sudan, Somalia, and Ethiopia. To my mind, sexual selection theory offers a powerful scientific rebuttal to the argument that we should accept female genital mutilation in such countries as part of "traditional tribal practice."

Just as the penis can be seen as a metaphor for the mind's sexually selected entertainment abilities, the clitoris can be seen as a metaphor for the mind's judgment and discrimination abilities. When we see a human perceptual or cognitive ability that looks curiously sensitive to stimulation yet resistant to satisfaction, we should not assume that it is a poorly designed information processing system. It may be part of a system for sexual or social discrimination. Consider humor. Some theories of humor have proposed that laughter evolved to promote group bonding, discharge nervous tension, or keep us healthy. The more laughter the better. Such theories predict that we should laugh at any joke, however stupid, however many times we have heard it before, yet we do not. A good sense of humor means a discriminating sense of humor, not a hyena-like shriek at every repetitive pratfall. Such discrimination is easy to understand if our sense of humor evolved in the service of sexual choice, to assess the joke-telling ability of others.

Breasts

By definition, all female mammals have mammary glands that

produce milk for feeding offspring. Any discussion about the evolution of breasts has to take this mammalian heritage as the starting point. Milk-substitute manufacturers have worked very hard for almost a century to convince women that they are not mammals and have no business breast-feeding. Even many science journalists support this view, as when some recent research was reported as showing that "breast-feeding raises IQ by five points," rather than "bottle-feeding reduces IQ by five points"—as if bottle-feeding was the biological norm. The popularity of bottle-feeding and breast implants should not mislead us into viewing breasts as nothing more than sexual ornaments.

During human evolution, female breasts would have been producing milk about half of the time between puberty and menopause. Babies probably nursed for at least a year or two, as they do in hunter-gatherer societies today. Without contraception, after a mother stopped nursing one baby she would typically have conceived the next baby within a few months. Assuming that the average female hominid produced at least 20 fluid ounces of milk per day when breast-feeding, and she spent a total of ten years breast-feeding in her life, the average hominid breast would have delivered over 35,000 fluid ounces (nearly 300 U.S. gallons) before menopause.

This high level of milk production does not itself explain why female humans breasts are so much larger than those of other apes. Most primate females are quite flat-chested, even when producing milk. Milk output depends on the amount of active glandular tissue in the breast, not the volume of fat. Human breasts have an unusually high ratio of fat to glandular tissue. They do not seem to be optimized for milk production. Most experts on breast-feeding claim there is no correlation between breast size before pregnancy and milk production ability after birth (though I know of no good data on this point). Milk output seems limited more by a woman's overall nutritional state than by her pre-pregnancy breast size. So, we have to distinguish between mammary glands, which evolved for milk production, and enlarged human breasts, which must have evolved for something else. It seems likely that sexual selection played a role. But how?

Perhaps breasts evolved as cues of sexual maturity. Human breasts enlarge at puberty, long before they are required for breast-feeding the first baby. Just as bipedal walking may have allowed female choice to focus more on the penis, bipedalism may have allowed male choice to focus on female breasts as a maturity cue. However, maturity cues do not have to be so dramatic. Males have evolutionary incentives to distinguish mature women from infertile girls, women have evolutionary incentives to advertise their fertility, and girls have evolutionary incentives to advertise their infertility. Given these shared interests, signals of sexual maturity could be very inconspicuous. Males of most other species have no trouble distinguishing mature from immature females using relatively subtle cues.

It seems likely that male choice shaped breasts not to distinguish girls from women, but to distinguish young women from older women. Here, the informative thing about breasts is the way they droop with the effects of age and gravity. There is a relatively narrow age window in which large breasts can appear pert before repeated cycles of pregnancy and breast-feeding cause them to sag. There were no bras or breast-lift operations in the Pleistocene. As we saw in the previous chapter, hominid males probably favored younger women for their higher fertility. Any indicator of youth, such as large, pert breasts, would tend to be favored by males. A male preference for size and pertness would spread at the expense of male preferences for droopiness and flatness, because the latter preferences would generally lead men to choose older, less fertile partners.

This argument sounds fine from the male point of view, but it takes a bit of thought to see why females should evolve youth indicators. The most informative cues of youth are also the most informative cues of age. Youth indicators might make women more attractive when they are truly young, but might make them less attractive when they are older. A mutation that caused an enlargement in breast size might benefit its carriers when they are in their teens and twenties, but impose high costs when they are in their thirties and forties. The question is whether the early benefits would outweigh the later costs. The answer is probably yes, because it is almost always better to have babies earlier than later in life. Females tend to be more fertile in youth, produce fewer birth defects, are in better shape to care for offspring, and are more likely to have living sisters and mothers to help with child-care. Also, fast breeders produce more generations per century, so can increase their population numbers faster than slow breeders. For these reasons an attractiveness benefit in youth can often outweigh an unattractiveness cost in older age. This is why it can be in the interest of females to evolve youth indicators such as large breasts that tend to droop, fine skin that tends to wrinkle, and buttocks that tend to develop stretch marks. This is one of the most counter-intuitive applications of Zahavi's handicap principle.

Breasts also make good fitness indicators because they come in symmetric pairs. I mentioned earlier that many bodily ornaments in many species advertise an aspect of fitness called developmental stability. When body traits grow in pairs, perfectly symmetric development of the pair indicates high fitness. The paired traits tend to grow large to make their symmetry more obvious during mate choice. Evolutionary psychologists John Manning and Randy Thornhill have shown that women with more symmetric breasts tend to be more fertile. It is possible that bipedalism made breasts a useful potential cue of developmental stability for male mate choice. Once men started paying attention to the symmetry of breast development, high-fitness women could better display the symmetry by evolving large breasts. The larger the breasts, the easier it is to notice asymmetries. Perhaps single mastectomies are so distressing to women because breast symmetry has been such an important fitness cue during human evolution. Large human breasts may have evolved to advertise fitness through their symmetry, not just youth through their pertness.

Finally, breasts are pretty good indicators of fat reserves. In the Pleistocene, starving was more of a problem than overeating. It was harder to have good fat reserves than to be extremely thin, because women had to use their own energy and intelligence to gather food from their environment. It would be possible to spread one's fat evenly over the whole body surface, like a porpoise, but that would make it hard for men to compare females, and it would give females too much insulation under the scorching African sun. Females who concentrated their fat-displays in breast and buttocks could attract male interest without overheating. Also, by not depositing too much fat on the abdomen (as males tend to), females could avoid appearing pregnant already—a sure sign of not being fertile at the moment, which might inhibit male sexual attention. Breasts appear to have evolved as highly condition-dependent indicators of a woman's nutritional state. Most women who have tried dieting know that breast size is the first thing to shrink when food intake is restricted.

The role of breasts as fitness indicators may help to explain why there is so much variation in breast size among women. If large breasts were critical for breast-feeding, which is one of the single most important stages in mammalian reproduction, all women would have large breasts. But as we have seen, fitness indicators do not tend to converge on a single size in a population. They maintain their variation indefinitely, due to the effects of genetic mutation and variation in condition. It has sometimes been argued that men's preferences for larger-than-average breasts must be an artifact of modern culture, because, if it were ancient, all women would have already have evolved large breasts. This argument is wrong if breasts evolved as fitness indicators. Bra manufacturers offer a range from A-cups to D-cups because evolution amplifies the variation in each fitness indicator rather than using it up.

However, even more important in explaining such variation is the fact that each sex assesses the other using a wide range of fitness indicators. This leads to surprising and subtle effects. Imagine that each indicator advertises a different aspect of physical or mental fitness. Because each indicator is costly (so it works according to the handicap principle), there are trade-offs between indicators. This allows scope for individuals to differ in their allocation of resources to different indicators. One individual may grow very tall and muscular; another may grow very symmetric breasts; yet another may grow very intelligent. Each may advertise the same general level of fitness, but may advertise it in a very different way. If height, breast symmetry, and intelligence are all fitness indicators, then—by definition—they must all correlate with fitness, so they must also be positively correlated with one another to some extent. However, such correlations might be quite modest. This implies that even if individuals select mates for their overall fitness, sexual selection may not have the power to drive every fitness indicator to its maximum value. Instead, sexual selection may produce a great diversity of strategies for allocating scarce bodily resources among different indicators. Variation in overall fitness level, combined with variation in these allocation strategies, may account for the rich human variation that we observe. It also explains why not all women have very large breasts—many women may be genetically programmed to prioritize other indicators of physical and mental fitness.

Like penises, breasts have given us some practical information about mate choice in the Pleistocene. The amplification of female human breast size beyond what was useful for milk production reveals the importance of male mate choice in human evolution. If males had not been picky about their sexual partners, female humans would be as flat-chested as chimpanzees. The clitoris does not yield evidence of male mate choice, but breasts do. This opens the door to the possibility of male mate choice influencing the evolution of female brains as well as bodies. Breasts seem to act simultaneously as indicators of youth, indicators of developmental stability, and indicators of foraging ability. We shall see that many of the human mind's most distinctive abilities seem to serve the same range of functions.

Buttocks and Waists

The emergence of upright walking put the buttocks of our ancestors in a new position—both posturally and evolutionarily.

Other great apes such as chimpanzees have small, hairy, flat rumps with tough skin patches on which they sit. But once our hominid ancestors started walking upright around 4.2 million years ago, the legs and buttocks were re-engineered. Much larger, stronger muscles evolved for powering the leg backwards so that it could propel the body forwards. These muscles are what give the human buttocks their basic rounded shape. Beyond this increased muscularity in both sexes, females evolved larger deposits of fat on the buttocks, hips, and upper thighs. Like breasts, these probably evolved through male mate choice as indicators of youth, adequate fat, and perhaps developmental stability.

We are the only species of primate with permanently protruding hemispherical buttocks, and the only species where this protrusion is permanently amplified in adult females by the addition of fat deposits. Buttock size and shape is a unique human feature and shows substantial sex differences. Buttocks are also age-specific, with almost no differences between the sexes before puberty, followed by a rapid accumulation of fat in female buttocks, hips, and thighs over a few years. Buttock size and protuberance normally peaks in young adulthood, around the time of peak female fertility, and then gradually diminishes relative to the rest of the body's fat reserves. Buttocks also show differences between human populations. In southern African Koi-San populations, female buttocks evolved through male mate choice to be especially prominent.

The sex difference in buttock size and shape is hard to explain through natural selection. Because female breasts and buttocks are composed primarily of fatty tissue, it has been suggested that they evolved to provide adequate fat reserves, to protect against the unpredictability and seasonality of food. However, almost all female vertebrates have evolved to store fat reserves, and only female humans have such an unusual distribution of fat. Gorilla females store plenty of fat inside their abdomens, as do human males. Why did human females deviate from this normal primate pattern to store fat below their hips?

Evolutionary psychologist Dev Singh has suggested that the female human distribution of fat evolved as an indicator of youth, health, and fertility. He found that men around the world generally prefer women who have a low "waist-to-hip ratio": a relatively narrow waist and relatively broad hips. Young, fertile women who are not pregnant have waist-to-hip ratios of around 0.7. This ratio would result from a waist circumference of 24 inches and a hip and buttock circumference of 36 inches, for example. Men almost always have a waist-to-hip ratio of at least 0.9, as do prepubescent girls and women past menopause. Obviously, pregnant women have even higher waist-to-hip ratios. Women with various health problems that impair fertility also tend to have higher than average waist-to-hip ratios. Indian temple sculptors have traditionally depicted Hindu goddesses with waist-to-hip ratios as low as 0.3, to symbolize their supernatural fertility and sexuality. In European fashion, corsets and bustle skirts have been used to lower waist-to-hip ratios deceptively. If male hominids have preferred low waist-to-hip ratios for many generations, this may explain why human females have such narrow waists, such broad hips, and such fleshy buttocks.

Women's breasts and buttocks did not evolve because hominid men happened to develop some arbitrary fixation on hemispheres as Platonic ideals of beauty. They evolved as reliable indicators of youth, health, fertility, symmetry, and adequate fat reserves. Starving, sickly women cannot maintain large breasts and buttocks. They need to burn up their fat reserves to stay alive, not keep them hanging around in the hope of attracting a mate. Because starving women tend to turn off ovulation, women without fleshy breasts and buttocks are usually women without fertility. Female long-distance runners, ballerinas, and anorexics who lose most of their body fat tend to have much smaller breasts and buttocks, and often stop menstruating and ovulating. Buttocks, like breasts, reveal the importance of male mate choice in human evolution.

Bodies, Faces, People, and Brains

Our four case studies—penis, clitoris, breasts, and buttocks—do not exhaust the body's complement of sexual ornaments. Because they are sexually differentiated, they are especially informative about male mate choice and female mate choice. However, they are relatively minor contributors to physical attractiveness compared with the face, and with overall body height, proportions, and condition. Our lack of body hair, our long head hair, and our sex differences in musculature are also important signs of sexual choice. Nancy Etcoff's *Survival of the Prettiest* and Desmond Morris's *Bodywatching* have discussed these charms in great detail.

However, I would like to note a few features of the human head that put the human brain in its bodily context. The head is a major target of sexual choice in both sexes. It is rich in fitness information because it is such a complicated piece of the body to grow, and so many things can go wrong. The front of the head has evolved a convoluted shape because evolution tends to pile sense organs up at the front of the body, where they are best placed to sample that part of the environment toward which we are headed, and from which signals can reach the brain quickly. This is why we have eyes, ears, noses, and tongues all huddled together, rather than spread around the body more evenly. The orifice for ingesting food also evolved to be near the brain so that we could efficiently control what we eat and how we chew. The result of evolution assembling the mouth and sense organs so close to the brain is called the face.

An alien biologist might consider such an unseemly concentration of organs on one tiny area of the body rather disgusting, so it is striking that we consider faces so crucial to physical beauty. If the alien did not understand fitness indicators, he or she (or it) might be puzzled that we pay so much attention to the one part of the body that is too complicated for anyone to grow in a perfect form. Wouldn't we find it easier to focus on thighs or backs, which are so easy to get right? Yes, it would be easier, but it would not give us the fitness information we want. Instead of averting our eyes from the unsightly front of another person's head, where harmful mutations show themselves most readily as unusual proportions and asymmetries, we are sometimes so rude as to stare at it, instead of their penis or their breasts. Have we no courtesy? Indeed, we pick the one part of the body where fitness differences are most manifest, and regard that as the seat of personhood. Where mutations show their effects most readily is where we direct our sexual judgment and social attention. A portrait of a human implies a representation of the face.

Much of this book applies the same fitness-indicator argument to the brain as well. Whereas we can perceive facial form visually, we can perceive a brain efficiency only indirectly, through a person's courtship behavior. Beauty is no longer skin-deep in our species. Sexual choice reached behind our faces to tinker with our minds. Mostly, it did so by connecting our brains in a unique way to our mouths, so that we could talk instead of just chewing and grunting. The attention we pay to faces and brains in sexual choice, our obsession with just those body parts that are most difficult to grow perfectly, is powerful evidence for the fitness-indicator view of sexual selection.

Weak Bodies, Strong Minds?

Now that we have seen a few examples of how sexual selection has shaped our bodies, we can step back and consider how the human body's evolution relates to the human mind's evolution. In the mid-20th century, many evolutionary theorists suggested that human bodies represent a degeneration from the wild, robust strength of other apes. They speculated that our supposed bodily weakness somehow forced our brains to become strong, so we could hold our own in the competitive ecology of prehistoric Africa. Reflecting this view, a persistent theme in Robert Heinlein's "Waldo" science fiction stories of the 1950s was that, as humans were allegedly ten times weaker and ten times smarter than chimpanzees, our space-faring, zero-gravity descendants will be ten times weaker and ten times smarter than us. Anthropologist Ashley Montagu influenced a whole generation of anthropologists with his view of neoteny: that the human body is weaker and more childlike than ape bodies, giving it a generality and flexibility uniquely suited for culture.

However, this compensatory view that our brains made up for our lack of brawn does not fit the fossil evidence. Since the rise of *Homo erectus* 1.7 million years ago, our ancestors were among the largest and strongest primates ever to have evolved. *Homo erectus* males seem to have averaged almost six feet tall, with robust skeletons suggestive of powerful muscles. When modern *Homo sapiens* lived as a hunter-gatherer in reasonably food-rich environments, they also grew tall and massive. While brain size was tripling in our ancestors, body size was increasing as well. We are two feet taller and twice as heavy as our earliest bipedal ancestors of 4.2 million years ago. They would be more immediately impressed by our astounding size and strength than by the little puffs of air we call language.

For the last 2 million years, our ancestors have been larger than any insect, amphibian, or bird, and larger and stronger than about 90 percent of reptiles and mammals (to a first approximation, most mammals are rodents and rabbits). Among more than 300 species of modern primates, only male gorillas (averaging around 350 pounds) are significantly larger than humans (around 150 pounds); female gorillas and male orangutans are slightly heavier than male humans, while male chimpanzees weigh up to 130 pounds, and bonobos up to 90 pounds, for both sexes. Our ancestors were the most powerful omnivores in Africa. There were some larger hoofed herbivores, a handful of larger carnivores, and the odd elephant, mastodon, hippopotamus, or rhinoceros. But once our ancestors evolved the ability to throw stones, to wave torches around, to attack in groups, and to run for long distances under the midday sun, they were probably the most terrifying animals in Africa. It is a wonder they bothered to evolve more intelligence at all.

Good Condition as the Evolutionary Norm

It is a mistake to envision our hominid ancestors as bedraggled, dirty, shuffling, sniffling, unhealthy cave-dwellers. They lived outside on a sort of perpetual camping trip, and got a lot of exercise. They had an excellent diet by modern standards, probably consuming about four pounds of fresh fruit and vegetables a day, and perhaps one pound of lean meat on good days (undomesticated game animals have very low body fat). They consumed hardly any salt or sugar, no chocolate, and no beer. They had no dairy products other than their mother's milk. They could not even eat pasta, bread, noodles, or oatmeal until cereal grains were domesticated around 10,000 years ago. The females would have been used to walking miles every day carrying infants and plant foods, and perhaps firewood and water. The males would have been used to chasing down wounded game, running for very long distances. Even our middle-aged ancestors would have remained in very good condition because they would still have made their livings as foragers.

Were we to be transported back 100,000 years in a time machine, we should not expect ancient humans of the opposite sex to fall on their knees and worship our god-like forms. If they were living in a reasonably food-rich habitat, they would probably have been as tall and healthy as us, and in considerably better shape. A week of living in the bush would have obliterated our initial cleanness and reduced our fine clothes to tatters. Any initial sexual interest we provoked would probably evaporate entirely after our total incompetence at hunting and gathering was revealed, and our cowardice in the face of wild baboons, leopards, snakes, elephants, and lions became the subject of jokes. Our bodies would, however, have provoked greater respect in any of the more recent pre-modern agricultural civilizations, in which nutrient-poor diets and communicable disease shrank average human stature by a foot and shortened human lifespans by decades.

Our ancestors would have considered most modern humans to be ridiculously fat, weak, breathless, unfit, and clumsy. They could not drive to the convenience store for a six-pack or a half-gallon of ice cream. They would not have been burdened by excess fat or by the excess muscle attained by modern body-builders by using weight machines, protein shakes, and steroids. Conan the Barbarian would have been too musclebound to run after and catch injured gazelles. Like modern human hunter-gatherers, our ancestors must have been relatively lithe, fit enough to run after game or away from predators, and strong enough to carry animal carcasses or infants long distances.

Sports as Fitness Indicators

This discussion of bodily condition brings us to our first example of a human mental ability that evolved through sexual selection: the capacity for sports. The ability to invent and appreciate new ways of displaying physical fitness is a distinctly human ability. The ritualized behaviors evolved by other animals to intimidate sexual rivals and attract mates almost always include costly, hard-to-fake indicators of physical condition. Male red deer roar at each other as loud as they can, showing off their size and energy. Usually, the weaker, quieter one gives up quickly. But sometimes the two are so closely matched that they roar for hours until endurance rather than strength decides the contest. As in other species, male humans participate much more often in competitive sports than females. But every human culture invents different sports. We inherit the physical capacities and motivations to learn sports, not the specific genes for football, skiing, or boxing.

Sports depend on rules. These prevent competitors from killing each other, as they might in ordinary sexual competition. Even a boxer must not take off his gloves, bite the opponent's ears, or hit below the belt where his opponent's genetic future hangs. Referees are supposed to stop athletic contests before injuries escalate into permanent debility or death. There are also rules for clearly determining who wins and who loses. Each sport could be viewed as a system for amplifying minor differences in physical fitness into easily perceivable status differences, to make sexual choice easier and more accurate. In this sense, sports are culturally invented indicators of physical fitness.

To a game theorist, many human sports look odd because the rules do not specify what the winner actually wins. In game theory, games are defined by a set of players, a set of possible strategies governed by rules, and a set of payoffs that specify what happens when somebody wins. Without specifying the payoffs, the game is meaningless. Modern professional sports offer monetary prizes. But almost no sport in traditional human cultures involves material or monetary prizes. One could say that the winners win "status," but what does that mean? Unless status translates into survival or reproductive benefits, it means nothing to evolution. I suspect that the rewards of winning were mostly reproductive during human evolution. Athletic ability is clearly valued in mate choice, and young people seeking mates are motivated to play competitive sports. This is why the payoff is left implicit for most sports. No referee could force a female to mate with a male winner. The sexual payoff could not be specified as part of the rules, because it still depended on individual mate choice. It was enough for sports competitors to understand that winners were more likely to attract high-quality sexual partners—as in the stereotype of the American high-school football captain dating the homecoming queen.

Sports rules are considered "fair" insofar as they produce the highest correlation between a competitor's fitness and his or her likelihood of winning. Fair rules make sports good fitness indicators; bad rules and rule violations undermine the correlation between winning and fitness. Boys learning to play sports argue endlessly about rules and their interpretation. Girls argue much less about rules, and tend to play less competitively, more often avoiding games with clear winners and losers. Adults playing sports care intensely about rule violations. If sports were just arbitrary cultural pastimes, why should competitors care so much about developing good rules? Fundamentally, I think they care about rules because they have a shared interest in presenting the sport as a good fitness-indicator to observers of the opposite sex. Obviously, competitors have conflicting interests in terms of who wins. But they all want their sport to be perceived as "cool," so that winning yields social status and sexual rewards. Cool sports like downhill skiing use good rules and clear outcomes to advertise major components of heritable fitness like strength, endurance, agility, and intelligence. Cool team sports such as volleyball also advertise the social intelligence abilities that allow a team to cooperate effectively. (Many modern sports are also cool in that they demand expensive equipment that makes them good wealth-indicators.) This obsession with rules and coolness reveals the importance of sexual choice in the evolution of sports.

In many tribal societies there is overlap between competitive sports, fighting, and warfare. All are ritualized and rule-governed to some degree. The rules usually emerge as social conventions for minimizing the risk of death from sexual competition over resources, territories, and status. In tribal warfare especially, there is always the temptation to violate the rules of engagement since dead enemies cannot report one's treachery to other tribes. But for competitions within a tribe, the rules governing fights can be enforced socially. Once we understand this continuum of sexual selection between male competitive sports and male fighting, it no longer seems so strange that men risk their lives and limbs in dangerous sports like motor racing, mountain climbing, and kickboxing. Males of all mammalian species risk their lives in ritualized sexual competition. We humans have invented thousands more ways of doing so, using our unique mental capacities to understand and follow the rules of sporting competition. As with other sexually selected behaviors, we do not need to know that sports evolved for a sexual display function in order to reap the reproductive benefits of manifest athletic skill.

There is almost no evolutionary psychology research on the mental adaptations underlying the human capacity for sports. For now, I can only make some guesses. In both sexes, there must be psychological adaptations for inventing, imitating, and participating in sports. Given children's high level of spontaneous motivation to learn and play sports, as distinct from learning to fight or play in other ways, I would assume that these adaptations are probably specific to sport, and not a side-effect of more general learning mechanisms. There must also be motivational systems for allocating energy and effort to athletic displays depending on who is watching and who else is playing. There must be cognitive systems that can invent the rules that govern sports, detect violations of those rules, and punish violators. We also seem to have a very flexible ability to make unconscious inferences about someone's physical fitness from their athletic displays, even when we have never seen a particular sport before. Such a general ability to make attributions about physical fitness given novel displays may explain why it was possible for so many different sports to emerge in different cultures.

Sports are the intersection of mind and body, nature and culture, competition and mate choice, physical fitness and evolutionary fitness. Sports advertise general aspects of bodily health and condition that are shared by both sexes, not just specific sexual ornaments like beards and breasts. An Olympic medal in swimming can be more sexually attractive than erotic dancing because swimming is a better fitness indicator. Sports evolved through sexual selection, but they are not crude sexual displays.

Sexual selection for the human body was not restricted to sexual ornaments. Once the capacity for sports evolved, sexual choice could favor fit bodies over unfit bodies much more directly. Evolutionary psychology needs to expand its analysis of physical beauty to embrace behavioral displays of physical fitness like sports. We need to be able to explain why women find champion sprinter Linford Christie's astounding speed and form attractive, even when they are used to run from one arbitrary place to another exactly 100 meters away, to no apparent biological purpose.

Sport Utility Vehicles?

Until recently, science and medicine have viewed the human body as a machine that evolved for its survival utility. In *The Selfish Gene*, Richard Dawkins proposed a radically evolutionary view of the body as a vehicle that carries its genes from one generation into the next. A sexual selection analysis views the body as an instrument for displaying physical fitness through costly displays like copulatory courtship and a huge variety of sports. Can we—playfully—combine these utility, vehicular, and sports views and consider the human body a sort of sport utility vehicle (SUV)?

The metaphor seems apt because SUVs make such a show of their rugged utility, all-terrain capability, enormous power, and absurd size. They pretended to be practical, but for most owners in America and Europe, they are just the latest form of conspicuous consumption. They are a status display that just happens to follow a utilitarian aesthetic. And, of course, they follow the handicap principle. Their huge size demonstrates the ability to incur a high initial cost, and their large engine demonstrates the ability to incur high running costs due to poor mileage. Although capable of transporting six adults across a mountain range, they are often used for nothing more demanding than driving one's toddler to and from day-care, through leafy suburbia. To some extent, their size looks like the outcome of a runaway arms race for vehicular safety. If everyone else is driving an SUV, one is no longer safe in an ordinary-sized car, so must buy an SUV oneself. But it would be a mistake to view the SUV phenomenon as simply an escalation of competitive crash-worthiness. Principally, their size is a wealth-indicator. The change from the original SUV utilitarian aesthetic into the recent SUV aesthetic of luxury ornamentation reveals that fact.

The human body seems to have evolved along similar lines. At first glance, it looks and acts like a utility vehicle evolved for survival. It looks as if it grew larger throughout the Pleistocene under the pressure of male sexual competition, because smaller males were not as safe for their genes to ride around in. But the proliferation of sexual ornamentation on our bodies suggests that sexual choice was also at work. This is especially clear for the male body. Its great size, fuel-hungry metabolism, and ability to burn energy in sports reveals a history of female choice for indicators of physical fitness. The demands of pregnancy and mothering did not permit the human female body to be quite so profligate, but women's bodies also show a set of fitness indicators that evolved through male mate choice. Our bodies evolved as sport utility vehicles for sexual display, not as the easiest way to carry the tools for hunting and fishing. Perhaps our minds evolved along the same pseudo-utilitarian lines. In the next chapter we shall see how sexual choice has given us the behavioral abilities and aesthetic tastes to extend our sexual ornamentation from our bodies to our works of art.

8

Arts of Seduction

Art has always been a puzzle for evolutionists. Michelangelo's *David* seems singularly resistant to the universal acid of Darwinism, which is otherwise so efficient at dissolving the cultural into the biological. Like any nouveau-riche connoisseur, we are both proud of our art and ashamed of our ancestry, and the two seem impossible to reconcile.

The evolution of art is hard to explain through survival selection, but is a pretty easy target for sexual selection. The production of useless ornamentation that looks mysteriously aesthetic is just what sexual selection is good at. Artistic ornamentation beyond the body is a natural extension of the penises, beards, breasts, and buttocks that adorn the body itself. We shall begin our tour of the human mind with a look at our artistic instincts for producing and appreciating aesthetic ornamentation that is made by the hands rather than grown on the body.

Our shift of art makes a turning point in this book. So far we have been considering generalities: sexual selection theory in general, and how sexual selection shaped the human mind and body in general. It is time to turn to specific mental adaptations to see whether the sexual choice theory can explain particular aspects of human psychology. The rest of this book is devoted to four human capacities: art, morality, language, and creativity. They will serve as case studies. Each has proven difficult to account for as a survival adaptation. We might make more progress by asking whether each may have evolved originally as a courtship adaptation. Of course, in modern life none are used exclusively for courtship, but they still show enough hallmarks of sexual selection for us to be able to trace their origin to the sexual choices made by our ancestors.

Art as an Adaptation

In her books *What Is Art For?* and *Homo Aestheticus*, anthropologist Ellen Dissanayake made one of the first serious attempts to analyze art as a human adaptation that must have evolved for an evolutionary purpose. She argued that human art shows three important features as a biological adaptation. First, it is ubiquitous across all human groups. Every culture creates and responds to clothing, carving, decorating and image-making. Second, the arts are sources of pleasure for both the artist and the viewer, and evolution tends to make pleasurable those behaviors that are adaptive. Finally, artistic production entails effort, and effort is rarely expended without some adaptive rationale. Art is ubiquitous, and costly, so is unlikely to be a biological accident.

Art fits most of the other criteria that evolutionary psychology has developed for distinguishing genuine human adaptations from non-adaptations. It is relatively fun and easy to learn. Given access to materials, children's painting and drawing abilities unfold spontaneously along a standard series of developmental stages. Humans are much better at producing and judging art than is any artificial intelligence program or any other primate. Of course, just as our universal human capacity for language allows us to learn distinct languages in different cultures, our universal capacity for art allows us to learn different techniques and styles of aesthetic display in different cultures. Like most human mental adaptations, the ability to produce and appreciate art is not present at birth. Very little of our psychology is "innate" in this sense, because human babies do not have to do very much. Our genetically evolved adaptations emerge when they are needed to deal with particular stages of survival and reproduction. They do not appear at birth just so psychologists can conveniently distinguish the evolved from the cultural. Beards have evolved, but they grow only after puberty, so are they "innate"? Is menopause "innate"? "Innateness" is a relatively useless concept that has little relevance in modern evolutionary theory or behavior genetics.

Some archeologists have argued that art only emerged 35,000 years ago in the Upper Paleolithic period, when the first cave paintings and Venus figurines were made in Europe. They follow archeologist John Pfeiffer's suggestion that this period marks a "creative explosion" when human art, language, burial ceremonies, religion, and creativity first emerged. This is a remarkably Eurocentric view. The Aborigines colonized Australia at least 50,000 years ago, and have apparently been making paintings on rock ever since. If art were an invention of the upper Paleolithic 35,000 years ago in Europe, how could art be a human universal? There is evidence from Africa of red ocher being used for body ornamentation over 100,000 years ago. This is about the latest possible time that art could have evolved, since it is around the time that modern *Homo sapiens* spread out from Africa. Had it evolved later, it is unclear how it could have become universal across human groups.

The Functions of Art

The aesthetic has often been defined in opposition to the pragmatic. If we view art as something that transcends our immediate material needs, it looks hard to explain in an evolutionary way. Selection is usually assumed to favor behaviors that promote survival, but almost no art theorist has ever proposed that art directly promotes survival. It costs too much time and energy and does too little. This problem was recognized very early in evolutionary theorizing about art. In his 1897 book *The Beginnings of Art,* Ernst Grosse commented on art's wastefulness, claiming that natural selection would "long ago have rejected the peoples which wasted their force in so purposeless a way, in favor of other peoples of practical talents; and art could not possibly have been developed so highly and richly as it has been." He struggled, like many after him, to find a hidden survival function for art.

To Darwin, high cost, apparent uselessness, and manifest beauty usually indicated that a behavior had a hidden courtship function. But to most art theorists, art's high cost and apparent uselessness has usually implied that a Darwinian approach is inappropriate, that art is uniquely exempt from selection's cost-cutting frugality. This has led to a large number of rather weak theories of art's biological functions. I shall briefly consider their difficulties before attempting to bring art back into the evolutionary framework.

Art for Art's Sake

Ever since the German Romanticism of Schiller and Goethe in the early 19th century, many have viewed art as a utopian escape from reality, a zone of selfless self-expression, a higher plane of being where genius sprouts lotus-like above the petty concerns of the world. This Romantic view opposes art to nature, but also opposes art to popular culture, art to market commodity, art to social convention, art to decoration, and art to practical design. It has often presented the artist as a male genius shunning the female temptresses that would sap the vital fluids that sustain his creativity. Thus, artistic success has also been seen as opposed to sexual reproduction.

Perhaps it is not surprising that many modern artists have adopted the ideology of these German philosophers. Romanticism makes excellent status-boosting rhetoric for artists. It presents them as simultaneously overcoming their instincts, avoiding banality, striving against capitalism, rebelling against society, and transcending the ornamental. The genius's need to shun sexual temptation also provides a ready excuse for avoiding sleeping with one's less attractive admirers. But this Romantic view makes no attempt to offer a scientific analysis of art—indeed, it actively rejects the possibility.

The kernel of truth in the Romantic view is that art is pleasurable to make and to look at, and this pleasure can seem a sufficient reason for art's existence. Its pleasure-giving power can seem to justify art despite its apparent uselessness. But from a Darwinian perspective, pleasure is usually an indication of biological significance. Subjectively, everything an animal does may appear to be done simply to experience pleasure or avoid pain. If we did not understand that animals need energy, we might say that they eat for the pleasure of eating. But we do understand that they need energy, so we say instead that they have evolved a mechanism called hunger that makes it feel pleasurable to eat. The Romantic view of art fails to take this step, to ask why we evolved a motivational system that makes it pleasurable to make and see good art. Pleasure explains nothing; it is what needs explaining.

Social Solidarity, Cultural Identity, and Religious Power

Many anthropologists view art, like ritual, religion, music, and dance, as a social glue that holds groups together. This hypothesis dates back to the early 20th century and the "functionalist" views of Emile Durkheim, Bronislaw Malinowski, A. R. Radcliffe-Brown, and Talcott Parsons. For them, a behavior's function meant its function in sustaining social order and cultural stability, rather than its function in propagating an individual's genes. The social functions postulated for art were usually along the lines of "expressing cultural identity," "reflecting cultural values," "merging the individual into the collective," "sustaining social cohesion," "creating a collective consciousness," and "socializing the young." It is not easy to be sure what any of these phrases really means, and in any case these putative social functions are not easy to relate to legitimate biological functions in evolution.

Primate groups work perfectly well without any of these mechanisms. Chimpanzees don't need to express their cultural identities or create a collective consciousness in order to live in groups. They need only a few social instincts to form dominance hierarchies, make peace after quarrels, and remember their relationships. Humans do not seem any worse at these things than chimpanzees, so there seems no reason why we should need art or ritual to help us "bond" into groups. Human groups may be larger than chimpanzee groups, but Robin Dunbar has argued convincingly that language is the principal way in which humans manage the more complex social relationships within our larger groups.

The view that art conveys cultural values and socializes the young seems plausible at first glance. It could be called the propaganda theory of art. The trouble with propaganda is that it is usually produced only by large institutions that can pay propagandists. In small prehistoric bands, who would have any incentive to spend the time and energy producing group propaganda? It would be an altruistic act in the technical biological sense: a behavior with high costs to the individual and diffuse benefits to the group. Such altruism is not usually favored by evolution. As we shall see in the chapters on morality and language, evolution can sometimes favor group-benefiting behaviors, if individuals can attain higher social and sexual status for producing them. But such opportunities are relatively rare, and one would have to show that art is well designed as a propaganda tool to create norms and ideals that benefit the group. Language is surely a much more efficient tool for telling people what to do and what not to do. The best commands are imperative sentences, not works of art.

A popular variant of the cultural-value idea is the hypothesis that most art during human evolution served a "religious function." Museum collections of art from primitive societies routinely label almost every item a fertility god, an ancestral figure, a fetish, or an altarpiece. Until recently, archeologists routinely described every Late Paleolithic statue of a naked woman as either a "goddess" or a "fertility symbol." Usually, there is no evidence supporting such an interpretation. It would be equally plausible to call them "Paleolithic pornography." The importance of church-commissioned art in European art history may have led archeologists to attribute religious content to most prehistoric art.

In any case, religious functions for art don't make much Darwinian sense. Some anthropologists have suggested that the principal function of art during human evolution was to appease gods and dead ancestors, and to put people in touch with animal spirits. In his textbook *The Anthropology of Art*, Robert Layton claimed that the function of Kalabari sculpture in Africa is "a pragmatic one of manipulating spiritual forces." This overlooks the possibility that gods, ancestral ghosts, and animal spirits may not really exist. If they do not exist, there is no survival or reproductive advantage to be gained from appeasing or contacting them. Some artists may believe that making a certain kind of statue will give them "spiritual powers." Scientifically, we have to take the view that they might be deluded. Their delusion, on its own, is not evolutionarily stable, because it costs them time and energy and the "spiritual powers" probably cannot deliver what is hoped for. However, if an individual's production or possession of a putatively religious object brings them higher social or sexual status, then it can be favored by evolution. A person can spend hours hacking at a piece of wood, making a fetish, and telling people about their extraordinary spiritual powers. If others grant the religiously imaginative individual higher status or reproductive opportunities, such behavior can be sustained by sexual selection.

The same argument applies to art that has the alleged function of curing disease, such as some Navajo sand-paintings. Navajo artists could speculate that the human capacity for making sand-paintings must have evolved through survival selection for curing diseases. If sand-paintings were proven medically effective in double-blind randomized clinical trials, they would have a good argument. But the sand-paintings probably have nothing more than a placebo effect. Like "appeasing the gods," "curing disease" works as an evolutionary explanation only if the trait in question actually does what is claimed.

Evolution is not a cultural relativist that shows equal respect for every ideological system. If an artistic image intended to control spirits or cure disease does not actually improve survival prospects, evolution has no way to favor its production except through sexual selection. Evolutionary psychologists should accept ideologies like religion and traditional medicine as human behavioral phenomena that need explaining somehow. This does not mean that we have to give them any credence as world-views. For scientists, science has epistemological priority.

There are important differences between the social functions of art (which may support religious, political, or military organizations), the conscious individual motivations for producing art (which may include making money, achieving social status, or going to heaven), and the unconscious biological functions of producing art (which must concern survival or reproduction). Darwinian theories of the origins of our capacity for art cannot hope to account for all of the social functions and various forms of art that happen to have emerged in diverse human cultures throughout history. Evolutionary psychology tries to answer only a tiny number of questions about human art, such as "What psychological adaptations have evolved for producing and appreciating art?" and "What selection pressures shaped those adaptations?" These are important questions, but they are by no means the only interesting ones. All the other questions about art will remain in the domain of art history and aesthetics, where a Darwinian perspective may offer some illumination, but never a complete explanation. We shall still need cultural, historical, and social explanations to account for the influences of Greek and Indian traditions on Gandhara sculpture, or the way in which Albert Hoffman's serendipitous discovery of LSD in 1943 led to the "happenings" organized by the Fluxus group in the 1960s. As we shall see, the human capacity for art is a particularly flexible and creative endowment, and identifying its evolutionary origins by no means undermines the delights of art history, or limits the range or richness of artistic expression.

A Bottom-Up View of Art

None of the standardly proposed "functions" of art are legitimate evolutionary functions that could actually shape a genetically inherited adaptation. As Steven Pinker has observed,

> Many writers have said that the "function" of the arts is to bring the community together, to help us see the world in new ways, to give us a sense of harmony with the cosmos, to allow us to

> appreciate the sublime, and so on. All these claims are true, but none is about adaptation in the technical sense . . .

If this is right, then what are we to do? The human capacity for art shows evidence of adaptive design, but its function remains obscure. Perhaps we need a broader view of art, inspired by more biologically relevant examples.

There are two strategies science can take in trying to understand the evolutionary origins of art: top-down or bottom-up. The top-down strategy focuses on the fine arts and their elite world of museums, galleries, auction houses, art history textbooks, and aesthetic theory. The bottom-up strategy surveys the visual ornamentation of other species, of diverse human societies, and of various subcultures within our society. In this broader view, the fine arts are a relatively unpopular and recent manifestation of a universal human instinct for making visual ornamentation. Most scientists, being anxious to display their cultural credentials as members of the educated middle class, feel obligated to take a top-down approach. There is a temptation to display one's familiarity with the canon of Great Art, to counter the stereotype that scientists are so obsessed with truth that they have forgotten beauty. One may even feel obliged to start with a hackneyed example of Italian Renaissance sculpture, as I have done in this chapter.

But what if we step back from the fine arts and ask ourselves what engagement ordinary humans have with visual ornamentation, once they step outside the dim museums of Florence and return to their real lives. Our opportunities to appreciate the fine arts typically arise during vacations and weekend trips to local museums. But visual ornamentation surrounds us every day. We wear clothing and jewelry. We buy the biggest, most beautiful houses we can afford. We decorate our homes with furniture, rugs, prints, and gardens. We drive finely designed, brightly colored automobiles, which we choose for their aesthetic appeal as much as their fuel efficiency. We may even paint the odd watercolor. This sort of everyday aesthetic behavior comes quite naturally, in every human culture and at every moment in history.

There is no clear line between fashion and art, between ornamenting our bodies and beautifying our lives. Body-painting, jewelry, and clothing were probably the first art forms, since they are the most common across cultures. Nor is there a clear line between art and craft—as William Morris argued when founding the Arts and Crafts movement in Victorian England. Fine art may be strictly useless in pragmatic terms, while good design merely makes beautiful that which is already useful. When we address the evolution of human art, we need to explain both the aesthetic made useless and the useful made aesthetic. We shall see that even apparently pragmatic tools like *Homo erectus* handaxes may have evolved in part through sexual selection as displays of manual skill.

In this chapter I take a bottom-up approach to analyzing the evolutionary origins of art, ornamentation, and aesthetics. This makes it easier to trace the adaptive function of these seemingly useless biological luxuries. As we have seen, most of the visual ornamentation in nature is a product of sexual selection. The peacock's tail is a natural work of art evolved through the aesthetic preferences of peahens. We have also seen that some of our bodily organs, including hair, faces, breasts, buttocks, penises, and muscles, evolved partly as visual ornaments. It seems reasonable to ask how far we can get with the simplest possible hypothesis for art: that it evolved, at least originally, to attract sexual partners by playing upon their senses and displaying one's fitness. To see how this idea could work, let's consider an example of sexual selection for art in another animal species.

Bowerbirds

Human ornamentation is distinctive because most of it is made consciously with our hands rather than grown unconsciously on our bodies. However, this does not mean that its original adaptive function was different. The only other animals that spend significant time and energy constructing purely aesthetic displays beyond their own bodies are the male bowerbirds of Australia and

New Guinea. Their displays are obvious products of female sexual choice.

Each of the 18 existing species constructs a different style of nest. They are constructed only by males, and only for courtship. Each male constructs his nest by himself, then tries to attract females to copulate with him inside it. Males that build superior bowers can mate up to ten times a day with different females. Once inseminated, the females go off, build their own small cup-shaped nests, lay their eggs, and raise their offspring by themselves with no male support, rather like Picasso's mistresses. By contrast, the male nests are enormous, sometimes large enough for David Attenborough to crawl inside. The golden bowerbird of northern Australia, though only nine inches long, builds a sort of roofed gazebo up to nine feet high. A hut built by a human male to similar proportions would top 70 feet and weigh several tons.

Males of most species decorate their bowers with mosses, ferns, orchids, snail shells, berries and bark. They fly around searching for the most brilliantly colored natural objects, bring them back to their bowers, and arrange them carefully in clusters of uniform color. When the orchids and berries lose their color, the males replace them with fresh material. Males often try to steal ornaments, especially blue feathers, from the bowers of other males. They also try to destroy the bowers of rivals. The strength to defend their delicate work is a precondition of their artistry. Females appear to favor bowers that are sturdy, symmetrical, and well-ornamented with color.

Regent and Satin Bowerbirds go an astonishing step further in their decorative efforts. They construct avenue-shaped bowers consisting of a walkway flanked by two long walls. Then they use bluish regurgitated fruit residues to paint the inner walls of their bowers, sometimes using a wad of leaves or bark held in the beak. This bower-painting is one of the few examples of tool use by birds under natural conditions. Presumably the females have favored the best male painters for many generations.

Sexual selection for ornamental bower-building has not replaced sexual selection for the more usual kinds of display.

Males of many bowerbird species are much more brightly colored than females, and they dance in front of the bowers when females arrive. They also sing, producing guttural wheezes and cries, and good imitations of the songs of other bird species. However, male bowerbirds are not nearly as spectacular as their relatives, the birds-of-paradise, the most gorgeous animals in the world. Somehow, having evolved from a drab crow-like form, the female ancestors of the bowerbirds and birds-of-paradise developed an incredible aesthetic sense. In the birds-of-paradise, their sexual choices resulted in an efflorescence of plumage in 40 species. In the bowerbirds, they resulted in a proliferation of ornamental nests in 18 species.

The bowerbirds create the closest thing to human art in a non-human species. Their art is a product of sexual selection through female choice. The males contribute nothing but their genes when breeding, and their art serves no survival or parental function outside courtship. The bowers' large size, symmetric form, and bright colors may reflect female sensory biases. However, the bowers also have high costs that make them good fitness indicators. It takes time, energy, and skill to construct the enormous bower, to gather the ornaments, to replace them when they fade, to defend them against theft and vandalism by rivals, and to attract female attention to them by singing and dancing. During the breeding season, males spend virtually all day, every day, building and maintaining their bowers.

If you could interview a male Satin Bowerbird for *Artforum* magazine, he might say something like "I find this implacable urge for self-expression, for playing with color and form for their own sake, quite inexplicable. I cannot remember when I first developed this raging thirst to present richly saturated color-fields within a monumental yet minimalist stage-set, but I feel connected to something beyond myself when I indulge these passions. When I see a beautiful orchid high in a tree, I simply must have it for my own. When I see a single shell out of place in my creation, I must put it right. Birds-of-paradise may grow lovely feathers, but there is no aesthetic mind at work there, only a body's brute instinct. It is a happy coincidence that females sometimes come to my gallery openings and appreciate my work, but it would be an insult to suggest that I create in order to procreate. We live in a post-Freudian, post-modernist era in which crude sexual meta-narratives are no longer credible as explanations of our artistic impulses."

Fortunately, bowerbirds cannot talk, so we are free to use sexual selection to explain their work, without them begging to differ. With human artists things are rather different. They usually view their drive to artistic self-expression not as something that demands an evolutionary explanation, but as an alternative to any such explanation. They resist a "biologically reductionist" view of art. Or they buy into a simplistic Freudian view of art as sublimated sexuality, as when Picasso repeated Renoir's quip that he painted with his penis. My sexual choice theory, however, is neither biologically nor psychologically reductionist. It views our aesthetic preferences and artistic abilities as complex psychological adaptations in their own right, not as side-effects of a sex drive. Bowerbirds have evolved instincts to construct bowers that are distinct from the instinct to copulate once a female approves of the bower. We humans have evolved instincts to create ornaments and works of art that are distinct from the sexual instincts behind copulatory courtship. Yet both types of instinct may have evolved through sexual selection.

Ornamentation and the Extended Phenotype

The bowerbirds show the evolutionary continuity between body ornamentation and art. They happen to construct their courtship displays out of twigs and orchids instead of growing them from feathers like their cousins, the birds-of-paradise. We happen to apply colored patterns to rock or canvas. Biologists no longer draw a boundary around the body and assume that anything beyond the body is beyond the reach of evolution. In *The Extended Phenotype*, Richard Dawkins argued that genes are often selected for effects that spread outside the body into the environment. It is meaningful to talk about genes for a spider's web, a termite's mound, and a beaver's dam. Some genes even reach into the brains of other individuals to influence their behavior for the genes' own benefit. All sexual ornaments do that, by reaching into the mate choice systems of other individuals. At the biochemical level, genes only make proteins, but at the level of evolutionary functions they can construct eyes, organize brains, activate behaviors, build bowers, and create status hierarchies. Whereas an organism's "phenotype" is just its body, its "extended phenotype" is the total reach of its genes into the environment.

In this extended-phenotype view, bipedalism freed our hands for making not just tools, but sexual ornaments and works of art. Some of our ornaments are worn on the body, while others may be quite distant, connected to us only by memory and reputation. We ornament the skin directly with ocher, pigments, tattoos, or scars. We apply makeup to the face. We braid, dye, or cut our hair. We drape the body with jewelry and clothing. We even borrow the sexual ornamentation of other species, killing birds for their feathers, mammals for their hides, and plants for their flowers. At a greater distance, we ornament our residences, be they caves, huts, or palaces. We make our useful objects with as much style and ornament as we can afford, and make useless objects with purely aesthetic appeal.

The Rise and Fall of Sexual Art

The idea that art emerged through sexual selection was fairly common a century ago, and seems to have fallen out of favor through neglect rather than disproof. Darwin viewed human ornamentation and clothing as natural outcomes of sexual selection. In *The Descent of Man* he cited the popularity across tribal peoples of nail colors, eyelid colors, hair dyes, hair cutting and braiding, head shaving, teeth staining, tooth removal, tattooing, scarification, skull deformations, and piercings of the nose, ears, and lips. Darwin observed that "self-adornment, vanity, and the admiration of others, seem to be the commonest motives" for self-ornamentation. He also noted that in most cultures men ornament themselves more than women, as sexual selection theory would predict. Anticipating the handicap principle, Darwin also stressed the pain costs of aesthetic mutilations such as scarification, and the time costs of acquiring rare pigments for body decoration. Finally, he argued against a cultural explanation of ornamentation, observing that "It is extremely improbable that these practices which are followed by so many distinct nations are due to tradition from any common source." Darwin believed the instinct for self-ornamentation to have evolved through sexual selection as a universal part of human nature, more often expressed by males than by females.

Throughout the late 1800s, Herbert Spencer argued that Darwin's sexual selection process accounts for most of what humans consider beautiful, including bird plumage and song, flowers, human bodies, and the aesthetic features of music, drama, fiction, and poetry. In his 1896 book *Paradoxes*, Max Nordau attributed sexual emotions and artistic productivity to a hypothetical part of the brain he called the generative center. Freud viewed art as sublimated sexuality.

However, these speculations did not lead very far because sexual selection theory was not very well developed at the beginning of the 20th century. By 1908, aesthetic theorist Felix Clay had grown weary of the facile equating of artistic production with reproduction. In *The Origin of the Sense of Beauty*, he complained:

> How the pleasure in some stately piece of beautifully proportioned architecture, the thrill produced by solemn music, or the calm sweetness of a summer landscape in the evening, is to be attributed to the feeling of sex only, it is hard to see; they have in common a pleasurable emotion, and that is all. That a very large part of art is directly inspired by erotic motives is perfectly true, and that various forms of art play an important part in love songs and courtship is obvious; but this is so because beauty produced by art has in itself the power of arousing emotion, and is therefore naturally made use of to heighten the total pleasure. That love has provided the opportunity and incentive to innumerable works of art, that it has added to the

> pleasure and enjoyment of countless beauties, need not be denied; but we cannot admit that it is due to the sex feeling that rhythm, symmetry, harmony, and beautiful colour are capable of giving us a pleasurable feeling.

In reading some of these century-old works, it is impressive how sophisticated and earnest their use of sexual selection theory was, and how favorably they compare to some current theories of art's evolution. Nevertheless, they repeat Freud's cardinal error, as Clay does here, of confusing sexual functions with sexual motivations. Art does not have to be about sex to serve the purposes of attracting a mate—it can be about anything at all, or about nothing, as in the geometric art of Islam, or Donald Judd's stainless-steel minimalist sculpture. As we saw with the bowerbirds, a sexually selected instinct for making ornamentation need not have any motivational or emotional connection with a sexually selected desire to copulate. The displayer does not need to keep track of the fact that beautiful displays often lead to successful reproduction. Evolution keeps track for us.

Great Artists of the Pleistocene

If art evolved through sexual choice, better artists must have attracted more sexual partners, or higher-fitness partners. How could that have happened? To appreciate the Pleistocene artist's reproductive advantages, we should not necessarily think of Modigliani's cocaine-fueled quest to have sex with every one of the hundreds of models he painted, or Gauguin's apparent drive to infect every girl in Polynesia with his syphilis. Perhaps it is better to remember how Picasso fathered one child by his first wife Olga Koklova, another by his mistress Marie-Thérèse Walter, and two more by his mistress Françoise Gilot. Picasso is not a bad example of the idea that artistic production serves as a fitness indicator. Before dying at age 91 and leaving an estate of $1 billion in 1973, he had produced 14,000 paintings, 34,000 book illustrations, and 100,000 prints and engravings. His tireless energy, prodigious output, and sexual appetite seem to have been tightly interwined, as he himself was aware. The old punk song was right about Picasso: "He was only 5 foot 3, but girls could not resist his stare."

Still, the extreme sexual success of modern professional painters like Modigliani, Gauguin, and Picasso would not have been the Pleistocene norm. It is unlikely that there were professionals of any sort during most of human evolution, since the division of labor was sexual, not vocational. The role of artistry in everyday life was more informal and ubiquitous. Everybody made things: tools, clothing, personal ornaments, shelters. Some individuals made things better than others. Making each object could serve as an occasion for demonstrating one's ornamental skills and aesthetic taste. Sometimes there was no time for such embellishments, but often there was.

For sexual choice to have favored good artistry, our ancestors needed only the opportunity to make sexual choices based on the extended phenotypes of potential mates, and the motive to pay attention to the extended phenotypes' aesthetic quality. It was not necessary for hominids to favor great artists over great hunters or great mothers. It was necessary only for them to favor those who showed taste and talent in their everyday self-ornamentation over those who did not, all else being equal.

Sexual Functions Versus Sexual Content

Prehistoric art had a lot of sexual content. Venus figurines are endowed with large breasts and buttocks. Rock-art often consists of nothing more than repeated motifs of female genitals. Ice Age Europeans carved phallic batons from bone and stone. One image from prehistoric Siberia appears to depict a man on skis attempting intercourse with an elk. This is all very interesting, but not very relevant to the sexual choice model for art's evolution.

Sexual selection for art need not imply that our ancestors favored hyper-sexual art in the style of Tantric Buddhism. They need not have gone around everywhere carving lingams (stylized phalluses) or yonis (stylized labia). Even if they did, that would reflect their interests without necessarily revealing the adaptive benefits of their art. Some bowerbirds make bowers that are tall and conical like a phallus, and some make avenue-shaped bowers that look like yoni, but that is a meaningless coincidence irrelevant to their evolution through sexual selection. Tantric myth does provide some lovely metaphors for evolution through sexual selection. Creation occurred through sexual play between an Originating Couple. Krishna seduced all the cow-girls of Brindaban with his blue skin, beauty, and flute music. The path to enlightenment lies in joyful copulation as a mutual escalation of consciousness. Nevertheless, the fact that Pleistocene art often looks Tantric is not very relevant to the sexual choice theory.

Darwinian Aesthetics

If we view art as an example of a biological signaling system, we can break it down into two complementary adaptations: capacities for producing art, and capacities for judging art. The second of these, our set of aesthetic preferences, seems more mysterious in some ways. If we assume a rich aesthetic sense to be part of human nature, we should not find it surprising that people figured out how to attract sexual partners and gain social status by producing things that others consider aesthetically pleasing. Neither, perhaps, should we find it surprising that sexually mature males have produced almost all of the publicly displayed art throughout human history. Given any set of human preferences about anything, males have more motivation to play upon those preferences to attract sexual partners. It seems reasonable to posit that our capacities for producing art are legitimate biological adaptations that evolved over thousands of generations, rather than cultural inventions. But our aesthetic sense seems a good place to focus our analysis to see how far the sexual choice theory can go.

Why is beauty so compelling? Why do we find some things more beautiful than others? As far as our subjective experience goes, these are the central mysteries of art. It seems hard to connect our experience of beauty to any evolutionary theory of aesthetics. Yet with every one of our pleasures and pains there is this lack of an explicit link. A burning sensation does not carry an intellectual message saying "By the way, this spinal reaction evolved to maximize the speed of withdrawing your extremities from local heat sources likely to cause permanent tissue damage injurious to your survival prospects." It just hurts, and the hand withdraws from the flame. Female sexual orgasm does not automatically create an intellectual appreciation of orgasm's role in promoting mate choice for good genes. No instinctive reaction to anything ever carries a special coded message saying why the reaction evolved. It doesn't have to—the reaction itself does the adaptive work of survival or reproduction.

Powerful reactions like aesthetic rapture are the footprint of powerful selection forces. Like our sexual preferences for certain faces and bodies, our aesthetic preferences may look capricious at first, but reveal a deeper logic on closer examination. If art evolved through sexual selection, our aesthetic preferences could be viewed as part of our mate choice system. They are not the same preferences we use to assess another individual's body, because, like most other animals, we already have rich sexual preferences about body form. Rather, they are the preferences we use in assessing someone's extended phenotype: the set of objects they made, acquired, and displayed around their bodies. To explain our aesthetic preferences, we should be able to use the same sexual selection theories that biologists use to explain mating preferences. As we saw in previous chapters, these boil down to three options: preferences that escalate through runaway effects, preferences that come from sensory biases, and preferences evolved to favor fitness indicators.

Runaway Beauty

Perhaps human aesthetics emerged through runaway sexual selection, with aesthetic tastes evolving as part of female mate choice. In this view, some female hominids just happened to have certain tastes concerning male ornaments. The artists best able to fulfill these tastes inseminated more aesthetic groupies and sired more offspring, who inherited both their artistic talent and their mothers' aesthetic tastes.

Something like this still happens among the Wodaabe people (also known as the Bororo), cattle-herding nomads who live in the deserts of Nigeria and Niger. At annual *geere wol* festivals, hundreds of people gather, and the young men spend hours painting their faces and ornamenting their bodies. The men also dance vigorously for seven full nights, showing off their health and endurance. Towards the end of the week-long ceremony, the men line up and display their beauty and charm to the young women. Each woman invites the man she finds most attractive for a sexual encounter. Wodaabe women usually prefer the tallest men with the whitest teeth, the largest eyes, the straightest nose, the most elaborate body-painting, and the most creative ornamentation. As a result, Wodaabe men have evolved to be significantly taller, white-toothed, larger-eyed, straighter-nosed, and better at self-decoration than men of neighboring tribes. This divergence probably happened within the last few hundred or few thousand years, illustrating runaway's speed. Journalists who know nothing of sexual selection often comment on the "reversal of sex roles" in Wodaabe beauty contests compared to European and American counterparts. But biologically, the Wodaabe are behaving perfectly normally, with males displaying and females choosing. The Miss America contests are the unusual ones.

As we saw with the runaway brain theory, runaway aesthetics would require polygamy and would result in large sex differences in artistic production. At first glance, it looks as if it should also produce large sex differences in aesthetic tastes, with females much more discriminating than males. If art were grown instead of made, that would be true. The peacock does not need the peahen's appreciation of a good tail—he needs only the tail itself. But for men to make good art, they must embody the same aesthetic discrimination as women. While decorating themselves, they must be able to access the same aesthetics that women will use in judging their decoration. Given this twist, the runaway aesthetic theory predicts sexual similarities in aesthetic taste, but much higher aesthetic output by males. That is roughly what we see in the history of art (although cultural and economic factors may have amplified the sex differences in artistic output over the last few millennia).

Yet the runaway theory cannot account for anything about human aesthetics other than their existence. It can explain why we find some things more beautiful than others, but it cannot explain any of the aesthetic criteria we use to make such judgments, because any standard of beauty can evolve through runaway. Runaway sexual selection is arbitrary, so it does not offer a very satisfying theory of aesthetics. It might still be the right explanation, but perhaps we can do better.

Aesthetic Tastes as Sensory Biases

Sensory bias theory seems ideal for explaining our aesthetic preferences. Whenever we encounter a human taste for a certain kind of aesthetic stimulation by identifying the brain circuits involved in perceiving that stimulation, we could show that they are optimally stimulated by just that stimulation. Perhaps we like stripes because our primary visual cortex happens to be most sensitive to stripe-like patterns. Perhaps we like highly saturated primary colors because our photo-receptors are most highly activated by such hues. Every time we find any brain mechanism underlying an aesthetic preference, we could just declare it an intrinsic sensory bias, and stop the analysis there.

This is a surprisingly venerable strategy for understanding human aesthetics, dating back to Hermann von Helmholtz, Gustav Fechner, and the earliest experimental days of 19th-century neurophysiology. It became integrated into the first wave of evolutionary psychology in the 1870s through the 1890s, as in Grant Allen's books *Physiological Aesthetics* and *The Color Sense.* By 1908, Felix Clay's *The Origin of the Sense of Beauty* could review dozens of theories about the evolution of human aesthetics—mostly forgotten now, but at least as good as many modern ideas.

More recently, Nancy Aiken took this physiological approach in her 1998 book *The Biological Origins of Art.* She tried to identify brain mechanisms that would favor certain colors, forms, patterns, and symbols. But she did not analyze the evolutionary costs and benefits of artistic behavior, or of having one set of aesthetic preferences rather than another. As we saw in the chapter on the ornamental mind, the sensory bias view is most useful when we can trace why our brain circuits evolved particular sensitivities. Sensory bias theorists do that by considering the relevant things that a particular species evolved to perceive under ancestral conditions. But physiological approaches like Aiken's do not usually take that next step of asking the evolutionary "why" questions. Why is our primary visual cortex most sensitive to stripes? Why does our color vision respond most strongly to highly saturated primary colors?

From an evolutionary viewpoint it is simply tautologous to say that humans have certain aesthetic preferences because our brains happen to have those aesthetic preferences. From a neuroscientist's viewpoint, we are our brains. It should come as no surprise that every one of our preferences is implemented somehow, somewhere in the brain. This is equally true of genetically evolved preferences and culturally acquired preferences. The identification of a brain mechanism may look as if it is providing evidence of an evolved adaptation, but it is not. Any culturally acquired behavior will be manifest in some brain mechanism too. Of course we will find neurochemicals and hormones and neural pathways that correspond to strong aesthetic emotions. So?

Sensory bias theory becomes more interesting when it is possible to show that the sensory bias evolved long before the relevant sexual ornamentation. For human aesthetic preferences, this would mean finding evidence for the preferences in other primates. So far most attempts to do that have failed. In the 1970s, Nicholas Humphrey tried very hard to find evidence of visual aesthetic preferences in rhesus monkeys. They preferred white light to red light, focused pictures to out-of-focus pictures, and pictures of monkeys to pictures of anything else. But they showed no sign of any aesthetic preferences for forms, patterns, symmetries, or compositions. Rhesus monkey visual systems are so similar to ours that they are often used by neuroscientists as experimental models for human vision. Yet they show no hint of the aesthetic preferences that we might expect as side-effects of having our sort of vision.

Other evidence against the sensory bias view comes from experiments on painting by chimpanzees. Desmond Morris's 1962 book *The Biology of Art* showed that, given paper, brushes, and paint, chimpanzees produced works resembling the abstract expressionist paintings that were in vogue at the time. Morris had been searching for evolutionary continuity between ape and human aesthetics, and thought he had found some evidence for a sense of pictorial composition and balance in apes. In appreciation of Morris's research, Salvador Dali declared that "The hand of the chimpanzee is quasi-human, the hand of Jackson Pollock is almost animal." However, later research suggested that chimps do not produce artworks according to a goal-oriented plan. They paint reactively in relation to the paper's edges and to any geometric forms already printed on the paper. If a human does not snatch away the paper in time, the chimp tends to cover it in a meaningless multicolored smear. Given paints and brushes in a more natural setting, chimps do not seek out a flat rectangular surface to make a picture—they just playfully paint the nearest bush or rock. Apes show few aesthetic preferences when given images, and show little patience for producing aesthetically structured images when given artistic materials. We should not expect to find any evidence in apes of human adaptations that probably evolved within the last million years, because our most common ancestor lived at least five million years ago.

The Beautiful, the Difficult, and the Costly

Runaway theory and sensory bias theory are not fully satisfying as explanations of human aesthetics. Runaway cannot explain why we have just the preferences that we do. Our sensory biases may be shared with other apes, but they show little evidence of our aesthetic tastes, so sensory biases do not appear to explain human aesthetics. Perhaps fitness indicator theory can do a better job of illuminating human aesthetics. According to this view, maybe our aesthetic preferences favor ornaments and works of art that could have been produced only by a high-fitness artist. Objects of art would then be displays of their creator's fitness, to be judged as such. As with the sexual ornaments on our bodies, perhaps beauty boils down to fitness.

To be reliable, fitness indicators must be difficult for low-fitness individuals to produce. Applied to human art, this suggests that beauty equals difficulty and high cost. We find attractive those things that could have been produced only by people with attractive, high-fitness qualities such as health, energy, endurance, hand–eye coordination, fine motor control, intelligence, creativity, access to rare materials, the ability to learn difficult skills, and lots of free time. Also, like bowerbirds, Pleistocene artists must have been physically strong enough to defend their delicate creations against theft and vandalism by sexual rivals.

The beauty of a work of art reveals the artist's virtuosity. This is a very old-fashioned view of aesthetics, but that does not make it wrong. Throughout most of human history, the perceived beauty of an object has depended very much on its cost. That cost could be measured in time, energy, skill, or money. Objects that were cheap and easy to produce were almost never considered beautiful. As Veblen pointed out in *The Theory of the Leisure Class*, "The marks of expensiveness come to be accepted as beautiful features of the expensive articles." Our sense of beauty was shaped by evolution to embody an awareness of what is difficult as opposed to easy, rare as opposed to common, costly as opposed to cheap, skillful as opposed to talentless, and fit as opposed to unfit.

In her books on the evolution of art, Ellen Dissanayake pointed out that the human arts depend on "making things special" to set them apart from ordinary, utilitarian functions. Making things special can be done in many ways: using special materials, special forms, special decorations, special sizes, special colors, or special styles. Indicator theory suggests that making things special means making them hard to do, so that they reveal something special about the maker. This explains why almost any object can be made aesthetically: anything can be made with special care that would be difficult to imitate by one who was not so careful. From an evolutionary point of view, the fundamental challenge facing artists is to demonstrate their fitness by making something that lower-fitness competitors could not make, thus proving themselves more socially and sexually attractive. This challenge arises not only in the visual arts, but also in music, storytelling, humor, and many other behaviors discussed throughout this book. The principles of fitness-display are similar across different display domains, and this is why so many aesthetic principles are similar.

Anthropologist Franz Boas insisted that in most cultures he studied, the artist's virtuosity was fundamental to artistic beauty. In *Primitive Art*, he observed that "The enjoyment of form may have an elevating effect upon the mind, but this is not its primary effect. Its source is in part the pleasure of the virtuoso who overcomes technical difficulties that baffle his cleverness." For Boas, works of art, were principally indicators of skill, valued as such in almost every culture. He added, "Among primitive peoples . . . goodness and beauty are the same." Whatever people make, they tend to ornament. He spent a good deal of *Primitive Art* trying to show that most of the aesthetic preferences of tribal peoples can be traced to the appreciation of patience, careful execution, and technical perfection. In his view, this thirst for virtuosity explains our preferences for regular form, symmetry, perfectly repeated decorative motifs, smooth surfaces, and uniform color fields. Art historian Ernst Gombrich made powerful arguments along similar lines in his book *The Sense of Order*, which viewed the decorative arts as displays of skill that play upon our perceptual biases.

Beauty conveys truth, but not the way we thought. Aesthetic significance does not deliver truth about the human condition in general: it delivers truth about the condition of a particular human, the artist. The aesthetic features of art make sense mainly as displays of the artist's skill and creativity, not as vehicles of transcendental enlightenment, religious inspiration, social commentary, psycho-analytic revelation, or political revolution. Plato and Hegel derogated art for failing to deliver the same sort of truth that they thought philosophy could produce.

They misunderstood the point of art. It is unfair to expect a medium that evolved to display biological fitness to be well adapted for communicating abstract philosophical truths.

This fitness indicator theory helps us to understand why "art" is an honorific term that connotes superiority, exclusiveness, and high achievement. When mathematicians talk about the "art" of theorem-proving, they are recognizing that good theorems are often beautiful theorems, and beautiful theorems are often the products of minds with high fitness. It is a claim for the social and sexual status of their favorite display medium. Likewise for the "arts" of warfare, chess, football, cooking, gardening, teaching, and sex itself. In each case, art implies that application of skill beyond the pragmatically necessary. Anyone who wishes to imply superiority in their particular line of work is apt to style themselves an artist. The imperatives of fitness display allow us to understand the passion with which people debate whether something is or is not an art. A claim that one's work is art is a claim for sexual and social status.

By this point in my argument, scowls may have crossed the faces of any readers who happen to have read Immanuel Kant's *Critique of Judgment* of 1790 on their last summer beach holiday. Didn't Kant argue that beauty cannot be reduced to utility, that aesthetic enjoyment must be disinterested, that "one possessed by longing or appetite is incapable of judging beauty"? Yes, but Kant recognized that in addition to "ideal beauty" (disinterested) there is "adherent beauty" (biologically relevant and personally interested). He pretended to have a philosophical proof that ideal, disinterested beauty exists. But it is hard to tell Kant's "proofs" from idealistic assertions about human psychology. If we can find an evolutionary function for an aesthetic taste, then it is "interested," and if we can find functions for all tastes, then ideal beauty was a figment of Kant's celibate imagination. If you want a philosopher who understood the biological functions of beauty, read Nietzsche instead.

But Is It Art?

This fitness display theory of aesthetics works much better for folk aesthetics than for elite aesthetics. Folk aesthetics concerns what ordinary people find beautiful; elite aesthetics concerns the objects of art that highly educated, rich elites learn are considered worthy of comment by their peers. With folk aesthetics, the focus is on the art-object as a display of the creator's craft. With elite aesthetics, the focus is on the viewer's response as a social display. In response to a landscape painting, folks might say, "Well, it's a pretty good picture of a cow, but it's a little smudgy," while elites might say, "How lovely to see Constable's ardent brushwork challenging the anodyne banality of the pastoral genre." The first response seems a natural expression of typical human aesthetic tastes concerning other people's artistic displays, and the second seems more of a verbal display in its own right.

Elite aesthetics follow the same signaling principles as sexual selection, but follow them in cultural direction specifically designed to contrast against folk aesthetics. Elites, free to enjoy all manner of costly and wasteful display, often try to distinguish themselves from the common run of humanity by replacing natural human tastes with artfully contrived preferences. Where ordinary folks prefer bright cheerful colors, elites may prefer monochromes, subtle pastels, and elusive off-whites. Where folks prefer good technique and manifest skill, elites may prefer expressiveness, randomness, psychoticism, or a childlike rejection of skill. Where folks prefer realism, elites prefer abstraction. With these preferences, elites can display their intelligence, learning ability, and sensitivity to emerging cultural norms. But to an evolutionary psychologist, the beauty that ordinary people find in ordinary ornamental and representational art says far more about art's origins.

The fitness indicator theory can explain some embarrassing questions that ordinary people ask when they are admitted to modern art museums. A common reaction to abstract expressionist painting is to dismiss it by saying "My child could have done that," "Any idiot could have done that," or "Even a monkey could have done that." Instead of condescending at such comments, we should ask what sort of aesthetic instincts they reveal. To say "My child could have done that" could mean "I cannot discern here any signs of learned skill that would distinguish an adult expert from an immature novice." The "Any idiot" comment could mean "I cannot judge the artist's general intelligence level from this work." The "Even a monkey" comment could mean "The work does not even include any evidence of cognitive or behavioral abilities unique to our species of primate."

Interpreted from a signaling theory viewpoint, such comments are not stupid. Most people want to be able to interpret works of art as indicators of the artist's skill and creativity. Certain styles of art make this difficult to do. People feel frustrated. They have efficient psychological adaptations for making attributions about the artist's fitness given their work, but some genres of modern art prevent those adaptations from working naturally. Having paid the museum's admission fee to see good art, they are instead confronted with works that seem specifically designed to undermine judgments about quality. Art historian Arthur Danto has observed that "We have entered a period of art so absolute in its freedom that art seems but a name for an infinite play with its own concept." This extreme artistic freedom makes it difficult for people to judge an artist's talent. This is not to say that all art should be easy, or that elite art is invalid, or that we should feel comfortable acting like philistines. The human tendency to regard works of art as fitness indicators is being used here as a clue to art's evolutionary origin—not as a prescription for how art should be made or viewed.

When we talk about the evolution of art, perhaps we are really talking about the evolution of a human tendency to make material objects into advertisements of our fitness. When we talk about aesthetics, perhaps we are really talking about human preferences that evolved to favor features of human-made objects that reliably indicate the artisan's fitness. This view suggests that aesthetics overlaps with social psychology. We possess a natural ability to see through the work of art to the artist's skill and intention. Seeing a beautiful work of art naturally leads us to respect the artist. We may not fall in love with the artist immediately. But if we meet them, we may well want to find out whether their actual phenotypes live up to their extended phenotypes.

The Work of Art Before the Age of Mechanical Reproduction

The Arts and Crafts movement of Victorian England raised a profound issue that still confronts aesthetics: the place of human skill in our age of mass production and mass media. During human evolution we had no machines capable of mechanically reproducing images, ornaments, or objects of art. Now we have machines that can do so exactly and cheaply. We are surrounded by mass-produced objects that display a perfection of form, surface, color, and detail that would astonish premodern artists.

Mechanical reproduction has undermined some of our traditional folk aesthetic tastes. Veblen observed that when spoons were made by hand, those with the most symmetrical form, the smoothest finish, and most intricate ornamentation were considered the most beautiful. But once spoons could be manufactured with perfect symmetry, finish, and detail, these features no longer indicated skilled artisanship: they now indicated cheap mass production. Aesthetical standards shifted. Now we favor conspicuously handmade spoons, with charming asymmetries, irregular finishes, and crude ornamentation, which would have shamed an 18th-century silversmith's apprentice. A modern artisan's ability to make any sort of spoon from raw metal is considered wondrous. Such low standards are not typical of premodern cultures. Drawing on his wide experience of tribal peoples in Oceania, Franz Boas observed in his book *Primitive Art* that "The appreciation of the esthetic value of technical perfection is not confined to civilized man. It is manifested in the forms of manufactured objects of all primitive peoples that are not contaminated by the pernicious effects of our civilization and its machine-made wares."

Likewise, the cultural theorist Walter Benjamin pointed out

that, before photography, accurate visual representations required enormous skill to draw or paint, so were considered beautiful indicators of painterly genius. But after the advent of photography, painters could no longer hope to compete in the business of visual realism. In response, painters invented new genres based on new, non-representational aesthetics: impressionism, cubism, expressionism, surrealism, abstraction. Signs of handmade authenticity became more important than representational skill. The brush-stroke became an end in itself, like the hammer-marks on a handmade spoon.

A similar crisis about the aesthetics of color was provoked by the development of cheap, bright aniline dyes, beginning with William Henry Perkins's synthesis of "mauve" in 1856. Before modern dyes and pigments were available, it was very difficult to obtain the materials necessary to produce large areas of saturated color, whether on textiles, paintings, or buildings. When Alexander the Great sacked the royal treasury of the Persian capital Susa in 331 B.C., its most valuable contents were a set of 200-year-old purple robes. By the 4th century A.D., cloth dyed with "purpura" (a purple dye obtained from the murex mollusk) cost about four times its weight in gold, and Emperor Theodosium of Byzantium forbade its use except by the Imperial family, on pain of death. Colorful objects were considered beautiful, not least because they reliably indicated resourcefulness—our ancestors faced the same problem of finding colorful ornaments as the bowerbirds. Nowadays, every middle-class family can paint their house turquoise, drive a metallic silver car, wear fluorescent orange jackets, collect reams of glossy color magazines, paint the cat crimson, and dye the dog blue. Color comes cheap now, but it was rare and costly to display in art and ornament during most of human evolution. Our ancestors did not live in a sepia-tint monochrome: they had their black skins, their red blood, the green hills of Africa, the blue night, and the silver moon. But they could not bring natural colors under their artistic control very easily. Those who could may have been respected for it.

Before the age of mechanical reproduction, ornaments and works of art could display their creator's fitness through the precision of ornament and the accuracy of representation. Modern technology has undermined this ancient signaling system by making precision and accuracy cheap, creating tension between evolved aesthetics and learned aesthetics. Our evolved folk aesthetics still value ornamental precision, representational accuracy, bright coloration, and other traditional fitness indicators. But we have learnt a new set of consumerist principles based on market values. Since handmade works are usually more expensive than machine-made products, we learn to value indicators of traditional craftsmanship even when such indicators (crude ornamentation, random errors, uneven surface, irregular form, incoherent design) conflict with our evolved preferences. Yet within the domain of manufactured goods, we still need to use our folk preferences to discern well-machined goods from poorly machined goods. This can lead to confusion.

For example, there was a famous case in 1926 when Constantin Brancusi sent his streamlined bronze sculpture "Bird in Space" from Europe to New York for an exhibition. A U.S. Customs official tried to impose a 40 percent import duty on the object, arguing that it did not resemble a real bird, so should be classed as a dutiable machine part rather than a duty-free work of art. Following months of testimony from artists and critics sympathetic to modernism, the judge ruled in favor of Brancusi, stating that the work "is beautiful, and while some difficulty might be encountered in associating it with a bird, it is nevertheless pleasing to look at." Although "Bird in Space" exhibited a perfection of form and finish that Pleistocene hominids would have worshiped, it was almost too perfect to count as art in our age.

Handaxes as Ornaments

Stone handaxes show that hominids did care about form and finish. Indeed, science writer Marek Kohn and archeologist Steven Mithen independently developed the theory that sexual selection favored symmetric handaxes as fitness indicators. If their arguments work, handaxes represent the first hominid works of art, and the first hard evidence of sexual selection shaping human material culture.

Two and a half million years ago, our small-brained ancestors evolved the ability to knock flakes from rocks to use as cutting edges. By doing so, they could also make the rocks themselves useful as choppers. This basic tool kit of flakes and choppers served the needs of hunting and gathering for a million years. Then, around 1.6 million years ago, a medium-brained African hominid (*Homo erectus*) evolved the ability to produce an extraordinary object that archeologists call a handaxe. A handaxe is a rock chipped into roughly the size and shape of a child's hand—flat with fingers together. There is a sharp edge all around and a point at the tip. The outline is midway between that of a pear and a triangle. The top and bottom faces are symmetrical (handaxes are also called "bifaces"), as are the right and left halves. Most were made of flint, some of quartzite or obsidian.

Handaxes proved enormously popular. They were made for over a million years, until about 200,000 years ago, by which time our ancestors had evolved into large-brained archaic *Homo sapiens*. Handaxes were made throughout Africa, Europe, and Asia, and in enormous numbers: sometimes hundreds are found at a single dig site. The persistence of a single design across such a span of space and time cannot be explained through cultural imitation. Designs passed down through mere imitation tend to deviate further and further from the original prototype, as languages do over hundreds of years. Handaxes must have been to hominids what bowers are to bowerbirds: part of their extended phenotype, a genetically inherited propensity to construct a certain type of object.

But why did the handaxe evolve? Handaxes are not particularly bad tools. They offer a fair amount of cutting edge for their weight, and they are somewhat safer and easier to use than flakes when butchering large animals. But the cutting edge all the way around the rim makes a handaxe rather difficult to hold, like a knife without a handle. For almost all practical purposes, sharp flakes and edged choppers would have been sufficient.

Perhaps handaxes were missiles rather than hand-held tools? H. G. Wells proposed in 1899 that handaxes may have been thrown at prey, but this "killer Frisbee hypothesis" has not fared well. In 1997, in a coal mine in Schöningen, Germany, archeologists found some well-preserved, six-foot-long sprucewood spears. They were almost as well engineered as modern javelins, quite lethal, and 400,000 years old. Given that such excellent missiles had been developed by that date, why did our ancestors keep making handaxes for another 200,000 years?

Some handaxes may have been practical tools, but Kohn and Mithen noted that many show evidence of skill, design, and symmetry far beyond the demands of utility. Some were made in large sizes too heavy and clumsy to use. The "Furze Platt Giant" handaxe is over a foot long, and seems designed to be held in both hands and admired. Others are under two inches long, too small to be of much use. Often they show far more exact symmetry than seems necessary, and, from a practical viewpoint, excessive attention to the regularity of form and finish. Handaxes were often made in very large numbers in the same place. Most importantly, many of the finest handaxes show no sign of use: no visible chips, and no evidence of edge wear under the electron microscope. Why were so many handaxes made so perfectly, with such care, and then discarded, apparently unused, still sharp enough to cut fingers a million years later?

In his book *As We Know It,* Marek Kohn argued that the handaxe "is a highly visible indicator of fitness, and so becomes a criterion of mate choice." Handaxes make good Zahavian handicaps. They impose high learning costs: it takes six months to acquire the basics of flint-knapping, and years more to perfect the skill. They take extra time to make. Modern experts with 25 years of flint-knapping experience take about 20 minutes to make a decent handaxe, whereas a simple edged tool can be made in just a couple of minutes. There are risks of injury: modern flint-knappers wear boots, leather aprons, and goggles to protect against flying rock shards, and they often get cuts on their hands. Expert handaxe production requires a combination of physical strength, hand–eye coordination, careful planning, conscientious patience, pain tolerance (to deal with the flying debris), and resistance to infection (to deal with the cuts)—as Kohn noted, "A handaxe is a measure of strength, skill and character." Their symmetry, like that of the peacock's tail and the human face, makes their perfection of form very easy to assess, but very hard to produce. In short, handaxes are reliable indicators of many physical and mental aspects of fitness. Kohn suggested that the normal, pragmatic handaxes may have been fashioned by females, while the very large, very small and very symmetrical ones were produced by males as sexual displays.

So, we have an object that looks like a practical survival tool at first glance, but that has been modified in important ways to function as a costly fitness indicator. Kohn and Mithen have made a fairly good case that the handaxe was often a work of art, and a sexual attractant. They suggested several ways to test their hypothesis further. If their radical idea proves correct, then handaxes may have been the first art-objects produced by our ancestors, and the best examples of sexual selection favoring the capacity for art. In one neat package, the handaxe combines instinct and learning, strength and skill, blood and flint, sex and survival, art and craft, familiarity and mystery. One might even view all of recorded art history as a footnote to the handaxe, which reigned a hundred times as long.

9

Virtues of Good Breeding

Murder, unkindness, rape, rudeness, failure to help the injured, fraud, racism, war crimes, driving on the wrong side of the road, failing to leave a tip in a restaurant, and cheating at sports. What do they have in common? A moral philosopher might say that they are all examples of immoral behavior. But they are also things we would not normally brag about on a first date, and things we would not wish an established sexual partner to find out that we had done. The philosopher's answer sounds serious and mine sounds flippant. But the philosopher's answer does not identify any selection pressure that could explain the evolution of human morality. Mine does: sexual choice.

Most evolutionary psychologists have viewed human morality as a question of altruism, and have tried to explain altruism as a side-effect of instincts for nepotism (kindness to blood relatives) or reciprocity (kindness to those who may reciprocate). I think human morality is much more likely to be a direct result of sexual selection. We have the capacity for moral behavior and moral judgments today because our ancestors favored sexual partners who were kind, generous, helpful, and fair. We still have the same preferences. David Buss's study of global sexual preferences found that "kindness" was the single most important feature desired in a sexual partner by both men and women in every one of the 37 cultures he studied. It ranked above intelligence, above beauty, and above status.

Oscar Wilde's play *An Ideal Husband* recognized the role of sexual choice in shaping human morality. The drama's theme is that men and women are under very strong pressure to make a credible show of high moral stature to their lovers and spouses. The drama centers on this: will the highly principled Lady Chiltern still love her husband after learning that he acquired his fortune by selling a government secret? Wilde put his finger on an evolutionary pressure for morality that has not yet received sufficient attention in evolutionary psychology: good moral character is sexually attractive and romantically inspiring. Conversely, liars and cheats are sexually repulsive—unless they have other charms that compensate for their flawed character. In the play, Sir Robert Chiltern retained his wife's affection only by making a parliamentary speech against an investment swindle—a public moral display which, due to the threat of blackmail by the swindlers, he believed would cost him his career.

As we have seen again and again in this book, sexual attractiveness alone is sufficient to explain the evolution of many traits. One does not always have to seek a survival function. Many theorists have tried and failed to trace human morality to various survival benefits for the individual or the group. I shall argue that some of our most valued moral virtues had no survival benefits, but they did have strong courtship benefits. Sexual selection enables us to explain a class of moral behaviors and moral judgments much broader than those considered by most philosophers and evolutionary psychologists. A sexual selection perspective allows us to explain sympathy, agreeableness, moral leadership, sexual fidelity, good parenting, charitable generosity, sportsmanship, and our ambitions to provide for the common good. The importance of sexual choice in the evolution of human morality, generosity, magnanimity, and leadership has also been analyzed by biologist Irwin Tessman, anthropologists Kristen Hawkes and James Boone, and primatologist Frans de Waal. I draw on many of their ideas in this chapter.

Human morality, in my view, includes any behavior that displays good moral character. It is not limited to altruism, which is the conferral of a benefit on someone else at an apparent cost to oneself. Displays of altruism can be among the most potent displays of moral character, but they are not the only such displays. As with most reliable fitness indicators, the point of moral displays is not so much the benefit conferred on others, but the cost imposed on oneself. Morality is a system of sexually selected handicaps—costly indicators that advertise our moral character.

Apathy as the Evolutionary Norm

According to a popular stereotype, evolutionary theory implies that organisms should engage in rampant, bloody, unrestrained competition. If we take any two animals from anywhere in the world and throw them in a pit, they should start tearing each other apart. Yet they do not. Does this imply that nature is more cooperative than evolution can explain?

No. Ecologists have long understood that the typical interaction between any two individuals or species is neither competition nor cooperation, but neutralism. Neutralism means apathy: the animals just ignore each other. If their paths threaten to cross, they get out of each other's way. Anything else usually takes too much energy. Being nasty has costs, and being nice has costs, and animals evolve to avoid costs whenever possible. This is why watching wild animals interact is usually like watching preoccupied commuters trying to get to work without bumping into one another, rather than watching a John Woo action film with a triple-figure body count.

Apathy is nature's norm. Predators ignore all but a few favored species of prey. Parasites usually focus their attention on just one species of favored host. Darwin pointed out that most of the violent competition happens within a species, because animals of the same species are competing for the same resources and the same mates. Evolutionary biology focuses on competition because competition between genes drives evolution. Nevertheless, animals usually tend to avoid competition as much as possible. In particular, evolution almost never favors spite, which means hurting a competitor at a net cost to oneself. The costs of spite are carried entirely by oneself and one's victim, while the benefits of hurting that victim are enjoyed by all of one's other competitors.

Since apathy is the default attitude of any one animal to any other, we need not seek any special explanation for human apathy towards other humans or other animals. The hard things to explain are costly behaviors that help others, and costly behaviors that hurt others. If we were typical animals, our attitudes to others would be dominated not by hate, exploitation, spite, competitiveness, or treachery, but by indifference. And so they are. Immanuel Kant suggested that we view people either as ends in themselves or as means to our ends. Neutralism suggests that we usually view them as neither—neither subject nor object, just an occasion for a blank stare and a lazy shrug. What evolution has to explain about human morality is why we ever do anything other than shrug when we see opportunities for care and generosity.

The Hidden Benefits of Kindness

The evolution of morality did not have to get us over some ethical hump to move us from spiteful animal to generous human. We started in the middle, already sitting on the ethical fence, neutral and apathetic. We just needed some kind of selection pressure capable of favoring kindness. Any good evolutionary theory of human morality must convert the apparent costs of helping others into a realistic benefit to one's genes, by turning material costs into survival or reproductive benefits. If it cannot do that, it cannot explain how moral behaviors like kindness or generosity could evolve. The rules of evolutionary biology demand that we find a hidden, genetically selfish benefit to our altruism.

Some philosophers, theologians, and journalists are unhappy with this hidden-benefit requirement. They wish to define morality as purely selfless altruism, untainted by any hidden benefit. In their view, only the morality of the celibate saint qualifies as worthy of evolutionary explanation. But to my way of thinking, a moral theory of saints explains little about human nature, because saints are rare. Of the 15 billion or so humans who have lived since the time of Jesus Christ, the Catholic Church has canonized only a few thousand. Saints are literally one in a million. They may be instructive as moral ideals, but they are statistically irrelevant as data about real human moral behavior. Moral philosophers are sometimes not clear about whether they are developing a descriptive explanation of human moral behavior as it is, or an ideal of saintly moral behavior as it should be. My interest here is in finding an evolutionary explanation of ordinary human kindness, not in accounting for the outer limits of saintly goodness.

One step down from theologians, but still high above the rest of us, sit the economists. They appear to explain human morality as they explain all behavior, in terms of rationally pursued preferences. If we are kind, we must have a taste for kindness, to which we attach some "subjective utility." If we give money to charity, that must be because the subjective utility we derive from giving exceeds the subjective utility that we would derive from holding on to the money. Most economists understand perfectly well that this "revealed preferences" principle is circular. It is a statement of the axioms that they use to prove theorems about the emergent effects of individual behavior in markets. It should not be confused with a psychological explanation of behavior, much less an evolutionary explanation.

Psychologists sometimes fail to understand how circular it is to "explain" moral behavior in terms of moral preferences. Of course, one can always say that we are kind because we choose to be kind, or it feels good to be kind, or we have brain circuits that reward us with endorphins when we are kind. Such responses beg the question of why those moral preferences, moral emotions, and moral brain circuits evolved to be standard parts of human nature. A costly behavior cannot evolve just because it happens to feel good. Feeling good must have evolved to motivate the behavior, which must have some hidden benefit.

Most evolutionary psychologists have agreed that kindness and generosity bring two major kinds of hidden benefit. One kind of benefit comes when the generosity is directed towards blood relatives. In such cases, the cost to one's own genes can be outweighed by benefits to copies of those genes in the bodies of relatives. This is the theory of kin selection, and it explains generosity toward kin. The other kind of benefit comes when generosity is directed toward individuals who are likely to reciprocate in the future. Today's altruism may be repaid tomorrow. This is the theory of reciprocal altruism, which, it has been claimed, explains most instances of kindness to non-relatives.

Kinship and reciprocity are certainly important in human affairs, and we have evolved many psychological adaptations to deal with them. They go a long way in explaining many aspects of human moral behavior. For example, kinship theory puts into an evolutionary context the Confucian virtues of family obligation, while reciprocity helps to explain prudence, loyalty, guilt, and revenge. However, there is a lot left over. Human morality includes a great variety of behaviors and judgments that are hard to explain through kinship and reciprocity. Let's have a closer look at their limitations, and then see whether there are any other hidden evolutionary benefits to kindness.

Kinship

Kin selection theory was developed by W. D. Hamilton in 1964. It pointed out that there is a hidden genetic benefit in being kind to one's offspring and close genetic relatives. A gene for kindness to relatives can prosper because it tends to help other copies of the same gene to prosper, copies that happen to be in bodies other than one's own. They are there because they were passed down from the same recent common ancestor to both oneself and one's relatives. Thus, generosity to blood relatives is actually genetic selfishness.

This can be hard to understand, so let's view evolution from a selfish gene's point of view. Genes are "selfish" in the sense that they evolve to generate as many copies of themselves as possible. They act as if they are trying to spread throughout a population. Often, they do so by constructing bodies and brains that act as self-interested individuals. But that is by no means the only way for a selfish gene to spread. All genes would profit from being able to recognize and help copies of themselves in other bodies, and would be able to spread themselves better if they could. But most of them cannot, because they help only to grow lungs, or livers, or some other blind organ incapable of generous behavior. Only a few kinds of genes have the power to be selectively generous to blood relatives. Some such genes are involved in growing a perceptual ability to distinguish kin from non-kin, or a behavioral ability to be kinder to kin. Others may grow nutrient-delivery systems such as wombs and breasts that nourish another individual that happens to be growing inside one's body, or clinging to oneself, which suggests they are probably one's own offspring. These are the main ways in which kindness-to-kin genes evolve.

Kin selection is really just a theory about how a particular kind of genetic mutation can spread. The only mutation that can assist other copies of itself in other bodies is a mutation for recognizing and favoring close genetic relatives. By favoring close relatives, who share a recent common ancestor, they are most likely to be favoring individuals who happen to carry a copy of this kin-favoring mutation. The whole edifice of family life, kinship, and parenting is based on a few very peculiar mutations that are directly responsible for kin recognition and nepotism. These kin recognition genes single-mindedly evolve better ways to recognize and assist one another, while all the other genes go about their business oblivious to kinship.

Kin selection predicts that we should be kinder to relatives who are more closely related to us: our altruism to kin should be in proportion to the likelihood of sharing the altruism gene with them. That likelihood is determined by how recently we shared a common ancestor, and how many common ancestors we shared. The likelihood of sharing the same kindness-to-kin gene is one half for siblings, parents, and offspring, but only a quarter for half-siblings, grandparents, grandchildren, nieces, nephews, aunts, and uncles. This is why we are usually kinder and closer to sisters than to nieces, even in societies where extended families live together.

A confusion commonly arises at this point. Kin selection theory is often misunderstood as saying we should be kind to other organisms in proportion to the true percentage of genes we share with them. Don't all humans share about 99 percent of our DNA? That sounds close to identical twins, who share 100 percent of their DNA. If we share so many of our genes with other humans, why should we discriminate between close relatives and distant relatives? And don't we share about half of our genes with other mammals, birds and even fish? We should treat all herring as brothers and all sloths as sisters. We should apply the golden rule to all primates, be true friends to all mammals, allies to termites and tapeworms, and just slightly grudging compatriots of baobab trees, stinging nettles, and Antarctic lichens. Universal peace, cooperation, and symbiosis should reign on our blessed planet, according to this genetic-similarity interpretation of kinship.

It would be a weird and wonderful world if there were an evolutionary process that could favor altruism in proportion to true genetic similarity. Racism, ethnocentrism, xenophobia, sexism, human competition, crime, warfare, deforestation, pollution, and cruelty to animals would all vanish. We would all behave like Jainists—members of the Indian religion who dare not eat or move for fear of injuring another living being. But there is no such evolutionary process. Kinship theory offers only a forgetful, myopic, fumbling imitation, in which we have evolved the delusion that only our extremely close relatives have anything genetic in common with us.

How much of human morality derives from kin selection? Opinions vary. W. D. Hamilton, E. O. Wilson, and many others have suggested that adaptations for kindness to kin may have been important building blocks for kindness toward non-kin. Kin selection does not require a brain, but having a brain helps. It was a major step for brains to evolve the abilities to recognize individual relatives, determine how much care they should receive based on cues of genetic similarity, and produce care behaviors that actually benefit them. It looks as though it should be fairly easy to modify such adaptations to recognize individual non-relatives, determine how much care they should receive based on other kinds of cues, and produce effective care behaviors.

But would evolution favor such modifications? The genes underlying kindness to kin could have evolved only if they discriminated against non-kin. Their success ever since they originated as rare little mutations has depended on them being selectively altruistic, not generally altruistic. Although psychologically it looks like a small step to extend kin-based altruism to non-kin, it is a huge evolutionary leap that violates the basic rationale of kin selection. Being able to imagine "all men are brothers" is a long way from acting as though they are. Kin recognition is widespread among mammals. If human morality evolved as a free-and-easy side-effect of kin-based altruism, we might expect most mammals to show human-like morality. They don't. Apparently, evolution ruthlessly eliminates all kinship genes that lose their discriminative abilities and start treating strangers as relatives. Kinship is powerful at explaining our kindness toward blood relatives, but it is hard to extend it beyond that.

Reciprocity

In the early 1970s, Robert Trivers pointed out that animals can benefit by being nice to one another if they interact often enough to build up trust. By keeping their promises and fulfilling their contracts, rather than opting for the short-term benefits of lying and cheating, they might obtain larger benefits over the longer term. Trivers's theory of "reciprocal altruism" suggested that many cases of apparent altruism are rationally selfish if viewed in their larger social context over the longer term. In reciprocity, there are three defining features: animals alternate giving and receiving benefits; each act has costs to the giver and benefits to the receiver; and giving is contingent on having received. As long as these three conditions hold, animals can trade benefits back and forth. Each act taken out of context may look altruistic, but the whole sequence is mutually beneficial. Trivers ingeniously showed how to connect the mathematics of reciprocity to the biology of altruism and the psychology of trust.

This logic of reciprocity was news to biologists, but not to economists. Trivers had rediscovered an economic principle called the "folk theorem of repeated games." It is called a folk theorem because it was discovered independently by so many different game theorists in the early 1950s that none has ended up with individual credit. The theorem says that repeated interactions can be as powerful as contract law in maintaining cooperation. Any mutual benefits that two individuals could agree to provide to each other through a formal contract can also be sustained if the individuals interact sufficiently often. This is why traditional Chinese-style business that builds trust through repeated interaction can work as well as American-style business based on contracts and litigation.

The folk theorem of repeated games clearly implied that cooperation depends on the threat of punishing cheaters who do not cooperate. With contracts, punishment implies litigation. With repeated interaction, punishment can consist of withdrawing from further interaction for a while, denying the cheat the benefits of cooperation. (If both individuals were not deriving benefits from cooperating, they would not be interacting at all.) For reciprocity to evolve according to the folk theorem, you do not need the concept of a contract, or the emotion of trust or betrayal, or a conceptual understanding of the future. All that is needed is a helpful behavior on your part if the other individual cooperated last time, a punishment routine you impose if he or she fails to cooperate, plus a capacity for telling the difference. Plants, flatworms, herring, and sloths could all evolve reciprocity if they evolved these three abilities.

The idea of reciprocal altruism promised to revolutionize the study of animal behavior. In the 1970s, biologists expected to find cooperation in thousands of species being sustained through repeated interactions. Unfortunately, three decades of intense research have produced almost no clear examples of reciprocity in animals other than primates. Evolution appears to avoid reciprocity whenever possible. The only decent non-primate example occurs in vampire bats. Biologist Gerald Wilkinson found that vampire bats that have drunk well on a particular night sometimes vomit surplus blood to hungry non-relatives. These non-relatives may vomit blood in return the next night, if they happen to have found a good vein. However, even in this often-cited case it is not clear whether generosity to particular individuals is truly contingent on their past behavior.

Social primates offer better examples of reciprocity. Primatologist Frans de Waal observed in his book *Good Natured* that chimps appear to show moral outrage if a long-term ally fails to support them in a fight. They seek out and attack the cowardly traitor. This looks like a punishment routine designed to sustain cooperation. De Waal also found good evidence of chimpanzees trading food for grooming in a truly contingent way. Higher levels of reciprocity in primates are not surprising, given that primates are good at recognizing individuals, forming social relationships with non-relatives, and giving one another social benefits such as grooming, food-sharing, and mutual defense.

Cheating for Status

In reciprocal altruism, one must be able to detect cheats who take without giving. Evolutionary psychologists Leda Cosmides and John Tooby reasoned that if humans evolved as reciprocal altruists, we must have moral capacities for detecting cheats. They have run many experiments demonstrating that the human mind is highly attuned to detecting situations where individuals take benefits without fulfilling a social requirement. Many of the situations described in their experiments do concern genuine reciprocity, in which mutual benefits are exchanged between two individuals.

However, some of their examples of detecting cheats seem more concerned with the reliability of sexual status indicators than with the maintenance of reciprocity. Consider their (fictional) example of "Big Kiku," which is often cited. A tribal chief called Big Kiku establishes the rule that an individual must have a special tattoo in order to eat cassava root, a local delicacy. When asked to identify various possible ways in which this rule could be violated by cheats, participants in the Cosmides experiments could easily see that individuals without tattoos might be cheats, and individuals eating cassava root might be too. Yet the participants' ability to reason correctly about this problem does not necessarily depend on their understanding reciprocal altruism in Trivers's sense. If I get a tattoo, it is not giving you some benefit that you reciprocate by allowing me to eat cassava root. The tattoo is simply a costly, painful signal of tribal sexual status which entitles me to enjoy the associated status display of eating cassava root.

The Cosmides experiments, often replicated and extended by other psychologists, are one of the best examples of empirical evolutionary psychology. They have revealed a specific human adaptation for detecting cheats that is distinct from general intelligence, social intelligence, or the comprehension of arbitrary social rules. But these experiments also reveal that reciprocity is not the only context in which we look for cheats. People seem to regard any status display as a benefit, and look for people who cheat by producing the display without deserving the status. They use the same mental adaptations to look for cheats who undermine fitness indicators (by pretending to a status they do not deserve) and cheats who violate reciprocity arrangements (failing to return a benefit to one who gave you a benefit).

Because the same cheater-detection module is apparently used to detect both status cheats and reciprocity cheats, evidence for cheat detection is not necessarily evidence of the importance of reciprocity in human evolution. The conventions of rank, privilege, and status are distinct from the conventions of reciprocity that yield mutual benefit. This was one of Karl Marx's key insights. A society could be based on status signals without reciprocity (a simple dominance hierarchy), or on reciprocity without status signals (an egalitarian utopia). In either, the ability to detect cheats would be useful. Our outrage against cheats is directed at those who display deceptive fitness indicators, not just those who fail to return a kindness.

There's More to Morality than Kinship and Reciprocity

Kinship and reciprocity are important, and explain a great deal of human behavior that looks initially puzzling from a survival-of-the-fittest viewpoint. Matt Ridley made a good case for their evolutionary, social, and economic significance in his book *The Origins of Virtue*. However, they hardly touch some of the moral virtues we consider most important. Parental solicitude and nepotism are widespread, adaptive, and important, but are not often praised as distinctly moral virtues. Reciprocity is certainly sensible, foresighted, and rational, but from the 1980s some scientists seem to have equated it with the whole of human morality.

For example, kinship and reciprocity have difficulty explaining charity to non-relatives. We know the difference between giving money to a nephew, lending money to a friend, and handing money to a beggar. Nor can kinship and reciprocity explain very satisfactorily other important virtues such as moral leadership, romantic generosity, sympathy, sexual fidelity, or sportsmanship. Moreover, sexual selection may cast new light on certain moral phenomena that were previously understood in terms of kinship and reciprocity.

Of course, it is possible to fit almost any human social behavior into the Procrustean bed of reciprocity, because many social interactions are repeated and many violations of social convention are frowned upon. But this does not mean that people are always giving benefits today in order to receive benefits tomorrow. Broadening evolutionary psychology's attention to aspects of morality other than kinship and reciprocity may lead to new research insights. It may also prove more appealing to those who believe that there is more to human virtue than nepotism and economic prudence.

Innate Depravity?

Some religions depict humans as born in sin and saved only through faith and good works. Some Darwinians have followed this line as well. T. H. Huxley's 1896 lecture *Evolution and Ethics* portrayed morality as a cultural invention, a sword to slay the dragon of our animal past and overcome our innate selfishness. In

Civilization and Its Discontents, Sigmund Freud took a similar line, arguing that society depends on the renunciation of animal passions and conformity to learned social norms. One of the few points of agreement between biologists and social scientists in the 20th century was that human morality must be taught, because it cannot be instinctive.

With the rise of selfish-gene thinking in 1960s and 1970s biology, the innate depravity view gained clarity and force. Biologists realized that all organisms must evolve to be evolutionary egoists in the sense of promoting the replication of their own genes at the expense of other genes. This inclined some to the view that organisms must usually be egoists in the vernacular sense as well: individually competitive, selfish, ungenerous, and ill-mannered. The evolutionary selfishness of the gene was seen as leading automatically to the selfishness of human individuals. Leading evolutionary theorists such as E. O. Wilson, George Williams, and Robert Trivers adopted this seemingly pessimistic view. In *The Selfish Gene*, Richard Dawkins followed Huxley's lead: "Be warned that if you wish, as I do, to build a society in which individuals cooperate generously and unselfishly towards a common good, you can expect little help from biological nature. Let us try to *teach* generosity and altruism, because we are born selfish."

Many critics reacted against the innate depravity view with moral outrage, sentimental anecdotes, and a failure to understand the power of the selfish-gene perspective. Confronted with an apparent conflict between modern evolutionary theory and human morality, some biologists such as Stephen Jay Gould dismissed the selfish-gene view of evolution. In reaction, selfish-gene biologists dismissed the critics as confused idealists. It has taken a couple of decades for scientists to get beyond this impasse, to accept that there are human moral instincts other than nepotism and reciprocity, and that they must have evolved somehow. Frans de Waal sounded this new note of optimism in his book *Good Natured*: "Humans and other animals have been endowed with a capacity for genuine love, sympathy, and care—

a fact that can and will one day be fully reconciled with the idea that genetic self-promotion drives the evolutionary process."

Evolutionary psychology is taking more seriously the evidence for human generosity, such as evidence that sympathy develops spontaneously in young children, and experimental economics research showing "irrationally" high levels of generosity between adults playing bargaining games. Economist Robert Frank's book *Passions Within Reason* was very important in putting true sympathy and generosity back on evolutionary psychology's agenda. He analyzed the evolution of human capacities for moral commitment, showing how apparently irrational tendencies to honor promises and punish cheats could bring hidden genetic benefits. He also showed that people are pretty good at predicting who will act generously and who will not when there is a temptation to be selfish: our moral character can be reliably judged. Philosopher Elliot Sober and biologist David Sloan Wilson have also insisted on the importance of bringing psychological evidence regarding sympathy and generosity into evolutionary discussions of morality. Some evolutionary economists have even turned to Adam Smith's *A Theory of Moral Sentiments*, which presents a much rosier picture of human generosity than his more famous *The Wealth of Nations*. Human kindness is becoming accepted as an adaptation to be explained rather than a myth to be ridiculed. The new Darwinian moral optimism is much more nuanced than either the innate selfishness view of the Catholic Church and sociobiology, or the innate goodness view of Rousseau and utopian socialism. It accepts our moral nature as we find it. But the goal remains: to find the hidden evolutionary benefits of human kindness.

Mating Well by Doing Good

Fortunately, kinship and reciprocity are not the only evolutionarily respectable ways to turn apparent altruistic costs into individual reproductive benefits. You may not be surprised to find me using the sturdy mule of mate choice to haul the cart of human nature up the mountain of morality. As we have seen before, sexual selection can explain things that few other evolutionary forces can. It can favor attractive, elaborate indicators that incur heavy costs in every domain other than reproduction. Could our moral acts be one class of such indicators? Do our moral judgments have some overlap with mate choice?

Immoral acts are mainly those we would be embarrassed by if our boyfriend or girlfriend found out about them. Why? Because they would then hold our character in lower esteem. The esteem of sexual partners sounds like a rather trivial basis for human morality. However, those who have been divorced for their moral failings may take a more respectful attitude towards mate choice as a shaper of moral instincts. As we have seen, David Buss's findings indicate that kindness is the most desired trait in a sexual partner around the world. Other research on human mate choice consistently confirms the attractiveness of kindness, generosity, sympathy, and tenderness.

In 1995, Irwin Tessman became the first to argue that sexual selection shapes morality. He pointed out that human generosity goes beyond the demands of kinship and reciprocity. Perhaps generosity works as a Zahavian handicap that displays fitness, and thus evolved through sexual selection. Amotz Zahavi has argued since the 1970s that apparent altruism could bring hidden reproductive benefits through the social status that it inspires. Anthropologist James Boone recently combined Zahavi's handicap theory and Veblen's conspicuous consumption theory to explain costly, conspicuous displays of magnanimity. While Tessman and I focus on direct mate choice for moralistic displays during courtship, Zahavi and Boone emphasize the indirect reproductive benefits of high status. Both effects were probably important during human evolution.

In theory, mate choice could be the single most powerful moral filter from one generation to the next. It could favor almost any degree of altruism or heroism, compensating for almost any risk to survival. If, for example, all females refused to mate with any males who ate meat, any genes predisposing individuals to vegetarianism (however indirectly) would spread like wildfire.

The species would turn vegetarian no matter what survival benefits were conferred by meat-eating, as long as the sexual selection pressure against meat-eating held. Natural selection for selfishness would be impotent against sexual selection for moral behavior.

Aristophanes' play *Lysistrata* of 411 B.C. illustrated the moral power of female sexual choice. Lysistrata convinced the other women of Athens to stop having sex with their men until the men stopped waging the Peloponnesian war. The women barricaded themselves in the Acropolis, while (in the original staging of the play) the sex-starved men wandered around with ever-larger leather phalluses, gradually realizing that military victory becomes meaningless without the prospect of sex. Although some women were also tempted to break the sex strike—one even tried to sneak off to a brothel—they outlasted the men. Lysistrata's sex-strike succeeded in forcing the Athenian men to make peace with the Spartans. Her strategy would have worked equally well over evolutionary time: female sexual preferences for peace-keepers could have reduced male belligerence and aggressiveness.

This "better morality through mate choice" hypothesis prompts several questions. Why would mate choice mechanisms evolve to favor displays of generosity, fair play, good manners, or heroism? Why do we consider such displays especially "moral," as compared with other courtship displays? Why do our judgments of different courtship displays feel so different? Bodily ornaments seem to provoke lust, artistic displays induce aesthetic feelings, and moralistic displays attract admiration. This chapter does not answer all these questions, but may chart some new territory in the evolution of human morality. We'll start with a simple example of how mate choice can favor costly behaviors that provide for the common good.

The Evolution of Hunting: An Altruistic Display of Athleticism?

Why did humans evolve to hunt relatively big game like eland and mammoth? The answer seems pretty obvious: you can eat their meat and survive better. The 1968 anthropological classic *Man the Hunter* took the view that hunting evolved through simple survival selection. Hungry hominid? No problem—go hunt.

It turns out not to be that simple. In the early 1980s, female anthropologists contributed to a corrective volume entitled *Woman the Gatherer*. They showed that in most hunter-gatherer societies women provide most of the sustenance, efficiently collecting plant foods and small game. The men often fail to bring any meat back from the hunt and often rely on their female partners for day-to-day sustenance. Trying to chase down large mammals that have evolved to run away from predators much faster than you is just not an efficient, reliable way to support yourself, much less your family. Anthropologist Kristen Hawkes found that in the tribe she was studying, men have only a 3 percent chance per day of successfully killing a large animal. That's 97 percent failure: not the stereotypical image of the cave-man bringing home the bacon. Data from other tribes shows slightly higher success rates, but they rarely exceed 10 percent each day.

To a female gatherer seeking a bit of meat on the side, the behavior of the males must be doubly annoying. If they must hunt to boost their egos, fine, but why must they try to catch really big animals? Men know very well that their hunting success is much higher when they go after smaller, slower, weaker animals. Usually, the smaller the prey they target, the more pounds of meat per day they bring home, and the less variable is the amount of meat from one week to the next. Also, the smaller the game, the more of its meat can be eaten before it goes rotten. When hunters really need to eat, they'll give up on the large game and catch the small. If hunting's function is to feed the hunter and his family, male human hunters look ridiculously overambitious. They aim for giraffes when they should be catching gophers.

Chimpanzee males hunt monkeys, but monkeys are little, so the chimps have more control over the distribution of their meat. The best predictor of male chimpanzee hunting effort is the number of females in the group that are currently in estrus, showing large red genital swellings. Males try to induce fertile females to mate with them by catching meat to give to them. Hunting in our closest living ape relatives apparently evolved through sexual selection. But male humans go after much bigger game than male chimps do, with a lower success rate and much less control over meat distribution.

Does meat from large game contain some special nutrient unavailable from small game or plants? If so, perhaps it makes sense for couples to split the work of feeding their families, for men to specialize in big-game hunting to get that precious nutrient, and for women to specialize in gathering the more dependable plant resources. In this vision of hunting's evolution, women demand meat in exchange for sex. Anthropologist Helen Fisher has even proposed that this was the first human contractual relationship, in her 1982 book *The Sex Contract.* Owen Lovejoy had a similar theory, that male hunting provided meat for sexual partners burdened by babies, whose gathering efficiency would suffer while they were breast-feeding. For a long time this sex-for-meat theory seemed reasonable. Many theorists even proposed that male hunting allowed humans to bear the nutritional burdens of evolving a larger brain: as long as men transferred enough protein to dependent offspring, those offspring could grow smarter. Note that even in this traditional theory, female choice drives the evolution of hunting. Women refuse sex to men who fail to bring home meat. They force men to invest paternal effort in their offspring, helping to bear the nutritional costs of raising their offspring.

There is another problem, though: even if men manage to kill a large animal, they cannot control how its meat is distributed. The bigger the kill, the harder it is for a hunter to make sure that the meat goes to his girlfriends and their babies. Anthropologists observe that in almost all tribal cultures, meat is shared very widely among tribe members. People come running when they hear of a successful kill or see the vultures circling. They demand their share, aggressively and insistently. Often the amount of meat the hunter gets is statistically indistinguishable from anyone else's share. After perhaps a month's hunting effort, the hunter gets around 10 percent of the carcass, around 20 to 30 pounds of meat that must be consumed within a few days before it rots. Within a week, he'll be hungry again. Good hunters are not just reciprocal altruists, because bad hunters will never manage to repay them for the meat they take, and reciprocity would favor hunting small game that was easier to defend from cheats.

Anthropologist Kristen Hawkes has argued that meat from large game is a "public good" in the technical economic sense: a resource that one cannot exclude others from consuming. When anthropologists considered meat a private good, with the hunter able to control its distribution and consumption, hunting seemed to make evolutionary sense as a way of supporting one's family. But meat as a public good seems to create a paradox. Hunting's costs are borne by the hunter: the time and energy spent learning how to hunt, making the weapons, tracking the animals, using the weapons, and running down wounded prey. The hunter also risks injury or death from an animal that is fighting for its life, when he is merely hunting for his dinner. Yet hunting's benefits are spread throughout the tribe, enjoyed by sexual competitors and unrelated offspring. Evolution cannot generally favor genetic tendencies to provide public goods at the expense of one's own genetic interests. Such a tendency would fit the definition of evolutionary altruism, which cannot evolve by any known natural process.

So we have a quandary. At first, hunting looked to be a simple matter of survival. Then it looked to be a simple case of sexual selection, a meat-for-sex exchange, a way for women to transform male courtship effort into paternal effort. Now it looks more like a risky, wasteful act of altruism, a way for males to feed their sexual competitors (and other members of their band) at high risk to themselves. All three of these views have some merit and some supporting evidence. Here I have focused on the apparently altruistic aspects of hunting, not because I am interested in hunting per se, but because it raises a more general issue: how could selfish genes possibly give rise to costly, seemingly altruistic forms of charity? We'll triangulate toward the answer from three directions: the analogy between hunting and sports, the behavior of birds called Arabian babblers, and the concept of equilibrium selection from game theory. These ideas will prove useful not only for explaining human morality, but also, later on, for explaining the evolution of language.

Blood Sports and Arabian Babblers

One perspective is that hunting should be regarded as just another competitive male sport, a contest in which winners can attract mates by demonstrating their athletic prowess. As we saw in Chapter 7, men spend huge amounts of time and energy doing useless sweaty things with one another: basketball, sumo, cricket, skiing, tae kwon do, mountaineering, boxing. To an evolutionist, male human sports are just another form of ritualized male contest in which males compete to display their fitness to females through physical dominance. From a female's point of view, sports are convenient because they make mate choice easier. She can tell which male is healthier, stronger, more coordinated, and more skillful by seeing who wins these ritualized contests. She doesn't need to weigh six hundred pounds to test a man's sumo ability herself; the other sumo wrestlers do it for her. Now that most of the cultural barriers against women participating in sports have fallen, men can equally assess a woman's physical fitness by observing her athletic abilities.

Now, consider two groups of hominids that evolve to prefer different sports. Suppose that one group prefers the club-fighting sport favored by the Yanomamo tribe of the Amazon: the males stand facing each other and take turns at bashing their opponent's head with a very long stick until one contestant gives up, faints, or drops dead. The females prefer mating with the winner, since he may have stronger arms, better aim, a thicker skull, or a pulse. Despite its wastefulness in terms of blood, death, and unsightly cranial scars, this is a perfectly good system of competitive courtship display, no worse than stags bashing their antlers together.

The second group develops a different sport: they compete to sneak up on big animals, throw spears at them, and chase the wounded animals until they drop dead. The females prefer mating with the successful animal-killers, since they are better at tracking, sneaking, spear-throwing, and running long distances. Here again, the competitive display system is wasteful: the males may spend all day, every day, chasing around after big animals, getting injured, getting tired, stumbling into thorn bushes, dropping spears, being gored by buffalo, and so forth. And yet, the hunting sport is not quite as wasteful as the club-fighting sport, because after a successful hunt there is this big carcass that a group can eat. Within each group, all individuals may be acting selfishly, competing to display their fitness, and choosing the highest-fitness mates they can. But the extra meat gives every gene and every individual in the hunting group a slight advantage over those in the club-fighting group. Over many generations, this advantage may lead to more groups hunting than club-fighting as the principal form of athletic display.

This process sounds like "group selection," which most biologists have rejected since the 1960s, but it is not quite the same. In traditional theories of group selection, competition between groups could supposedly lead individuals to sacrifice some of their own survival and reproduction prospects for the greater good of their group. In these theories, there was assumed to be a direct conflict between individual self-interest and group interest. But in this example of hunting versus club-fighting there is no such conflict. In both groups, all individuals are selfishly trying to attain the highest sexual status they can through ritualized sports; it just happens that one sport yields a higher group-level payoff than the other sport.

Another perspective on the provisioning of public goods comes from some songbirds that live in Israel. The birds are called Arabian babblers. They weigh three ounces, they live in big groups, and they are the stars of *The Handicap Principle* by Amotz and Avishag Zahavi. The Zahavis have studied babblers for three decades, and report that the birds behave in several ways that look altruistic. Some act as sentinels for the group, giving alarm calls when predators approach. If a predator approaches, they mob the intruder, trying to drive it away. They share food with non-relatives. They practice communal nest care, taking care of babies that are not their offspring, kibbutz-style. They look like paragons of avian virtue, with altruism as conspicuous as a peacock's tail.

What is going on? Kin selection can't explain it, because the birds are kind to non-relatives. Reciprocity theory would predict that the birds would try to cheat, reaping the group benefits without paying the individual costs of being sentinels, mobbing predators, sharing food, or caring for nestlings. Instead, the birds do the opposite: they compete to perform the apparently altruistic behaviors. The Zahavis report that dominant animals, upon seeing a subordinate trying to act as a sentinel, will attack and drive off the subordinate, taking over the sentinel role. The birds try to stuff food down the throats of well-fed non-relatives. The Zahavis propose that the birds are using these altruistic acts as handicaps to display their fitness, thereby attaining higher social status in the group and improving their reproductive prospects. Only the birds in the best condition with the highest fitness can afford to act altruistically. Individuals seeking a mate can find good genes by finding a good altruist. That is how altruism apparently evolved in babblers. Most bird species do not appear to display their fitness by carrying out such pro-social good works. But those that do may have significant advantages, both as individuals and as groups.

John Nash Versus the Taxi Drivers of Bangalore

The altruistic human hunters and altruistic babbler birds are two outcomes of a very important evolutionary process called "equilibrium selection." It is an intimidating term, not widely understood even by biologists who have read some game theory. But I think the idea can clarify many mysteries, not only in evolution but in human culture.

To understand equilibrium selection, we first have to understand a little about equilibria and game theory. Game theory is the study of strategic decision-making, where your payoff for doing something depends not only on what you do, but on what other people do. A "game" is any social situation in which there are incentives to pick one's own strategy in anticipation of the strategies favored by others—but where their strategies will in turn depend on their anticipations of your own behavior. This sounds like an infinite regress: I anticipate that you anticipate that I anticipate that you anticipate . . . How can game theory make any progress in predicting human behavior in such games, when games seem like hopeless muddles?

Around 1950, the economist John Nash cut through this Gordian knot by developing the idea of an "equilibrium" (now known as a Nash Equilibrium). An equilibrium is a set of strategies, one for each player, that has a simple property. The property is that no player has an incentive to switch to a different strategy, given what the other players are already doing. An equilibrium tends to keep players playing the same strategies. The idea of an equilibrium is the foundation of modern game theory, and therefore of modern economics, business strategy, and military strategy. For his insight, Nash received a share of the 1994 Nobel Prize for Economics.

Driving on the left side of the road is a good example of an equilibrium. If everybody else is already driving on the left, as in Britain, no rational individual has a good reason to start driving on the right—such rebels against convention would quickly be eliminated from the population of drivers. But driving on the right side of the road is also an equilibrium, apparently favored by some former British colonies in North America as a mark of their independence. There is a third equilibrium in the driving game, which consists of driving on the left 50 percent of the time and on the right 50 percent of the time. If everybody is already doing that, you might as well too. This randomized equilibrium seems to be favored in Britain's former colonies in south Asia, especially by the taxi drivers of Bangalore. Nash realized that in most realistic games there are many equilibria. We cannot necessarily predict which equilibrium will be played, but we can predict that players will coordinate their behavior on one of the equilibria. In the driving game, different countries play different equilibria.

Equilibrium selection is the gradual process by which an equilibrium becomes established for a particular game. Imagine an anarchic country without cars that suddenly starts importing cars. People would start driving without knowing which side of the road other drivers will favor. Some would pick the left consistently (the British equilibrium), others would pick the right consistently (the American equilibrium), and still others would toss a coin every day to decide (the Bangalore equilibrium). Now we have a process of competition between three strategies that would each produce a different equilibrium. Suppose that every head-on collision kills both drivers involved. If left-driver meets left-driver, they both survive. If right meets right, they both survive. If Bangalore meets Bangalore, they both die half the time. If right-driver meets left-driver, they both die. There is no rational basis for predicting which equilibrium will become established. Every equilibrium is equally "rational" in the sense that every individual is doing as well as possible given what everyone else is already doing. Although rationality cannot select between equilibria, the contingencies of history can. We can be virtually certain that within several weeks, either the drive-left equilibrium or the drive-right equilibrium will win out. Which of them wins will be due to chance, but one of them will win. (There is only a very small chance that the Bangalore equilibrium will win.)

In this example, the equilibrium selection problem is solved not by rational logic but by historical contingency. When species evolve to play one equilibrium rather than another in the game of courtship, evolutionary contingency can play the role of historical chance. It is easy to simulate this process in a computer, as Brian Skyrms did in his wonderfully lucid 1997 book *Evolution of the Social Contract*. The same equilibrium selection processes must happen all the time in real biological evolution. Most interactions between animals can be interpreted in strategic terms, and so can be modelled using game theory. But for most realistically complex games, there are vast numbers of equilibria: not just three equilibria as in the driving game, but hundreds or thousands of possible equilibria. For realistic games with many equilibria, equilibrium selection processes become absolutely crucial to understanding and predicting behavior.

In our sports example, we considered two possible equilibria in the game of displaying athletic fitness: club-fighting and hunting. If everyone is already club-fighting, you can attract a mate only by club-fighting too, so you have no reason to do anything else, and that makes club-fighting an equilibrium. But if everyone is already hunting, you can only attract a mate by hunting well, so hunting is an equilibrium too. The mate preferences that favor good hunters or good fighters tend to be genetically and culturally conservative, and this sexual conservatism maintains the equilibrium.

Club-fighting and hunting are equally rational from the individual point of view, but hunting is the equilibrium with the higher payoff for everyone. With the Arabian babblers, we saw that altruistic behaviors such as food-sharing and alarm-calling could work as an equilibrium in the game of displaying fitness. The general point is that courtship games have many possible equilibria, and some of them will include a lot of apparently altruistic behavior. Most of them do not, because most ways of wasting energy to display one's fitness do not transfer any benefits to others. The peacock's tail simply wastes one peacock's energy to display his fitness, without transferring that energy to any other peacocks or peahens. But in some species, such as Arabian babblers and humans, our costly courtship displays actually bring some benefits to others.

Anthropologist James Boone described how equilibrium selection can favor altruistic displays in his 1998 paper "The Evolution of Magnanimity." He envisioned different groups playing different equilibria in the game of conspicuous display:

> Now imagine that, in some of these groups, elites signal their power by piling up their year's agricultural surplus in the plaza and burning it up in front of their subordinates. In other groups, elites engage in status displays by staging elaborate

> feasts and handing out gifts to their subjects. After several generations of intense warfare, which type of display behavior is likely to survive in the population? One might expect that the "feasters" would be much more successful at attracting supporters than the "burners."

Competition between groups would favor a magnanimous equilibrium over a wasteful equilibrium. Yet this would not be "group selection," as traditionally defined by biologists, in which individuals incur an individual cost to produce a group benefit. In this case, every individual is acting selfishly and rationally in trying to gain high status and sexual attractiveness through their costly display. The individual sexual benefits, not the group benefits, maintain the equilibrium: group competition merely picks between equilibria. Anthropologists Robert Boyd and Peter Richerson have argued that this sort of interaction between equilibrium selection and group competition is extremely important, not only in genetic evolution but in cultural history. Their ideas offer a new foundation for the comparative analysis of human cultures and social institutions, and I wish I had more space to discuss them further here.

In summary, evolution sometimes favors courtship equilibria in which animals are very generous to others. This does not mean that evolution favors truly selfless altruism, simply that the hidden benefit of generosity is reproductive rather than nepotistic or reciprocal. In principle, evolution could sustain very high levels of altruism by rewarding the altruistic with high social status and improved mating opportunities. Without sexual selection, generosity to unrelated individuals unable to reciprocate would be very unlikely to evolve. With sexual selection, such generosity can evolve easily as long as the capacity for generosity reveals the giver's fitness. In our species, the fact that we find kindness and generosity so appealing in sexual partners suggests that our ancestors converged on a rare and wonderful equilibrium in the game of courtship.

Leadership

High status among chimpanzees and gorillas does not depend only on physical dominance. It also depends on an individual's ability to prevent fights among other group members, to mediate conflicts, to initiate reconciliations, and to punish transgressors. Frans de Waal observed that one of the chimpanzees in Arnhem Zoo named Yeroen sustained his high status late into life by being good at this sort of moral leadership. Yeroen had the social intelligence to notice when trouble was developing between group members, and the social skills to intervene in just the right way to defuse tension and maintain group harmony. He was remarkably impartial, not allowing his own social relationship and consortships to bias his peacekeeping. Other individual males could beat Yeroen in a fight, but his high status was maintained through popular support and respect.

Chimpanzees have apparently transformed the ancient tradition of primate dominance hierarchies into a status system based on moral leadership. We used to imagine that this was a distinctively human achievement, but it is not. If chimpanzees and gorillas respect peace-keeping and policing ability, and modern humans do too, then it is likely that our common ancestor five million years ago did as well. Status based on moral leadership is a legacy of the great apes. For at least five million years, our ancestors have been striving to attain status through their moral leadership, rather than just through their physical strength.

But what exactly does "high status" mean? In primates, it generally brings greater reproductive success, which depends on greater sexual attractiveness. Status is not a piece of territory that can be taken by force. It must be granted by others, based on their likes and their dislikes, their respect and their disrespect. "Status" is a statistical abstraction across the social and sexual preferences of the members of one's group. If our ancestors attained high status through moral leadership, that meant moral leadership was socially and sexually attractive. It was favored by social choice and sexual choice. Because sexual choices have so much more evolutionary power than social choices about friends, grooming partners, and food-sharers, we come to this conclusion: moral leadership evolved through sexual choice in both chimpanzees and humans.

Leadership is like hunting in this respect: it provides a common good that looks purely altruistic until one considers the behavior's sexual attractiveness. Sexual selection could have favored the opposite of moral leadership, but that preference would tend to go extinct along with its tense, bickering, exhausted groups. One could imagine a primate species in which females happened to develop a runaway sexual preference for hair-trigger psychopaths who randomly pick fights. Males could obligingly evolve into violent bullies. But groups playing that psychopathic equilibrium would go extinct in competition with efficient, peaceful groups playing the good-leadership equilibrium. As with hunting versus club-fighting, this is an example of equilibrium selection. It is not an example of the discredited group selection process in which individuals pay an individual cost for a common benefit. The sexual rewards of moral leadership mean that good leaders obtain a net individual benefit from behavior that provides for the common good.

Where chimpanzees evolved moral leadership, humans evolved the more advanced capacity of moral vision, including the passionate articulation of social ideals concerning justice, freedom, and equality. Moral vision is sexually attractive, and may have been generated by sexual selection. It takes the impartiality of the peacekeeping primate to a more conscious, principled level. In discussing such an important human capacity, we must be especially careful to distinguish evolutionary function from human motivation. When Malcolm X used his verbal genius and moral charisma to forge a vision of a Muslim society free of racism, he was motivated by moral instincts, not "sexual instincts." His moral instincts happened to attract a beautiful young woman named Betty Shabazz to become his wife, as they had evolved to do through sexual selection. Likewise for Martin Luther, whose Protestant vision attracted the ex-nun Katharina von Bora to marry him and raise six children. The peacock's tail is no less beautiful when we understand its sexual function. Nor should the validity of human moral vision be reduced when we understand its origin in sexual choice.

Why Scrooge Was Single: The Evolution of Charity

Survival of the fittest was supposed to make us act selfishly. Like Charles Dickens's Scrooge, the traditional Darwinian account depicts humans as mean and miserly, perhaps with a little nepotism toward close relatives and some prudent loan-sharking to those who might pay us back. This is a convenient myth for educators, priests, and politicians, because it presents us as badly in need of socialization through schools, churches, and prisons. Supposedly, we need these character-improving regimes as our Ghosts of Christmas Past, Present, and Future, to transform our selfish, biological, pre-Christmas Scrooge into our generous, cultured, post-Christmas Scrooge.

The Dickens story makes a poor parable for human evolution, though. Scrooge survived well enough, but he was single and childless. In Victorian London, his manifest selfishness exiled him from the mating market. No self-respecting Englishwoman would pay him the slightest notice. As far as his genes are concerned, his miserliness was not self-interested, but self-castrating.

Without sexual selection, the human proclivity for charity might remain an evolutionary enigma. It is hard to imagine how instincts for giving resources away to strangers would benefit the giver. Usually, evolutionary psychologists explain charity as a side-effect of humans having evolved in small tribal groups, in which any kindness would probably be reciprocated. There would have been no such thing as a stranger in Pleistocene Africa. Yet the psychology of charity is different from the psychology of reciprocity, and several important features of charity cannot be explained as side-effects of reciprocity instincts. I shall take examples from the charitable behavior of people in modern societies. Pleistocene generosity was not the same as modern charity, but it may not have been so different either. We have already seen that traditional hunting was a time-consuming way to attain higher social, sexual, and moral status by providing a public good. Hunting was charity work.

One puzzle is why many people care so little about the efficiency of charities in transferring resources from givers to receivers. If charity derives from reciprocity, and if reciprocity favors the efficient trading of resources, then we should care deeply about maximizing the benefits of charity to the receiver. This is because, if the beneficiaries of charity ever found themselves able to repay what they had received, they might well feel that it was fair to give back only what had been given to them, and not be asked in addition to meet the high overhead costs incurred by the giver. Yet many contributors show an odd lack of interest in the efficiency of charities. Some of the largest charities have high administrative overheads, a large proportion of donations going to pay the salaries of their administrators and fundraisers. Some French cancer charities were notorious for distributing less than 10 percent of their revenues to actual research. Many "charity events" are luxurious parties at which donors can meet other donors while drinking champagne. Within two weeks of Princess Diana's death in 1997, British people had donated over £1 billion to the Princess of Wales charity, long before the newly established charity had any idea what the donations would be used for, or what its administrative overheads would be. Only a minority of donors seek out the really efficient charities such as Oxfam, which transfers about 80 percent of donations to the needy, while spending only 3 percent on administration. Charities vary enormously in their efficiency, but most donors do not bother to get good charitable value for their money. This attitude contrasts starkly with our concern for government efficiency when we pay taxes that support the ill, the elderly, and the arms dealers.

The phenomenon of "charity work" also reveals how generosity is used as an inefficient fitness display rather than an efficient resource-transfer device. If the wealthy really wanted to help people, they should make as much money as they can doing what they are trained to do, and hand it over to a lower income group who are trained to help people. The division of labor is economically efficient, in charity as in business. Instead, in most modern cities of the world, we can observe highly trained lawyers, doctors, and their husbands and wives giving up their time to work in soup kitchens for the homeless or to deliver meals to the elderly. Their time may be worth a hundred times the standard hourly rates for kitchen workers or delivery drivers. For every hour they spend serving soup, they could have donated an hour's salary to pay for somebody else to serve soup for two weeks. The same argument applies not only to lawyers, but to everyone with an above-average wage who donates time instead of money. So why do they donate their time? Here again the handicap principle applies. For most working people, their most limited resource is time, not money. By donating time, they help the needy much less efficiently, but show their generosity and kindness much more credibly.

Another feature of human charity is that givers must usually be given tokens of appreciation, which they can display publicly. In the United States, donors to the Public Broadcasting Service (PBS) are rewarded with PBS tote bags, PBS umbrellas, and PBS T-shirts. In Britain, charities offer donors red paper poppies for buttonholes, red plastic clown noses, or red tomato T-shirts. Blood drives usually give donors buttons saying something like "I gave blood today," which essentially proclaim "I am altruistic, not anemic, and HIV-negative." Major benefactors of universities or hospitals usually expect buildings to be named after them. There is the phenomenon of the "anonymous donor," but we should not take the term at face value. A London socialite once remarked to me that she knew many anonymous donors. They were well known within their social circle—the set of people whose opinion matters—even though their names may not have been splashed across the newspapers. I suspect that few male millionaires keep their charitable donations secret from their wives and mistresses.

A final oddity is that people usually avoid giving to charities that nobody else has heard of, however worthy the cause. The result is something approaching a winner-takes-all contest, with the charities that grow large and well-known attracting ever larger proportions of donations. Charities must spend a large proportion of revenue on "fundraising." This sounds like the pragmatic solicitation of donations. But it often turns out to mean the costly creation of a strong brand identity for the charity, hiring advertising firms to promote the charity in the same way that any other luxury good is marketed. Fundraisers know that when a new charity is launched, it is important to attract a few major donors, so their rivals feel obliged to top those donations with larger ones. The charity's goal is to provoke a donation arms race between local millionaires. From the viewpoint of efficiently transferring resources from the wealthy to the needy, such arms races look pathological. They result in overfunding a few salient diseases in the developed world. They lead to the neglect of more cost-effective programs in the developed world, such as drilling for clean water wells, anti-malaria programs, pro-breast-feeding campaigns, elementary school education, and capital for women's small businesses. If charity really resulted from altruistic instincts for solving other people's problems, we should expect people to take more time to research which charities are most cost-effective and most likely to produce immediate, measurable improvements. This would result in money being spread around much more widely, ameliorating more of the world's avoidable misery. Instead, most donors spend less time researching their charities than they do picking which video to rent. This results in charity fashion cycles, and over-giving to this season's stylish causes.

How can we explain these peculiar features of human charity? They cannot be traced to nepotism or reciprocity. They do not seem to result from socialization for genuine altruism. Instead, they often look like just another form of wasteful, showy display. If the point of charity is to incur the cost of giving rather than to bring real benefits to others, we can understand why people do not care much about the efficiency of charities, and why they donate time when they should be donating money. If charitable donations must be advertised to be effective as signals, we can understand why donors receive little badges to indicate their generosity, and why charities spend so much fundraising money creating a strong brand identity. If donations are signals subject to the usual demands of recognition and memorability, we can understand why people give to famous, oversubscribed causes rather than obscure, worthier ones. Donations as courtship displays would also explain the charity fashion cycles, which are especially apparent among young, single donors. For most of us, our charities are cosmetic.

This is not to say that people giving to charity are "trying to get more sex." They are simply trying to be generous. That is their motivation. My question is why the motivation evolved. Their genuine instincts for generosity just happen to have many of the showy, fashion-conscious features common to other products of sexual selection.

Understanding charity's origin as a sexual display should not undermine its social status. As Robert Frank argued in *Luxury Fever*, we may have evolved instincts for achieving higher social status through conspicuous display, but as rational and moral beings we can still choose conspicuous charity over conspicuous consumption. Every hundred dollars we spend on luxuries could probably have saved a sick child from death somewhere in the developing world if we had donated it to the appropriate charity. The ten-thousand-dollar premium that distinguishes a sport utility vehicle from an ordinary automobile probably cost India a hundred dead children. We may pretend that it did not, but our self-justifications are no comfort to the dead. Perhaps if we imagined a hundred hungry ghosts haunting every luxury vehicle, runaway consumerism would lose some of its sexual appeal. While designer labels advertise only our wealth, the badges of charity advertise both our wealth and our kindness. As it is, the car manufacturers can afford better advertising than the needy children, which is why our instincts for display have been directed more toward consumerism than toward charity.

Why Men Tip Better than Women

Waitresses know more about human generosity than most moral

philosophers. Their incomes depend on tips. To economists, leaving tips in restaurants is the classic example of "irrational" human kindness: tips are voluntary donations to non-relatives who are unlikely to reciprocate. According to standard Darwinian models, we should all be very bad tippers. But that is not what we observe. Instead, most waitresses report that groups of men leave much better tips than groups of women, and men on dates with women leave especially good tips if they pay for the meal. That is consistent with sexual selection favoring displays of generosity. One might argue that men leave bigger tips because they have more money to spare. But that is an economically naive argument, because selfish men could have eaten in a slightly more expensive restaurant, or ordered more expensive wine, and left a smaller tip. It is also an evolutionarily fallacious argument in a more interesting sense: it begs the question of why the men bothered to make more money in the first place.

When Ted Turner announced in 1997 that he would donate $1 billion to the United Nations, his wife Jane Fonda did not. We could explain this in two ways. We might say that he could afford it because his personal wealth was over $4 billion, and she could not because hers was only several hundred million. We could take the sex difference in earnings as a given, and use it to explain the sex difference in charity. On the other hand we could ask, from a Darwinian viewpoint, why men should bother acquiring more resources if they just end up giving them away. One clue emerged in a Larry King interview. Turner revealed that when he told his wife of his intended gift, she broke down in tears of joy, crying, "I'm so proud to be married to you. I never felt better in my life." At least in this case, charity inspired sexual adoration.

One of the most extreme examples of male acquisitiveness in the service of charity was John D. Rockefeller, Sr., the 19th-century oil magnate. In business he was a ruthless monopolist, but in private, he was a devout Baptist committed to good works right from adolescence. Even during his first year of work as an assistant accountant at age 15, he gave 6 percent of his paltry annual salary to charity. This rose to 10 percent by age 20 in 1859, when he raised $2,000 to save his church from bankruptcy by paying off its mortgage, and contributed to a fund for an African-American man in Cincinnati to buy his wife out of slavery. His magnanimity did not go unnoticed: one young woman from his congregation reported of the young Rockefeller that, though not especially handsome, "He was thought much of by these spiritual minded young women because of his goodness, his religious fervor, his earnestness and willingness in the church, and his apparent sincerity and honesty of purpose." Even after he was earning $10 million a year in dividends from his Standard Oil monopoly by age 40, he avoided the ostentation of other Gilded Age magnates, preferring to spend his money creating institutions such as the Rockefeller Institute for Medical Research and the University of Chicago (which incidentally appointed Thorstein Veblen as one of its first faculty members). After age 50, Rockefeller spent much more time researching his charitable efforts than minding his business, and he managed to give much of his billion-dollar fortune away to intelligently chosen causes before dying at age 93. The Rockefeller Foundation was his peacock's tail.

Male Generosity in Courtship

Traditional evolutionary theories of morality have trouble explaining unreciprocated generosity toward non-relatives. They worry about trivial cases like tipping, while ignoring the case where male generosity is most apparent—during sexual courtship. During courtship, males incur very high costs in terms of time, energy, risk, and resources. Some of these costs, like those of bird song, evaporate into thin air, yielding no benefit to the female other than information about male fitness. Other male courtship efforts bring wider social benefits to a whole community, like the legendary knights who slew dragons to win the hand of a princess, or Pleistocene hunters killing mammoths. A few cases even bring benefits to the female, like the prey offered by male scorpionflies.

Some researchers such as Helen Fisher and Camilla Power have viewed human courtship as a social contract where a male offers resources like meat in exchange for sex. One might caricature human courtship as men using gifts to buy the reproductive potential of women, using the same reciprocity instincts that sustain human trade. In this view, prostitution was the oldest profession, and marriage is a form of prostitution. Economist Gary Becker won his 1992 Nobel Prize for Economics in part for analyzing human marriage in similar contractual terms. In the modern world where every thing becomes commodified and every relationship becomes contractual, the reciprocity theory of courtship seems plausible. However, the reciprocity theory collapses on closer inspection.

Gentlemen and feminists understand the difference between contractual prostitution and male courtship gifts. When a man buys a woman dinner, she is emphatically not obliged to have sex with him. He would be a sexist cad for suggesting that she was. He cannot take her to a small claims court if she says, "Thank you for a lovely meal, but I do not believe we are suited to one another." Of course, an amorous male may be frustrated and resentful if his courtship fails, but that is not to say that the female has cheated him according to the terms of some implicit contract. It means that she has rejected him. It is her power of sexual choice that determines whether the relationship will escalate to intercourse, not his imposition of a gift. (However, in some cultures, if a couple continues to date and a woman accepts an escalating series of gifts over a long time period, this may create an implicit sexual contract.)

What's more, most males could not possibly afford to buy a woman's reproductive potential if courtship were a simple economic exchange. What would be an appropriate market price for a nine-month pregnancy, the pain of childbirth, the exhaustion of breast-feeding, and twenty years of maternal care? At least half a million dollars at a basic salary of $25,000, one would think. How much do men spend on courtship in the first few months? Perhaps a tenth of 1 percent of the proper market price. Their generosity might continue after a baby arrives, but it might not. One could do the same sort of analysis for hunter-gatherers, in terms of the calorie cost of pregnancy and maternal care versus the calorie value of the meat that males offer. Are women just undervaluing themselves by a factor of a thousand? It seems unlikely that evolution could have produced such low female self-esteem, if the reciprocity theory is correct. Mutant women who demanded more should have replaced those who demanded so little, since their offspring would materially benefit.

Finally, male generosity during courtship is relatively inefficient as a way of transferring resources to females. It is like charity: we don't seem to care about the efficiency, only the cost of donation and the good intention. Efficient benefit-transfer is extremely unromantic. If human courtship evolved under the reciprocity model, it would be very, very simple. Today, women would auction their reproductive potential on the Internet, accepting wire transfers of bank funds from all male suitors, awarding their favors to the highest donor, and keeping all the money. Women would have emotions well adapted to falling in love with the most generous bidder—even though there were no interbank electronic transfers during the Pleistocene. The fact that we find this scenario so unappealing is psychological evidence against the reciprocity model.

Romantic gifts are those that are most useless to the women and most expensive to the man. Flowers that fade, candles that burn, overpriced dinners, and walks on exotic beaches are the stuff of modern romance. They do not increase a woman's survival prospects as much as they reduce a man's bank account. One might say that these things bring pleasure, but, as we have seen, pleasure is what evolution must explain. How could evolution possibly have favored humans who fall in love with individuals who provide them with useless luxuries that bring no survival benefit? The fact that a diamond engagement ring happens to be made out of durable matter does not make it a biologically relevant material benefit to a woman. If she wanted the diamond as a purely material benefit, she should not mind if her suitor bought it on sale from a mail-order catalog. But in reality, she wants him to pay the full retail price at Tiffany's, because that is more "romantic," which is to say, costly. Moral philosophers might not consider male courtship generosity very "moral" behavior. But to a woman receiving a romantic gift, it is a capital virtue.

Sexual Selection for Sympathy

Empathy is the mental capacity to understand the suffering of others, while sympathy is the emotional capacity to care about that suffering. Much of human courtship consists of sympathy displays. We show kindness to children; we listen to sexual prospects enumerating their past sufferings. The development of emotional intimacy could be viewed as the mutual display of capacities for extremely high levels of sympathy.

When we favor kindness in courtship, we are favoring a real personality trait that has been measured and dissected by psychologists. In the leading "5-factor" model of personality, one of the factors is "agreeableness." People who score high on this trait are compassionate, loving, sincere, trustworthy, and altruistic. Empirical research shows that these personality features really do cluster together, and are fairly independent of other personality traits like conscientiousness, extroversion, and intelligence. When people are asked to rate personality features as positive or negative, the agreeableness feature always tops the charts. The worst-rated adjectives describe the opposite of agreeableness: dishonest, cruel, mean, phony. Also, agreeableness appears to be moderately heritable, so agreeable parents tend to produce agreeable children.

By looking for sympathy during courtship, people may also be trying to avoid psychopaths. True psychopathy ("antisocial personality disorder") is very rare, occurring in less than 1 percent of the population, but psychopaths account for a very high proportion of murders, rapes, assaults, and other serious crimes. Psychopathy is basically the absence of sympathy. There are fewer female psychopaths, perhaps because female psychopaths would not have shown sufficient sympathy to their babies in past generations. But an absence of parental sympathy did not consign male psychopaths to reproductive oblivion. On the contrary, male psychopaths tend to seduce women, get them pregnant, and abandon them. Women prefer to avoid this, so psychopaths know they must feign agreeableness during courtship. Very few psychopaths flaunt their lack of sympathy like Hannibal Lecter, because very few of them are glamorous, urbane geniuses. Mostly, they are just ordinary creeps who beat their girlfriends, stab guys in bars for no reason, get caught, and then apply for parole four times as often as non-psychopaths because they don't think they've done anything wrong.

If sexual preferences evolved to avoid anything, they should have evolved to avoid psychopaths. During human evolution there may have been a three-way arms race: females developed better tests for male sympathy, male psychopaths developed better ways to fake sympathy, and male non-psychopaths developed sympathy-displays that were harder and harder to fake. Just as fitness indicators evolved to advertise freedom from harmful mutations, perhaps sympathy indicators evolved to advertise freedom from psychopathy.

The psychologist Hans Eysenck argued that, apart from true psychopathy, there is a much more common personality trait of psychoticism in which people are aggressive, cold, egocentric, impersonal, antisocial, unempathetic, and tough-minded. Like psychopathy itself, these features are not generally favored in sexual relationships, though they may bring advantages in dominance contests. In our current context, the interesting thing about psychoticism is that the innate depravity view of human morality mistook one extreme of this personality dimension for the whole of human nature. People with extreme psychoticism are perfectly capable of nepotism and strategic reciprocity when it suits them—they just lack the sympathy and agreeableness that average people have. If one equates evolutionary egoism with psychological egoism, it looks as if all humans should be psychopaths. That prediction is wrong, because there is a hidden sexual-selective advantage to sympathy.

Our sexually selected instincts for displaying sympathy tend to affect our belief systems, not just our charity and courtship behavior. When individuals espouse ideological positions, we typically interpret their beliefs as signs of good or bad moral character. Individuals feel social pressure to adopt the beliefs that are conventionally accepted as indicating a "good heart," even when those beliefs are not rational. We may even find ourselves saying, "His ideas may be right, but his heart is clearly not in the right place." Political correctness is one outcome of such attributions. For example, if a scientist says, "I have evidence that human intelligence is genetically heritable," that is usually misinterpreted as proclaiming, "I am a disagreeable psychopath unworthy of love." The arbiters of ideological correctness can create the impression that belief A must indicate personality trait X. If X is considered sexually and socially repulsive, then belief A becomes taboo. In this way our sexually selected instincts for moralistic self-advertisement become subverted into ideological dogmas. I think that human rationality consists largely of separating intellectual argument from personality attributions about moral character. Our difficulty in making this separation suggests that political, religious, and pseudo-scientific ideologies have been part of moralistic self-display for a very long time.

Sexual Fidelity and Romantic Love

Sleep around with too many people, and your lover will probably leave you. Sexual choice is not just the power to initiate relationships, but the power to end them. Our capacity for sexual fidelity, imperfect though it may be, is a result of our ancestors favoring the faithful by breaking up with the unfaithful. As David Buss has emphasized, humans have evolved specialized emotions for detecting and punishing infidelity in sexual relationships, distinct from our instincts for detecting cheats in reciprocity relationships.

Evolutionary psychology has rightly stressed how pervasive human sexual infidelity is when compared with the cultural ideal of monogamous commitment. The greater tendency of males to philander is certainly consistent with the predictions of sexual selection theory. However, I also find it astonishing how faithful most humans are when compared with other mammals. Some male birds are relatively faithful, but most male primates never turn down an opportunity to copulate with a willing female. Other female primates show no sense of sexual commitment to a particular male. If a better male comes along and a female is not too afraid of a jealous beating (as can often happen to female chimpanzees), she may switch partners. Humans are different. We value sexual fidelity in others, and have the capacity to inhibit our own courtship and copulation behavior, even in the face of awesome temptation.

Fidelity could be viewed as an example of reciprocity, insofar as cheating by one individual tends to provoke punishment by the other. However, the punishment is usually implemented by sexual choice. The "punisher" ends the relationship, denies further sexual access, or chooses to have sex with someone else. It may not matter whether we view this as sexual choice in the service of reciprocity, or reciprocity in the service of sexual choice. In either case, sexual preferences favor the virtue.

Sexual selection produced a sort of two-stage defense against sexual infidelity: romantic love, and then companionate sexual commitment. Romantic love powerfully focuses all courtship effort on a single individual to the exclusion of others. For at least a few weeks or months, it inhibits infidelity. Needless to say, romantic love is sexually attractive. It may not increase the appeal of an otherwise unattractive individual enough to provoke mating, but, all else being equal, it is clearly valued in mate choice. Love evolved through sexual selection, not least because it signaled fidelity.

However, passionate sexual love, "being in love," rarely lasts more than a couple of years. That is not nearly long enough to keep a couple together to raise a toddler, which they may have a shared interest in doing. Much more important over the long term is the feeling of friendly, mutually respectful sexual commitment. This does not work by shutting off all sexual attraction to others, but by managing that attraction through flirtation and sexual fantasy. The human capacity for flirtation (sexually inhibited pseudo-courtship) is one of the modern world's most underrated virtues, the principal spice of adult social life throughout history. Equally important is sexual fantasy, the spice of adult mental life. It permits sexual infidelity in the virtual reality of the imagination, without offending one's real sexual partner as much as a real affair would.

Our sexual fidelity evolved as a compromise between two selection pressures. On the one hand there was sexual selection favoring high fidelity through romantic love and sexual commitment. On the other hand there were the potential reproductive benefits of philandering. Especially for males, those potential benefits made it maladaptive to completely turn off their sexual attraction to everyone other than their partners. Flirtation and fantasy sometimes escalated into real affairs, and those affairs sometimes gave our ancestors net reproductive benefits. Sexual choice could not reach into our minds and totally eliminate polygamous desires. It could only punish observable infidelity and blatantly wandering eyes. We are not always sexually faithful, but that does not mean that our capacity for fidelity is a flawed adaptation. It may be perfectly adapted to a Pleistocene world in which the highest reproductive success went to those who were almost always faithful, except when a significantly more attractive option arose.

Virtues of Good Fathers

Male courtship generosity does not usually end after the first copulation, or even after the first baby arrives. As we saw in Chapter 7, many men are fairly good and generous fathers, even to their step-children. They do nowhere near as much hands-on child care as mothers, of course, but do vastly more than most male primates. We saw how fatherly solicitude could be interpreted as courtship effort rather than parental effort. Women may break up with bad fathers and continue to sleep with good fathers, and that would have been sufficient sexual selection to favor good fathers. There is not much more to say about that here, other than to put the virtues of fatherhood in this chapter's moral context.

The generosity of step-fathers in particular cannot be explained by nepotism or reciprocity. The step-children will probably never reciprocate, and to say that their mother "reciprocates" with sex just trivializes her mate choice. Of course, good mothering is a virtue as well, but natural selection has already been favoring it for 200 million years of mammalian evolution. Indeed, the maternal virtues of female mammals, including their capacity for milk production, were a major factor in the success of mammals. Beyond that mammalian legacy, male mate choice may have favored some indicators of mothering ability in women, such as a conspicuous interest in unrelated babies and children, and a verbose pride in the achievements of one's own children. Sarah Blaffer Hrdy has recently analyzed these human maternal virtues in her book *Mother Nature*.

Sportsmanship

Cheating during an athletic competition provokes a particular sort of moral outrage. It leads to finger-pointing, name-calling, and arguing. And yet, it is not the same resentment that we feel when someone fails to reciprocate a kindness. If someone cheats during social reciprocity, we initiate our punishment routine: we sulk and withdraw from further social contact. If someone cheats during sports, we complain loudly and publicly, and then go back to playing against them. Why should we do this?

Sportsmanship is not a matter of being altruistic, but of ritualizing one's intense sexual and social competitiveness in a particular, restrained manner. We are not normally playing against our kin, so kin selection cannot explain the restraint. Reciprocity looks relevant, but only in the sense that it always looks relevant to any social interactions that continue over time, include costs and benefits, and offer the possibility of cheating.

Turn-taking and rule-following in sport is not the same as reciprocity. When Pete Sampras and Andre Agassi hit a tennis ball back and forth, they are not reciprocating costs and benefits;

they are trying to win points. Of course, sports contestants play only because the expected average benefits of play exceed those of not playing. But the benefits from play are not transferred from one participant to the other, as in economic trade. The benefits come from the spectators, in the social status they confer on good players. Of course, the spectators do not need to be physically present to award status to the winners, they only need to hear about the result through gossip. The claim that a winner cheated during an athletic contest is a potent sexual insult, because it undermines their claim to status, which is one of the most valuable currencies in the sexual marketplace.

In competitive sports, games and contests, cheating means anything that interferes with a meritocratic outcome. The ideal is for the best player to win. Anything that significantly undermines the correlation between the players' ability and the outcome of the contest is viewed with suspicion, whether or not it is a violation of some explicit rule. In fact, the rules of sports are often changed to maintain the link between "true ability" and outcome, such as when professional sports banned steroids. We value a "level playing-field" not only in sports but in other kinds of meritocratic competition. Fair competition maximizes the information that winning carries about the relative fitness of the winner. The result is to maximize the efficiency of sexual choice based on sporting results. We saw in Chapter 8 that our mental capacities for sports may have evolved through sexual selection. But our sporting instincts also include some powerful moral judgments.

A concern for meritocracy—or fairness—pervades human social life, extending far beyond sport. What does "meritocracy" mean? It seems to imply maximizing the information about "merit" (fitness) carried by social status. At first glance, it looks as if instincts for meritocratic moral intuitions could not evolve biologically. By definition, a person of average merit and average fitness should not win a meritocratic contest. Why should they care whether contests are meritocratic or not? I think people prefer meritocracy because they want to be able to choose the best mate they can, so they favor meritocratic competition among the opposite sex. While no one wants to appear inferior to individuals of higher merit, everyone wants to appear superior to those of lower merit. Winning the number one position is not the only thing that matters. Every increment of apparent fitness and apparent social status matters. So, people have powerful shared interests in setting up meritocratic competitions to make mate choice more efficient. It is likely that our concerns for fairness evolved in this context.

There is the distinction between equality of opportunity (meritocracy) and equality of outcome (egalitarianism). Hunter-gatherer tribes are intensely egalitarian about certain issues like sharing meat equally, articulating their views during tribal discussions, and preventing anyone from becoming a tribal "chief." Yet they are often meritocratic about sexual reproduction. This is because mate choice makes it impossible to impose equality of outcome at the level of reproductive competition. The discriminatory nature of sexual choice undermines all egalitarian utopias. Women might like the idea of all men being able to have equal amounts of sex, but no individual woman would be willing to forgo her power of sexual choice to allow an unattractive, unfit man to copulate with her. In the realm of human sexuality, no one would agree to the maxim "from each according to his abilities; to each according to his needs." While tribes have shared interests in meritocratic reproductive competition, they have no such shared interests in equalizing reproductive success across individuals by violating mate choice.

Sexual Selection and Nietzsche

The emphasis on reciprocity has led evolutionary psychology to concentrate on what Friedrich Nietzsche called the morality of the herd: prudence, humility, fairness, conscience, dependability, equality, submission to social norms, and the cult of altruism. In *The Genealogy of Morals*, Nietzsche argued that many human cultures attributed moral value to other virtues: bravery, skill, beauty, fertility, strength, pride, leadership, stoicism, sacrifice, tolerance, mercy, joy, humor, grace, good manners, and the creation of social norms. In *The Will to Power*, he listed the core elements of these pagan virtues: "(1) virtue as force, (2) virtue as seduction, (3) virtue as [court] etiquette." What is striking here is that Nietzsche's virtues sound remarkably like sexually selected fitness indicators.

More than any other moral philospher, Nietzsche inquired into the biological origins of our moral judgments, trying to understand how they could serve the needs of organic life. He wrote of virtue as "a luxury of the first order" which shows "the charm of rareness, inimitableness, exceptionalness, and unaverageness." By their luxuriant excess, virtues reveal "processes of physiological prosperity or failure." For Nietzsche, virtue was what the strong and healthy could afford to display.

Of course, we should remember the butler Jeeves's response to Bertie Wooster's asking whether Nietzsche was worth reading: "I would not recommend him sir; he is fundamentally unsound." Nietzsche read Darwin but did not understand him. Nietzsche intuited that sexuality and power lay at the heart of human perceptions, judgments, values, ideologies, and knowledge, but he did not understand sexual selection. Like Alfred Russel Wallace, he often used fallacious "surplus-energy" arguments to explain costly displays that had no apparent survival function.

Nietzsche's name remains taboo in polite society because of his misappropriation by the Nazis. But perhaps it is worth considering his argument that Christian values, which he called the morality of the herd, may not be the only human values worth analyzing from a biological and psychological viewpoint. The Nietzschean virtues do not raise the same evolutionary-theoretical problems as the Christian virtues, because they are not so altruistic. But our analysis of human morality should not be limited to behaviors that raise intriguing theoretical issues. Some aspects of human morality may have direct, unproblematic survival value. Other aspects, such as the Nietzschean virtues, may reflect evolved adaptations for certain kinds of costly display, just like other sexually selected handicaps.

Science could benefit by broadening its attention to the full range of human virtues that have been considered worthy of praise in various cultures. As individuals, we may find some of those virtues no longer praiseworthy. Military heroism, stoicism, and etiquette are distinctly out of fashion at the moment; there may even be good philosophical or practical reasons why they should stay out of fashion. But that is no reason for scientists to ignore them. Moral philosophers consider only a tiny fraction of human virtues and moral judgments worthy of analysis. But scientists must consider them all.

What's So Funny About Peace, Love, and Understanding?

In this chapter we have found one of the hidden evolutionary benefits to human kindness: the reproductive advantages it brings through mate choice. Our ancestors favored kind, fair, brave, well-mannered individuals who had the ability and generosity to help their sexual partners, children, step-children, and other members of their tribe. They were sexually unattracted to cheats, cowards, liars, and psychopaths. Is that really so hard to believe? Darwinians have searched so hard for the selfish survival benefits of morality that we have forgotten its romantic appeal.

Does this reduce our noblest ideals to a crude sex drive? Emphatically not. When our ancestors were favoring kindness, they were not looking for fake kindness, strategic kindness, or short-term kindness. They were looking for the real thing—genuine concern for others. Because of the power of sexual choice, they had the power to evolve it. Human altruism is not an evolutionary paradox. It is a sexual ornament.

Clearly, sexual choice does not account for all of morality. Kinship and reciprocity, too, were very important. And I have barely alluded to many other virtues, such as prudence, temperance, justice, courage, faith, hope, mercy, compassion, friendship, gratitude, patience, and humility. Sexual choice may have favored some of them, but other forms of social selection were undoubtedly powerful as well. Different selection pressures probably interacted in different ways to produce each moral adaptation, and it will take decades to sort them all out.

Some may be unhappy with attributing a sexual function to human morality. But we must remember that a sexual function is not a sexual motivation. This theory does not claim that we are only virtuous when we want sex; rather, it suggests that moral emotions, judgments, and reasoning were favored during courtship between our ancestors. Their sexual choices were not satisfied with a few tokens of romantic generosity. They selected instincts to provide for the common good even at high personal risk. They selected principled moral leadership capable of keeping peace, resolving conflict, and punishing crime. They selected unprecedented levels of sexual fidelity, good parenting, fair play, and charitable generosity. They helped to shape the human capacity for sympathy. They helped to make us reasonably agreeable, sincere, and socially responsive. It is a remarkable achievement for an evolutionary process that began with amoral bacteria, and unfolded through pure genetic self-interest right up to the moment when each of us was conceived.

10

Cyrano and Scheherazade

A classic symptom of paranoid schizophrenia is the belief that alien beings sometimes transmit their thoughts to us through invisible waves that influence our behavior. But every professor of linguistics knows that all ordinary people routinely transmit their thoughts to us through invisible waves that influence our behavior. The linguistics professors sound even more paranoid than the schizophrenics, but they simply have a greater respect for language. Most schizophrenics, like most other people, take language for granted, whereas language researchers recognize it as a signaling system of almost miraculous power and efficiency.

To other animals, we must seem a species endowed with telepathic powers. Consider things from a mammoth's perspective, a hundred thousand years ago. You are peacefully browsing somewhere in Eurasia when you spot a previously unknown type of two-legged primate. The creature watches you for a few minutes, then runs off. A few hours later you see a few of the creatures loping toward your vicinity, carrying pointy little tree-branches. How would a bunch of them suddenly know you were here? Must be a coincidence. Anyway, they don't look big enough to hurt you, since you stand ten feet at the shoulder and weigh about 14,000 pounds. But one of the creatures suddenly makes some strange squeaky sounds, and instantly all of the horrid little things start trying to stab you with their pointy branches. How annoying! You lumber away from them, but they make more squeaks, and a few seconds later another band of them springs up from a hiding place in front of you. Another coincidence? The ones in front have somehow set the grass on fire, not in one place the way lightning would, but all at once, creating an impassable wall of crackling heat. You must turn back. Yet the creatures behind you are still there, looking more confident, like the pack-hunting carnivores you feared as a youngster. Time to deploy your defense against pack-hunters: charge one until it's injured, then another, until selfish fear breaks down their coordination. Your tusks manage to injure a few, but every time you charge one, the others try to stick their pointy branches into you, all at once. Their coordination just will not break, and they continue that infernal squeaking as your stab wounds accumulate. Worse, as you weaken, one of them points to your head and squeaks loudly, and then all of the pointy branches are being aimed at your eyes. Within minutes you are blind, and charging blindly, but the stab wounds come more quickly now. New, higher-pitched voices are now audible: perhaps their females and young already calling for your meat to be pulled from your bones. Your last thought before you bleed to death is: I am extinguished by a bunch of little bodies that weave themselves, through that odd squeaking, into one great body with dozens of eyes, dozens of arms, and one lethal will.

This Pleistocene fantasy could be criticized on many counts. It may overestimate the awareness of mammoths, though I doubt it, since their brains were five times the size of ours. It may over-estimate the hunting ability of our recent ancestors, though I doubt that too, since there is fairly good evidence that they hunted many species of mammoths, mastodons, and elephants to extinction in the last hundred thousand years. The real problem with most fantasies like this is that they show the telepathy-like power of language being used only in the struggle for survival. Doubtless language was useful in coordinating hunting, as it was in many survival activities. But language was also sure to be useful in courtship. In this chapter I shall put survival selection to one side, and consider how our ancestors developed the ability to fall in love by talking to each other.

Forget Chomsky and Kanzi!

The history of research on the evolution of language resembles the history of sexual selection theory. Darwin had some good ideas, then scientists got distracted for a century by the wrong questions, and only recently took up where Darwin left off. In *The Descent of Man*, Darwin proposed that language evolved gradually through sexual selection, as an instinct to acquire a particular method of verbal display similar to music. He recognized that language, like conceptual intelligence and principled morality, was an unusual human adaptation deserving a serious evolutionary analysis. Yet after Darwin, there followed a century of speculations about language origins which focused on tangential issues like "ape language" and the "innateness" of language. Only recently have we come back to Darwin's viewpoint, where we can once again ask what the adaptive functions of language may have been.

The ape language controversy was unenlightening because we already knew that chimpanzees do not naturally talk. The fact that they do not suggests that the last common ancestor we shared with chimpanzees, five million years ago, did not talk either. Language therefore evolved in the last five million years. If a human adaptation clearly evolved after the split from our last common ancestor with chimps, there is no more reason to look for rudiments of language in chimps than in baboons, beavers, or birds. The trained use of visual symbols by very clever individual apes like the famous Kanzi is marginal to understanding the evolution of human language.

The situation would have been very different if the other species of hominid had not all gone extinct. We could potentially learn a great deal about language evolution if there were still living descendants of *Australopithecus robustus* (a small-brained, strong-jawed bipedal hominid), Asian *Homo erectus* (a medium-brained hominid offshoot), and European Neanderthals (a large-brained, near-human species). As it stands, to discover whether Neanderthals talked, we would have to identify a lot more of the genes underlying human language, and then test the scraps of Neanderthal DNA that we can recover from their bones to see if they share the same genes. That might take another couple of decades. The presence of language in Neanderthals would tell us much more about the evolution of human language than the absence of language in chimpanzees does.

The other 20th-century controversy about language concerned its "innateness." The language theorist Noam Chomsky and other language "nativists" fought hard against the social science dogma that all human mental abilities are products of learning. It was a heroic fight, but for our purposes all we need to know is that the nativists won. Steven Pinker's excellent book *The Language Instinct* reviewed why they won. Pinker listed the features of language that mark it as a proper biological adaptation: "Language is a complex, specialized skill, which develops in the child spontaneously, without conscious effort or formal instruction, is deployed without awareness of its underlying logic, is qualitatively the same in every individual, and is distinct from more general abilities to process information or behave intelligently." These features show that language really is a human instinct, a mental adaptation. But they are common to all of our mental adaptations. Our capacities for language, depth perception, face recognition, sexual attraction, autobiographical memory, and social planning are all specialized skills—spontaneously learned, unconsciously deployed, and universally enjoyed. These features do not help to identify exactly what adaptive functions were served by language. They show that it evolved, but not why it evolved.

Chomsky's own research had the same limitation. He offered convincing arguments that children could not possibly learn the fundamental syntactic principles of language through parental feedback or formal instruction. This demonstration undermined the 1950s behaviorist view of language as a learned cultural invention. But Chomsky's demonstration that language depended on innate genetic capacities failed to give him any useful insights into how it evolved. In fact, Chomsky has rejected the possibility that language evolved through normal Darwinian processes.

This is a common reaction. Sometimes researchers can get so caught up in demonstrating an adaptation's complexity, elegance, and innateness that they can no longer imagine how the adaptation could have evolved through normal Darwinian processes. Alfred Russel Wallace fell into this trap when he analyzed human rationality, morality, and musical ability. When healthy respect for an adaptation tips over into awe, it becomes impossible to make any progress in understanding the selection pressures that shaped the adaptation. Like Chomsky, many researchers interested in the evolution of language suffer from this awe-of-language syndrome. Chomsky has even speculated that any sufficiently large brain (like that of a mammoth?) might automatically develop the capacity for language as a mysterious side-effect of packing 100 billion nerve cells into a small volume of space. To avoid the intellectual paralysis that the awe-of-language syndrome sometimes produces, I shall not review here the evidence for language's power and complexity—Steven Pinker has already done an excellent job of that in *The Language Instinct.*

More has been written about language evolution than about the evolution of any other specific human mental ability. However, very little of this writing has been genuinely adaptationist in the sense of assessing particular fitness benefits that could have driven the evolution of language. Very few "theories of language evolution" identify particular selection pressures that could favor the gradual accumulation of genetic mutations necessary to evolve a complex new mental capacity that has costs as well as benefits.

The current debate no longer concerns whether language is an adaptation, but what it is an adaptation for. It seems so easy to imagine survival functions for language that its possible sexual functions have been overlooked. Postulating survival functions has the appeal of the exotic, because we can daydream about mammoth hunts, tribal wars, and flint-knapping from the comfort of our armchairs. Verbal courtship is less fun to think about, perhaps because it may remind us of failed attempts at self-introduction, disastrous first dates, ardent self-revelations that met with cold, pitying stares, broken promises of fidelity, and relationship-terminating arguments. From the viewpoint of any normal living individual, all of one's past survival attempts have succeeded, whereas most of one's past courtship attempts have failed. (If most of your courtship attempts have succeeded, you must be a very attractive and charming person who has been aiming too low.) This, I think, is a useful clue: it is easier to live with language than to court with language.

Selfish Language: Communication, Manipulation, or Display?

The trouble with language is its apparent altruism. Most speech, except for commands and questions, appears to transfer potentially useful information from speaker to listener. Speaking costs the speaker time and energy, and brings information benefits to the listener, so it looks altruistic. But, as we saw in the last chapter, evolution tends to avoid altruistic behavior.

Fifty years ago, altruistic communication did not seem such a problem. The animal behavior researcher Konrad Lorenz supposed that communication was for the good of the species. Animals could save their species lots of time and energy by evolving signals that reveal their intentions and motivations, especially in combat and courtship. This would reduce the deaths from combat and the confusions of courtship. Ritualized threats such as a dog's growling were supposed to convey accurate information about the dog's level of aggression and willingness to fight over a resource. If a growly dog meets a non-growly dog, the non-growly dog should back down, saving the species a wasteful dogfight. For several decades, the biologists' dogma was that animal signaling meant communication, communication revealed emotions and intentions, and communication evolved to make a species work more efficiently.

The rise of selfish-gene thinking in the 1970s shattered this idyllic view of animal signaling. Traits did not evolve for the good of the species. In their seminal 1978 paper, Richard Dawkins and John Krebs argued that animals should evolve to produce signals only when signaling gives them a net fitness benefit that helps their own genes replicate at the expense of other genes. Evolution cannot favor altruistic information-sharing any more than it can favor altruistic food-sharing. Therefore, most animals' signals must have evolved to manipulate the behavior of another animal for the signaler's own benefit. Dogs growl because it was easier for them to intimidate a rival than to fight. Smaller dogs could be intimidated by deep growls because a deep growler is probably a larger dog that would beat them in a fight anyway. Both the growl and the growl-sensitive ears evolved for selfish reasons.

The modern theory of animal signaling grew from this insight. Signals don't usually convey information about the world, because signalers have so many reasons to lie about the world. The theory suggests that animals usually evolve to ignore the signals from other animals that may be attempting to manipulate them. There are only a few exceptions. Predators listen to signals from prey that reliably say "You can't catch me," or "I'm poisonous." (Animals hiding from predators also evolve camouflage, the purpose of which is to hide signals of existence rather than to broadcast them.) Relatives listen to signals from other relatives that reliably say "Watch out for that predator!" Animals competing for a resource listen to signals that reliably say "I could kill you." And animals looking for a good mate listen to signals that say "I have good genes." Basically, that's it. Except for the warning signals about poison and predators, these signals are all fitness indicators. Any other kind of signal that evolved in nature would probably be pure manipulation, making the listener vulnerable to lies, sweet talk, and propaganda.

The handicap principle can make fitness indicators reliable. It can do so because the signal's cost is in the same currency—the currency of biological fitness—as the signal's information. This can work not only for fitness indicators that advertise good condition to potential mates, but for signals of desperation that advertise poor condition to relatives. For example, the handicap principle may also account for the effectiveness of a baby bird's gaping-mouth hunger display. Desperation signals also work with the currency of fitness: the animal reliably shows how much a desired resource would improve its fitness. Basically, fitness indicators advertise good condition and desperation indicators advertise poor condition. Signals between unrelated animals can convey information only about the signaler's own condition, broadly construed. There are no credible models showing that evolution can favor signals that carry any other kind of information, as long as there are incentives for deception.

This is a crippling problem for almost every existing theory of language evolution, but the problem is not widely understood. The handicap principle is not a magic wand that makes all communication truthful just because a speaker has paid a fitness cost. It cannot guarantee that a sentence conveys valid information. For example, just because someone accepted the pain and risk of infection necessary to get a tattoo does not make the tattoo's message valid. It just implies that the tattooed person is stoical and healthy.

Anthropologist Chris Knight has emphasized that human language is especially vulnerable to deception because it depends so much on "displaced reference"—referring to things that are distant in time or space. To a person dying of thirst, we can say, "There's a river over that hill." But displaced reference is hard to verify. We might be lying about the river, and the thirsty person might die if he goes over the hill expecting to reach a river and finds a desert instead. In fact, there are no theories of animal signaling in which reliable displaced reference could evolve, given significant conflicts of interest between signaler and receiver. Bee dances use displaced reference to indicate the direction and distance of food, but the bees are sisters from the same hive, so they have common interests. Between our Pleistocene ancestors there were always conflicts of interest, so it is very hard to see how reliable displaced reference could have evolved. If displaced reference was not reliable, listeners would not have bothered to listen, so speakers would not have bothered to speak.

This brings us back to the altruism problem. At first it sounds plausible to suggest that "language evolved to convey propositional information from one mind to another." But that raises the question of why the speaker should altruistically give away information to an evolutionary competitor. Truthful communication is rare in nature because altruism is rare. As we saw in the previous chapter, naive altruism theories cannot explain human morality. Why should we invoke them to explain human language?

To explain language evolution, then, we need to do the same things we did for morality: find a hidden survival or reproductive benefit in the apparently altruistic act of speaking. As with morality, there are three basic options for the hidden benefit: kinship, reciprocity, or sexual selection. The fitness benefits of speaking must have come from giving useful information to a relative, sustaining a mutually beneficial information-trading relationship, or attracting a mate. I am sure all three were important, and I am not going to claim that sexual choice was the only selection pressure that shaped human language. However, I do want to highlight some features of how people talk that are not very consistent with the kinship and reciprocity theories.

Language Through Kinship and Reciprocity?

Shared information is multiplied, whereas shared food is divided. By giving you a useful fact, I do not automatically lose the benefits of knowing it. Potentially, this information-sharing effect could have made it rather easy for language to evolve through kin selection and reciprocal altruism. Our ancestors lived in small, semi-stable groups full of relatives and friends. By evolving the ability to share information with them, our genes and our social relationships would have benefited.

This sounds useful, and it is probably mostly right. However, there are still conflicts of interest. Relatives do not share all of the same genes, so do not have identical evolutionary interests. Likewise for friends in a reciprocity situation: there is always the temptation to cheat by receiving more than one gives. Given these conflicts of interest, we can look at the costs and benefits of language to see whether people's real behavior follows the predictions of kinship and reciprocity models.

As long as language is viewed purely in terms of information transmission, it will be seen as bringing more benefits to the listener than to the speaker. The speaker already knows the information being conveyed, and learns nothing new by sharing it, but the listener does gain information by listening. Information is still like food in this sense: it is better to receive than to give. In the bare-bones kinship and reciprocity theories, the principal benefit of language must be to the listener. This leads to an interesting prediction: we should be a species of extremely good listeners and very reluctant talkers. We should view silent, attentive listening as a selfish indulgence, and non-stop talking as a saintly act of altruism. People should pay huge amounts of money to engage in the vice of being psychotherapists, who get to hear people's innermost secrets while having to reveal little of themselves.

This does not describe the human species as I know it. Watch any group of people conversing, and you will see the exact opposite of the behavior predicted by the kinship and reciprocity theories of language. People compete to say things. They strive to be heard. When they appear to be listening, they are often mentally rehearsing their next contribution to the discourse rather than absorbing what was just said by others. Those who fail to yield the floor to their colleagues are considered selfish, not altruistic. Turn-taking rules have emerged to regulate not who gets to listen, but who gets to talk. Scientists compete for the chance to give talks at conferences, not for the chance to listen. For psychotherapists to use the "non-directive" methods advocated by Carl Rogers—in which the therapist says nothing back to the client except paraphrases of what they have heard—requires an almost superhuman inhibition of our will to talk.

Nor do the kinship and reciprocity theories predict our anatomy very accurately. If talking were the cost and listening were the benefit of language, then our speaking apparatus, which bears the cost of our information-altruism, should have remained rudimentary and conservative, capable only of grudging whispers and inarticulate mumbling. Our ears, which enjoy the benefits of information-acquisition, should have evolved into enormous ear-trumpets that can be swivelled in any direction to soak up all the valuable intelligence reluctantly offered by our peers. Again, this is the opposite of what we observe. Our hearing apparatus remains evolutionarily conservative, very similar to that of other apes, while our speaking apparatus has been dramatically re-engineered. The burden of adaptation has fallen on speaking rather than listening. Like our conversational behavior, this anatomical evidence suggests that speaking somehow brought greater hidden evolutionary benefits than listening.

Verbal Courtship

Much of human courtship is verbal courtship: "boy meets girl" usually means boy and girl talk. At every stage of courtship, language is displayed, and language is subject to mate choice. Teenagers agonize over the words they will use when they telephone someone to ask for a date. Stuttering, sudden changes in voice pitch, awkward grammar, poor word choice, and uninteresting content are usually considered such fatal errors by their perpetrators that they often hang up in shame, assuming that they will remain sexual failures forever. Things are not so different a little later in life. Adults in singles bars nervously rehearse their pickup lines, and mentally outline their conversational gambits.

After basic greetings, verbal courtship intensifies, progressing through self-introduction, observations concerning immediate social surroundings, compliments, and offers of minor favors. If mutual interest is displayed, people go on to trade more personal information, searching for mutual acquaintances, shared interests, and ideological common ground. If there is no common language or if accents are mutually unintelligible, courtship usually breaks down. At each stage, either person may break off courtship or attempt to escalate intimacy, but usually at least several hours of conversation precede even minor physical contact, and at least several separate conversations over several encounters precede real sex. This verbal courtship is the heart of human sexual selection. Although people may be physically attracted before a word is spoken, even the most ardent suitors will offer at least a few minutes of verbal intercourse before seeking physical intercourse.

All of this is quite obvious to any adult human with a modicum of social experience. But whereas toddlers can learn to speak reasonably well within three years of birth, it usually takes at least a decade of practice before young adults are comfortable with the basics of verbal courtship. To an evolutionist interested in sexual selection, adolescence is fascinating. The 19th-century biologist Ernst Haeckel's claim that "ontogeny recapitulates phylogeny" is often misleading, but there are cases, especially in sexual selection, where stages of life-cycle development may reflect past stages of evolutionary history. The awkward, uneven, sometimes witty verbal courtship of teenagers may not be such a bad model for the verbal courtship of our ancestors during the evolution of language. There must have been some similarities: poor vocal control, small vocabulary, uncertainty about conversational conventions, difficulty in finding phrases to express thoughts. As every parent of a teenage boy knows, the sudden transition from early-adolescent minimalist grunting to late-adolescent verbal fluency seems to coincide with the self-confidence necessary for dating girls. The boy's same-sex friends seem to demand little more than quiet, cryptic, grammatically degenerate mumbling, even when playing complex computer games or arguing the relative merits of various actresses and models. Girls seem to demand much more volume, expressiveness, complexity, fluency, and creativity. If natural selection had shaped human language for the efficient, cooperative communication of useful information, we would all speak this sort of "Early Adolescent Mumbled Dialect." At least in males, only with the demands of verbal courtship do we witness the development of recognizably human-level language.

Computer pioneer Alan Turing alluded to the importance of verbal courtship for testing someone's mental capacities in the original 1950 version of his "imitation game," which has come to be known as the "Turing test." In the imitation game, an interrogator tries to determine whether he is interacting with a real woman or a computer program that imitates a woman. Turing was more interested in intelligence than female flesh, so he eliminated the physical cues of womanhood, and limited the interrogator to typing questions on a terminal, and receiving answers on a screen. The questions can be as challenging as the interrogator likes, such as "Please write me a sonnet on the subject of the Forth Bridge." In Turing's view, if a computer can successfully lead an interrogator to believe that he is interacting with a real woman, it should be considered intelligent. Turing emphasized that the computer must be capable of credibly demonstrating a very wide range of behaviors—his list included being kind, using words properly, having a sense of humor, catching us by surprise, claiming to enjoy strawberries and cream, falling in love, and making someone fall in love with it. (Strikingly, many of these behaviors overlap with the courtship adaptations we have considered in previous chapters.)

After Turing, philosophers of artificial intelligence dismissed the sexual aspect of the imitation game as a confusing distraction, and stripped it away from modern versions of the Turing test. However, Turing's original version subtly pointed to the special challenges of demonstrating human intelligence during courtship. Even a very simple 1970s computer program like ELIZA can fool people into thinking that they are interacting with a real psychotherapist—but no one has fallen in love with ELIZA, as far as I know. Turing's more sexualized imitation game offered a key insight: human intelligence can be demonstrated very effectively through verbal courtship, and any machine capable of effective verbal courtship should be considered genuinely intelligent.

The idea that language evolved for verbal courtship solves the altruism problem by identifying a sexual payoff for speaking well. Once the rudiments of language started to evolve, for whatever reason, our sexually motivated ancestors would probably have used their heritable language abilities in courtship. Language complexity could have evolved through a combination of runaway sexual selection, mental biases in favor of well-articulated thoughts, and fitness indicator effects.

Language Displays and Social Status

This verbal courtship theory fits nicely with some ideas developed by three other language evolution theorists—Robbins Burling, John Locke, and Jean-Louis Dessalles. They are not as well known as Noam Chomsky or Steven Pinker, but they share my belief that a good theory of how language evolved must show how selfish genes can derive hidden benefits from the apparently altruistic act of speaking. In an important paper published in 1986, anthropologist Robbins Burling advanced arguments similar to mine. He contrasted the excessiveness of our baroque syntax and enormous vocabulary with the sufficiency of simple pidgin languages for trade, hunting, and tool making, and considered this alongside the problem of language's apparent altruism. He proposed that complex human language evolved through male orators competing for social status by speaking eloquently, since high status would give them reproductive advantages. Burling cited anthropological evidence of the links in tribal societies between verbal skill, social status, and reproductive success. As long as those links held true during human evolution, language could have evolved ever greater complexity. As Burling noted, "All that is needed for the mechanism I suggest to be effective is that the average leader in the average society have slightly more verbal facility and slightly more children than other men." Although he emphasized verbal leadership more than verbal courtship, he did acknowledge that "We need our very best language for winning a lover." I think Burling's sexual selection model of language evolution deserves much more attention than it has received, and complements my ideas about verbal courtship.

Cambridge linguist John Locke has extended Burling's social-status model with more linguistic evidence, paying more attention to the role of "verbal plumage" in human sexual mate choice. He quoted from a study in which a young African-American man from Los Angeles patiently explained the sexual-competitive functions of language to a visiting linguist: "Yo' rap is your thing . . . like your personality. Like you kin style on some dude by rappin' better 'n he do. Show 'im up. Outdo him conversation-wise. Or you can rap to a young lady, you tryin' to impress her, catch her action—you know—get wid her sex-wise." In a few concise phrases, this teenager alluded to both classic processes of sexual selection: male competition for status, and female choice for male displays.

Along similar lines, language researcher Jean-Louis Dessalles has pointed out that listeners award higher social status to speakers who make relevant, interesting points in conversation. Language may have evolved through social selection to permit these "relevance" displays. This is why people compete to offer good ideas and insights when talking in groups. While Burling and Locke focused on dramatic public displays of oratorical prowess, Dessalles focused on social competition to say interesting things in ordinary small-group conversation.

Burling, Locke, and Dessalles have all identified important selection pressures that have been neglected in previous theorizing about language evolution. They have shown how language's hidden status and sexual benefits could have driven its evolution. In their theories, sexual attractiveness depends on social status, which in turn depends on verbal ability displayed in large or small groups. In my verbal courtship theory, sexual choice favored verbal ability more directly through one-to-one conversation. Sexual selection probably shaped human language in both ways: directly, through mate choice, and indirectly, through social status. Here I focus on verbal courtship only because it has received less attention so far.

A Million Words of Courtship

Verbal courtship can be quantified. Conception of a baby is the evolutionarily relevant threshold for success in courtship. Without contraception, it takes an average couple about three months of regular sex before a pregnancy occurs. If we assume two hours of talk per day in the early stages of sexual relationship, and three words spoken per second (an average rate), each member of a couple would have uttered about a million words before they conceived any offspring. Each would have talked enough to fill six books the length of this one. In modern societies, the surprising thing is not that couples run out of things to say to each other, but that they do not run out much sooner.

From the first greeting to the millionth word, much can go wrong. Personalities clash. Arguments go unresolved. Incompatibilities arise. Jokes fall flat. Boredom ensues. Both individuals must clear the million-word hurdle before they contribute to the next generation. When language first evolved, it may have been a ten-word hurdle, or a thousand-word hurdle. But at each step, both individuals were trying to extract, by using the language available to them, as much information as they could. The more talking they did, the more of their minds they revealed. The more verbal courtship revealed, the greater effect sexual selection could have.

This courtship theory has been mocked as the "chat-up theory" of language evolution. It is all too easy to describe in salacious terms. One could write about nimble tongues playing across strong columns of warm air, the syncopated breath of lovers tickling those most sensitive surfaces of the human body—the eardrums—and conversation as minds dancing together in a tango of frenzied cognitive foreplay. But there is no reason to make sexual selection sound so lubricious. Human verbal courtship is the least superficial form of courtship that evolution has ever produced. A million words give a panoramic view of someone's personality, past, plans, hopes, fears, and ideals. It would be misleading to make our verbal courtship sound like second-rate erotica, or to focus on the risible chat-up lines sometimes heard in singles bars. Verbal courtship continues for months after people first meet, and it becomes the bedrock of human intimacy and love.

Public Speech as Covert Courtship

Verbal courtship can be viewed narrowly as face-to-face flirtation, or broadly as anything we say in public that might increase our social status or personal attractiveness in the eyes of potential mates. Sexual flirtation during early courtship accounts for only a small percentage of language use, but it is the percentage with the most important evolutionary effects. This is the time when the most important reproductive decisions are made, when individuals are accepted or rejected as sexual partners on the basis of what they say. Yet, if language evolved only for face-to-face flirtation, we would talk much less than we do. Why do we bother altruistically giving away information when we are not directly courting a particular individual?

Verbal courtship in the broader sense explains why we compete to say interesting, relevant things in groups. Sexual choice permeates human social life, because anything that raises social status tends to improve mating prospects. If a man gains a reputation as an incisive thinker who consistently clarifies group decision-making and mediates social conflicts, his social status and sexual attractiveness increase. If a woman gains a reputation as a great wit and an inventive storyteller, her status and attractiveness increase as well. Public speaking and debate allow individuals to advertise their knowledge, clear thinking, social tact, good judgment, wit, experience, morality, imagination, and self-confidence. Under Pleistocene conditions, the sexual incentives for advertising such qualities would have persisted throughout adult life, in almost every social situation. Language puts minds on public display, where sexual choice could see them clearly for the first time in evolutionary history.

Form and Content

If language evolved for sexual display, shouldn't we go around trying to say the most difficult possible tongue twisters? Shouldn't human sexual competition follow the style of Cyrano de Bergerac, who demonstrated his physical and mental fitness to the beautiful Roxane by improvising a ballad of rhyming alexandrines, including three eight-line stanzas and one quatrain, while sword-fighting his sexual rival the Vicomte de Valvert, all timed perfectly so that Cyrano's last word coincided with the Vicomte's death? That would be impressive. But it is not what sexual selection demanded.

What we say is generally more important than how we say it.

The formal structure of language evolved principally as a medium for conveying ideas and feelings, which tend to attract sexual partners by revealing our personalities and minds. Sexual selection shapes language's content more than its form. Or rather, the form evolved in the service of the sexually selected content, rather than as a sexual display in its own right, as bird song did. Some of us prefer sexual partners with deep thoughts expressed succinctly to partners with many words but no thoughts. Sexual selection need not favor the superficial chatterbox over the Zen master who utters an enlightening and memorable 17-syllable haiku once a day. If it had, we would all resemble people with Williams syndrome, who tend to produce fluent, grammatical, large-vocabulary streams of relatively trite speech.

Nevertheless, there are some hints of sexual ornamentation in the human voice's pitch and timbre, the size of our vocabularies, the complexity of our grammar, and the narrative conventions of storytelling. For example, adult human males have deeper voices than children or women, which may reflect female choice favoring a low-pitched voice as an indicator of large male body size. (A deep voice does not have to correlate perfectly with large body size in order to work as an indicator.) Female frogs prefer lower-pitched male frog calls, and women generally find the deep, resonant voice of Isaac Hayes more sexually attractive than those of the Vienna Boys' Choir. Even in the television show *South Park*, the sexual charisma of Hayes's voice shows through in his school chef character, who, despite his low job status, credibly says lines indicative of sexual desirability, like "Damn, woman, I just gave you sweet lovin' five minutes ago!" On the other hand, low pitch could also have evolved through male competition as a threat display, as when the actor James Earl Jones provided the terrifying voice of Darth Vader in *Star Wars*.

Apart from examples like this, there is not much evidence of sex differences or sexual selection in the details of language form. In analyzing these details, linguistics made reasonable progress by assuming that language evolved as a cooperative system for the transmission of information. The acoustics of speaking and listening can be modeled fairly well by optimal information-transmission models where it is assumed that speakers and listeners are trying to minimize the joint costs of such transmission. Speakers pronounce words just clearly enough to be understood, but not so clearly that their jaws and tongues get exhausted; listeners work pretty hard to understand what is said, but not so hard that their auditory cortex evolves to enormous size. Likewise, the cooperative model has helped language researchers to understand grammar (syntax), word structure (morphology), and word meaning (semantics). These aspects of language make it look like a system designed for efficient information transmission.

However, the same cooperative model would work reasonably well in analyzing many details of peacock courtship displays. If one assumed that peacock courtship evolved for the efficient, cooperative transmission of iridescence patterns from peacock to peahen, one could successfully describe most of the anatomy of the peacock's tail and the physiology of the peahen's visual system. His tail works pretty hard to produce iridescence, but her eyes work pretty hard to perceive it. Her eyes may be optimally attuned to the wavelengths of light reflected by his tail, just as our ears are optimally attuned to the sound spectra produced by speech. The movement patterns of his tail may be optimally adjusted to produce maximum iridescence-transfer to her eyes under most lighting conditions. And so forth. At the level of signal transmission and reception, peacock courtship may have the appearance of a cooperative system.

The details of signal production and perception cannot usually distinguish cooperative communication from courtship display. The differences emerge more at the level of signal cost, signal content, receiver attitude, and overall pattern of social interaction—aspects of language not typically studied in linguistics. Courtship displays usually have high costs and high degrees of difficulty, taking into account everything relevant to display effectiveness. At first glance, human language looks like a very cheap and easy form of signaling. Once your species has evolved language, and you have learned language, and you are fit and healthy, and you have something to say, and you have the attention of a potential mate, it does not take much time, energy, or effort to say it. The hard part, of course, is having something interesting to say. The difficulty of effective verbal courtship is not the cost of moving your jaw and tongue, but the cost of thinking of something verbally expressible that will impress another human. This cost depends entirely on the listener's threshold for being excited: intelligent listeners demand intelligent utterances, and these are difficult to produce.

In cooperative communication, the receiver may be mildly skeptical about the information conveyed. In courtship, the receiver is extremely judgmental not only about the information, but about the signaler. When listening, we automatically evaluate whether what is being said makes sense, whether it is congruent with what we know and believe, whether it is novel and interesting, and whether we can draw intriguing inferences from it. But we also use all of these judgments to form an impression of the speaker's intelligence, creativity, knowledge, status, and personality. We assess the information content of utterances, not just to make inferences about the world, but to make attributions about the speaker.

This is why perfectly grammatical, well-spoken, true sentences can fail as conversational gambits. Consider the old English nursery rhyme:

> Tommy Snookes and Bessy Brookes
> Were walking out one Sunday;
> Says Tommy Snookes to Bessy Brookes,
> "Tomorrow will be Monday."

As a sentence evaluated according to traditional linguistic standards, Tommy's utterance is perfectly successful. It passes the tests of grammaticality. But as a social act of courtship, Bessy will not be impressed. Tommy's comment is too obvious. It is true, but irrelevant. It provokes no further thought or response. Bessy may suspect Tommy of low intelligence, social laziness, or nervousness.

In real human social life, conversational failures like that of Tommy Snookes are relatively rare. This is not because everyone is good at verbal display, but because those who are not learn to keep relatively quiet. People tend to socialize with friends and sexual partners who show roughly their own verbal ability level—their verbal compatibility has already determined which social relationships were formed. The majority of human conversation occurs between sexual partners and long-term friends. They have already chosen each other as mates or friends precisely because their first few conversations were mutually interesting, evoking mutual respect and attraction. Ordinary talk between old friends and lovers still includes sufficient verbal display to maintain mutual respect, but may not include the same verbal fireworks as the first few conversations did. That is why conspicuous verbal display plays only a minor role in everyday speech. Thus the costs of effective display and the risks of display failure look low. But this is an illusion: meet someone new, and these costs and risks surge back into salience.

Many language researchers remain preoccupied with studying the principles of syntax, by inviting native speakers of a language to tell them which sentences follow the language's grammatical rules and which do not. These decisions are called "grammaticality judgments." From an evolutionary perspective, it seems peculiar for linguistics to focus on this very narrow sort of normative judgment. People often speak ungrammatically in real conversation, but such rule violations are almost always ignored. People are much more interested in normative judgments about whether a speaker is truthful, relevant, interesting, tactful, intelligent, and sympathetic. Traditional linguistics has exiled all such questions to the subdiscipline of "sociolinguistics," which concerns how people use and judge language in real social interactions. Sociolinguistics is the evolutionarily crucial level of analysis, where all the social and sexual pressures that could have shaped language show themselves. But modern sociolinguistics is a small, underfunded social science that has proved highly skeptical of evolutionary psychology.

We have a quandary: the syntax theorists who study grammaticality judgments dominate the conferences on language evolution, while the sociolinguists who study more evolutionarily important social judgments about speakers will not talk to evolutionary psychologists. Grammaticality judgments are extremely useful scientific data for analyzing the principles of syntax, but social judgments about the intelligence, personality, and attractiveness of speakers are much more potent as selection pressures. (Of course, social judgments of a speaker's intelligence may rely, in part, on judging their grammar, along with what they have to say, their voice quality, their social tact, their verbal self-confidence, and so forth.) We need an evolutionary sociolinguistics that can finally test evolutionary theories of the social and sexual benefits of language against data on the social and sexual uses of language in different cultures. From the standpoint of traditional linguistics, syntax, morphology, and semantics are the core of human language—but from a Darwinian viewpoint they are just the incredibly complex design details of a signaling adaptation centered upon social functions and social content.

Life Stories

Verbal courtship allows individuals to tell their life stories quickly and verifiably. Humans can learn more about each other in an hour than mute animals can in months. Within minutes of boy meets girl, boy and girl typically know each other's names, geographical origins, and occupations. In the first heady hours of chatter, they usually learn about each other's families, past and current sexual relationships, children, friendships, work colleagues, adventures, travels, ideological convictions, hobbies, interests, ambitions, and plans. By the time a sexual relationship has lasted a few months, lovers usually have a pretty good idea of each other's lives from childhood onwards. By contrast, chimpanzees can never gain direct information about one another's past experiences or long-term plans. They can only make a few rough inferences about personality from social behavior, so what they see in mate choice is pretty much all they get. Language lets us learn about potential mates much more efficiently and interactively than any other species can.

Are these life stories reliable as indicators of anything? Who has not been tempted, when sitting next to a stranger on an airplane, to make up an utterly fictional account of oneself, inventing a new name, origin, and occupation? But, as every undercover police officer knows, false autobiography is vulnerable to logical inconsistency, to claims being proven wrong, to insufficient background knowledge, and to accidental revelation of true identity by one's actual acquaintances. The life stories that we reveal over days and weeks of courtship are kept reasonably reliable by logical, empirical, and social pressures.

Of course, we present our lives in the best possible light. We mention our successes rather than our failures, impressive relatives more than wastrels, dramatic trips more than solitary depressions, and palatable beliefs more than secret bigotries. Our life stories present us as the heroes of the grand adventures that are our lives, rather than the Rosencrantz or Guildenstern to someone else's Hamlet. Nevertheless, because most people distort their life stories to more or less the same degree, they remain a valid basis for mate choice. Initially at least, our life stories will be compared not to the truth, but to the equally distorted life stories of our sexual competitors. You might effuse about your package holidays to Bermuda, while your rival reminisces about his or her space shuttle flights as mission commander. Even if you both hide your tendency to periods of indigence and self-doubt, a potential mate can still judge that flying a billion-dollar spaceship at 17,000 miles per hour is a better fitness indicator than a weekend of immoderate drinking at Club Med.

Our ancestors could not brag of orbiting the Earth, but neither can most of us. Our lives are generally safe and sedentary compared to theirs, so our life stories are probably less dramatic, and less informative about our ability to handle challenges and emergencies. By the time they reached sexual maturity, our ancestors would have had plenty of close encounters with dangerous wild animals, some experience of physical violence, a great many travel stories concerning diverse places, and encounters with potentially hostile members of other tribes. By middle age, they would have seen death and injury, lost many relatives, and experienced sickness and starvation. Surviving males would have killed very dangerous animals, and perhaps killed another human. Surviving females would have suffered miscarriages, difficult births, the death of infants, sexual harassment, stalking by unwanted men, and perhaps rape. Our ancestors had plenty of life to fill their stories.

When life stories became important in verbal courtship, our ancestors began to judge one another's past experiences, not just their present appearance. Language made each individual's entire history a part of their "extended phenotype" in courtship. Like our body ornaments, our pasts became part of our sexual displays. We dragged them around after us, into every new relationship. As a result, sexual selection could favor any mental trait that tended to produce an attractive past. It sounds like a time-travel paradox, but it is not. It just means that sexual selection could have favored genes for a good autobiographical memory, a tendency to have risky adventures, or a credibly restrained sex life without too many infidelities. The handicap principle suggests that sexual selection could even have favored a masochistic taste for memorable discomfort, since the ability to survive hardship reveals fitness. Even in the carnage of mechanized warfare or the intellectual bloodbath of an academic job interview, one can always think, "This will make a hell of a story someday." Through memory and language, we can transform a pure fitness cost in the past (such as a physical wound or a social rejection) into a reliable fitness indicator in the present (a story about our ability to heal without disability, or to overcome depression).

Introspective, Articulate Ape Seeks Same

Sexual selection for verbal courtship may have re-engineered our minds in other ways, favoring abilities to articulate a wider range of our mental processes. Before language evolved, there may have been little reason for animals to introspect about their thoughts and feelings. If introspection does not lead to adaptive behavior, it cannot be favored by evolution. However, once verbal courtship became important, sexual selection pressures could have increased the incentives for being able to consciously experience more of the thoughts and feelings that guide our behavior, and being able to report those experiences verbally.

Lovers sometimes say, "Words cannot express what I feel about you," but this attention-getting device usually precedes hours of impassioned chatter or lovemaking. Articulate people can articulate anything that they consciously experience. Insofar as sexual choice favored verbal self-disclosure, it may have favored an expansion of conscious experience itself. The result is the effortless, fluid way we can translate from perceived objects through consciously attended qualities into spoken observations. We can walk with a lover through Kew Gardens, notice a rose, describe its distinctive color and fragrance, and perhaps even whisper a relevant quote from Shakespeare's sonnet fifteen, observing

> Where wasteful Time debateth with Decay,
> To change your day of youth to sullied night;
> And all in war with Time for love of you,
> As he takes from you, I engraft you new.

This high-bandwidth channel, from perception through consciousness and memory to articulate communication, seems unique to humans. Only when sexual choice favored the reportability of our subjective experiences—with the emergence of the mental clearing-house we call consciousness—did our strangely promiscuous introspection abilities emerge, such that we seem to have instant conscious access to such a range of impressions, ideas, and feelings. This may explain why philosophical writing about consciousness so often sounds like love poetry—philosophers of mind, like lovesick teenagers, dwell upon the redness of the rose, the emotional urgency of music, the soft warmth of skin, and the existential loneliness of the self. The philosophers wonder why such subjective experiences exist, given that they seem irrelevant to our survival prospects, while the lovesick teenagers know perfectly well that their romantic success depends, in part, on making a credible show of aesthetic sensitivity to their own conscious pleasures.

Such evolutionary pressures to report our conscious experiences may have even influenced how we perceive and categorize things. Psychologist Jennifer Freyd has argued that some of our cognitive processes have become adapted to the demands of verbal "shareability." For example, we may tend to perceive some naturally continuous phenomena in discrete ways, just because it is easier to give verbal labels to discrete categories than to points on fuzzy continua. Applied to verbal courtship, Freyd's shareability idea suggests that sexual selection may have made human mental processes well adapted for producing romantically attractive language, not just effective survival behavior.

Gossip: Social Information, Entertainment, or Indicator?

Apart from ourselves, we mostly talk about other people—language is mostly gossip. Evolutionary psychologist Robin Dunbar has proposed that gossip helped our ancestors to keep track of a larger number of social relationships than they could by direct observation and direct interaction. Talking proved more efficient than grooming as a way of servicing our friendships. This view of gossip as "social grooming" explains why gossip includes so many sympathy displays. The idea that gossip helps to manage large numbers of relationships clarifies why gossip sometimes sounds like a fairly methodical review of the state of every social relationship known to both speakers.

However, gossip has other features that may be better explained as status displays, and sometimes even courtship displays. Jean-Louis Dessalles has pointed out that a speaker's utterance must seem relevant to listeners if it is to attract their attention. If language's content was shaped by the psychological biases of our ancestors, what subject matter would seem most relevant to a highly social primate? The answer, of course, is social content. If our ancestors were already spending most of their conscious lives thinking about one another, and worrying about their relationships, they would have a psychological bias to favor social content in their conversations. Gossip would fill their hunger for social information. If we had evolved from solitary spiders, our language would be as dominated by webs and flies as were our spidery minds. The social content of human speech may have no direct social function: it may simply reflect the optimal way to excite a mind already geared to social information, as a form of socially and sexually attractive entertainment. The better entertainers benefit by attracting better friends and mates. Gossip may exploit the social obsessions of the human mind as much as soap operas and romantic films do.

Yet there may be more to gossip than the passive appeal of soap operas with fictional characters. Beyond the psychological bias view of sexual selection, there is indicator theory. As with all courtship displays, we can ask what information about the displayer might be revealed in their display. To be worth listening to, gossip must be novel, but credible and interesting, which generally means that it must be new, verifiable information about mutual acquaintances. We have little interest in old information about old friends, or new information about total strangers. It is not easy to consistently produce new, verifiable information about mutual acquaintances. Since the object of gossip is a mutual acquaintance, then, all else being equal, the listener is as likely to know the news as the gossiper. If the gossiper usually knows some news that the listener does not know, the gossiper may have privileged access to secrets, or a wider social network, or a better social memory, or friends who themselves have privileged access to social information. That is, the gossiper must have high social status, and high social intelligence. This is how gossip can function as a reliable indicator of social status and social skills. Gossip may have evolved as a status display, favored by sexual selection and other forms of social selection.

Dunbar tested his gossip theory in a 1997 paper by analyzing the content of ordinary human conversations between British adults. His results appear to support a mixture of his gossip-as-grooming theory, and my gossip-as-courtship theory. Across all conversations analyzed, social topics such as personal relationships accounted for about 55 percent of male conversation time and about 67 percent of female time. That high proportion is generally consistent with both theories. Of the time spent discussing any kind of social relationship, talking about one's own relationships accounted for 65 percent of male speech and only 42 percent of female speech. Males appeared more motivated to display the quality and number of their relationships. Also, males tend to talk more about intellectual topics such as cultural, political, or academic matters, particularly when females are present. Dunbar observed that:

> Female conversations can be seen to be directed mainly towards social networking (ensuring the smooth running of a social group), whereas males' conversations are more concerned with self-promotion in what has all the characteristics of a mating lek. This is particularly striking in the two university samples where academic matters and culture/politics, respectively, suddenly become topics of intense interest to males when females are present.

For males, verbal self-advertisement appears to be a fairly constant function of speech, while for females, it may be an occasional function, more limited to one-on-one conversations with desired mates. A complete theory of the evolution of language will probably have to combine sexual selection and social selection, integrating the gossip-as-courtship theory with the gossip-as-grooming theory.

Supercalifragilisticexpialidocious

Given that the word "blue" exists, why does the word "azure" exist? They are nearly identical in meaning. It is hard to envision a situation in which natural selection would favor the hominid who could say, "The sky on the other side of that mountain was azure" over one who could say "It was blue." Perhaps poetic words like "azure" were invented for some special ritual or religious function. But why then do we also need "cobalt," "sapphire," "ultramarine," "cerulean," and "indigo"?

Human vocabulary sizes seem to have rocketed out of control. The average adult human English-speaker knows 60,000 words. The average primate knows only about 5 to 20 distinct calls. The largest bird song repertoires are estimated at about a thousand, though their songs do not have distinct symbolic meanings. Unusually intelligent bonobos such as Kanzi can be taught about 200 visual symbols in ape language experiments. No other animal has a signal repertoire with distinct meanings that comes anywhere near the human vocabulary size.

In this section I look at vocabulary size as an example of how sexual selection may have shaped language evolution. If language evolved in part through sexual choice as an ornament or indicator, it should be costly, excessive, luxuriant beyond the demands of pragmatic communication. How could we measure whether language is excessive? Vocabulary is convenient to study because we can count how many words people know, whereas we do not yet know how to measure the complexity of grammar or the social strategies of conversation. More importantly, we can count how many words people would need to know for pragmatic purposes, and see whether our vocabularies are excessive.

We acquire our vocabularies with such speed that we must have evolved special adaptations for learning word meanings. To build an adult vocabulary of 60,000 words, children must learn an average of 10 to 20 words per day between the ages of 18 months and 18 years. Often these words are learned through a single exposure: an adult points to a bassoon and says "that's a bassoon" just once, and the child knows the word forever after. Human children are word-sponges. By contrast, even the brightest chimps in ape language experiments require at least 20 to 40 exposures before they learn the meaning of a visual sign. One has to do the equivalent of saying "bassoon, bassoon, bassoon" over and over until it loses all meaning to a human and acquires one for the chimp. In humans, word meaning appears to be stored in special brain areas, and damage to these areas through injury or stroke produces vocabulary deficits.

Is a vocabulary of 60,000 words excessive? Most of it is not used very often. The most frequent 100 words account for about 60 percent of all conversation; the most frequent 4,000 words account for about 98 percent of conversation. This sort of "power law" distribution is common: the 100 most successful movie actors probably account for 70 percent of all money paid to all actors; the 100 most popular Internet sites probably handle a similar proportion of Internet traffic; and so forth. It is not surprising that vocabulary use follows a power law, but it is surprising that our average vocabulary is so large, given how rarely we use most of the words that we know. It could easily have been that just 40 words account for 98 percent of speech (as it does for many two-year-olds), instead of 4,000 (as it does for most adults). As it is, any of the words we know is likely to be used on average about once in every million words we speak. When was the last time you actually spoke the word "cerulean"? Why do we bother to learn so many rare words that have practically the same meanings as common words, if language evolved to be practical?

To see whether our large vocabularies evolved as ornamental luxuries, we can compare them with artificial languages and "pidgin" languages specifically created for pragmatic communication. Artificial languages can work with very small vocabularies. In the 1920s, the Oxford philosopher I. A. Richards and collaborator C. K. Ogden developed a stripped-down English vocabulary of just 850 words that they called Basic English. Their motive was to promote international peace and understanding by making it easier for non-native speakers to acquire a minimal, functional version of English, which they recognized as the emerging planetary language.

Basic English works with ordinary English grammar. Despite it having a vocabulary only 1 per cent as large as normal, Richards wrote that "it is possible to say in Basic English anything needed for the general purposes of everyday existence—in business, trade, industry, science, medical work—and in all the arts of living, in all the exchanges of knowledge, desires, beliefs, opinions, and news which are the chief work of a language." Indeed, Richards wrote this passage using Basic English. Richards and Ogden also found that they could easily define any other English word using just the Basic vocabulary: their *General Basic English Dictionary* did this for 20,000 non-Basic words. Basic is really quite simple: it gets by with just 18 verbs, which Richards called his "willing, serviceable little workers . . . less impressive than the more literary verbs, but handier and safer." Basic is not quite as compact as ordinary English—it takes perhaps 20 percent more words to state a given idea—but it is vastly easier to learn, and easier to understand by a wider range of people. A slightly expanded Basic even works for expressing scientific ideas: the Basic Scientific Library series in the 1930s included introductory textbooks on astronomy and biology.

Like Basic English, "pidgin" languages illustrate how useful even small vocabularies can be. Pidgins arise when people speaking mutually unintelligible languages are thrown together in a situation, such as a slave plantation, that forces some means of communication to develop. Most pidgins have small vocabularies, like Basic English, and minimal grammar. Yet they suffice for trade, cooperative work, and ordinary survival functions. However, children brought up learning a small-vocabulary pidgin tend to transform it into a larger-vocabulary "creole," which is a full-sized language. Language researchers take "creolization" as evidence that small-vocabulary pidgins must have been insufficient for pragmatic communication in some respect. But that implies that all complexity must be due to pragmatic demands. A different view is possible: perhaps creoles, like language itself, arose as better verbal ornaments and better indicators of verbal intelligence.

If Basic English and pidgins allow people to communicate, trade, cooperate, and live together using very small vocabularies, why do all mature, natural human languages have a hundred times as many words? An analogy to bird song may be useful here. Most bird song evolves under sexual selection through mate choice. Most birds produce a fairly small repertoire of courtship songs, but in a few bird species, such as marsh warblers and nightingales, the number of distinct songs seems to have undergone some sort of explosive evolution, resulting in repertoires of over a thousand distinct songs. In these species repertoire size itself became a criterion for mate choice, with males who sing more songs being perceived as more attractive. Above-average repertoires may work as reliable indicators of a bird's age, learning ability, intelligence, brain size, brain efficiency, or general fitness. Males with larger repertoires appear to sire healthier offspring, suggesting that repertoire size may be an indicator of heritable fitness.

Although particular bird songs do not have any meaning, their overall repertoire size does; it indicates heritable fitness. Human words do have meaning, but perhaps our overall vocabulary size has the same meaning as their song repertoires. A large vocabulary may be a good fitness indicator. Large vocabularies may have been favored in mate choice, and may have evolved through sexual selection.

Obviously, vocabulary size differs enormously between people, so it could be a useful cue in mate choice. The American Scholastic Achievement Test includes plenty of vocabulary questions because vocabulary knowledge varies enough to be a reasonable indicator of intelligence and general learning ability. Evidence shows that vocabulary size is at least 60 percent genetically heritable, and has about an 80 percent correlation with general intelligence. (The correlation with intelligence is not 100 percent, of course—people with Williams syndrome, for example, have lower than average general intelligence, but delight in unusual words such as "diplodocus," and develop fairly large vocabularies.)

Since words are learned, it may seem odd that overall vocabulary size should be heritable, but that is what behavior-genetic studies find. Identical twins reared apart (who have the same genes but different family environments) correlate about 75 percent for their vocabulary size. By contrast, the environmental effect of parenting accounts for only a small proportion of the variation in the vocabulary size of children, and just about 0 percent of the variation in adult vocabulary size. If you have a large vocabulary, that is because your parents gave you genes for learning lots of words quickly, not because they happened to teach you lots of words. Actually, most of vocabulary's heritability is carried by the link between vocabulary learning ability and general intelligence, which in turn is highly heritable.

This link between vocabulary and intelligence may extend all the way to biological fitness. Perhaps general intelligence itself, or what intelligence researchers call "the *g* factor," is a fitness indicator. One study has shown that intelligence correlates about 20 percent with body symmetry, which is a known fitness indicator. Thus, vocabulary size could indirectly advertise fitness. Our ancestors would have benefited by favoring sexual partners with large vocabularies. If vocabulary was a criterion for mate choice, they would also have benefited by evolving larger vocabularies, just as peacocks evolved larger tails.

Few will admit to—or even be aware of—a sexual preference for a large vocabulary. It would be unusual to see a personal advertisement that ran "Single female seeking man who knows fifty thousand useless synonyms." However, couples in long-term relationships tend to have vocabularies of similar sizes, and the strength of this assortative mating for vocabulary size is higher than for most other traits. Although one may not consciously prefer a date who uses "azure" instead of "blue," one may shudder if a date uses "azure" as if it meant teal, mauve, or vermilion.

So how would one display a large vocabulary size in courtship? Consider vocabulary as an intelligence-indicator. We know from intelligence-test research that there tend to be minimum IQ thresholds for producing and comprehending certain words.

According to the widely used WAIS-R intelligence test, for example, English-speaking adults with an IQ of 80 typically know the words "fabric," "enormous," and "conceal," but not the words "sentence," "consume," or "commerce." IQ 90 speakers typically know "sentence," "consume," and "commerce," but not "designate," "ponder," or "reluctant." If you are flirting with someone, and they say they would like to "consume" your body in a passionate embrace, but they do not understand when you say you are "reluctant," you can probably infer they have an IQ between 80 and 90. We make these sorts of inferences quite automatically and unconsciously, of course.

We may not realize that we use vocabulary as an intelligence-indicator. Yet, what we do not admit, wise nannies may understand. In the film *Mary Poppins*, the song "Supercalifragilisticexpialidocious" celebrated the power of unusual words to advertise intelligence, attract mates, and make friends with maharajas.

Near the end of the song, Mary suggested that when you cannot find the right word to express your thoughts, the "super"-word can fill the gap. However, she also warned that its life-changing power must be used with caution. At that point, her back-up drummer interjected a personal example: he once uttered the word to his girlfriend, and it led straight from his verbal courtship to their marriage. Mary's song captures a key feature of the verbal courtship theory—words can work as reliable indicators of intelligence (and articulation ability), even when, like birdsong, they have no meaning whatsoever.

To test this verbal courtship theory of vocabulary properly, we would have to find out much more about human verbal behavior than language researchers know at present. We don't know the size of typical ancestral or tribal vocabularies. We don't know whether people use more impressively obscure words during courtship. We don't know whether large vocabularies are valued directly in human mate choice. We don't know how vocabulary sizes correlate with brain size, physical health, physical attractiveness, fertility, or general fitness. Sex differences in the distribution of vocabulary sizes are rarely reported in the scientific literature (though they are perfectly well known to the Educational Testing Service that administers the SAT).

Words appear to have evolved for symbolic reference. This appears to set them apart from other forms of animal signaling. My point in this section has been that words can also evolve as indicators. The servicable little vocabularies of Basic English and pidgins suggest that we learn and display many more words than we really need to communicate: our huge vocabularies make no sense as pragmatic adaptations for survival. Human vocabulary size may have evolved through the same sexual selection process that favored enormous song repertoires in some bird species. But whereas only male birds sing, both men and women use large vocabularies during courtship, because courtship and choice are mutual, and because unusual words work as reliable displays only if their meanings are understood.

Why Do Women Have Higher Verbal Ability than Men, if Language Was Sexually Selected?

When sex differences do show up in human mental abilities, women typically show higher average verbal ability, while men show higher average spatial and mathematical ability. For example, women comprehend more words on average, and this sex difference accounts for almost 5 percent of the individual variation in vocabulary size. But sexual selection normally predicts that males evolve larger ornaments. If language evolved as a sexual ornament, it seems that males should have much higher average verbal abilities. Is this a fatal problem?

The standard predictions of sexual selection are hard to apply because language is used for both speaking and listening—both verbal display and the judgment of verbal displays by others. Normally, sexual selection makes males better display-producers and females better display-discriminators. Peacocks can grow bigger tails, but peahens may be better at seeing and judging tails. Most tests of human verbal abilities are tests of language comprehension, not tests of language production. Given a strict male-display, female-choice mating system, we should expect female superiority in language comprehension and male superiority in language production.

For example, females should recognize more words, but males should use a larger proportion of their vocabulary in courtship, biasing their speech towards rarer, more exotic words. In this simple picture, more women might understand what "azure" really means (so they can accurately judge male word use), but more men might actually speak the word "azure" in conversation (even if they think it means "vermilion"). Standard vocabulary tests measure only comprehension of word meaning, not the ability to produce impressive synonyms during courtship. Reading comprehension questions are more common than creative writing tests. Women are faster readers and buy more books, but most books are written by men.

But the male-display, female-choice system is not an accurate model of human conversation anyway. Throughout this book I have stressed the importance of mutual mate choice in human evolution. Human courtship means, above all, men and women talking to one another. It is not restricted to men standing up and pouring forth a stream-of-consciousness verbal display to anyone who will listen. Such male verbal broadcasts can be observed in churches, parliaments, and scientific conferences, but human speech is typically more private and more interactive. The interactiveness of conversation makes terms like signaler and receiver problematic. All humans are both. As with other mental abilities, mutual display and mutual choice tend to produce sexual equality in the display ability.

How should we interpret the female superiority on language comprehension tests, given the male motivation to produce public verbal displays? The latter has not been so well quantified yet, but it is still obvious. Men write more books. Men give more lectures. Men ask more questions after lectures. Men dominate mixed-sex committee discussions. Men post more e-mail to Internet discussion groups. To say this is due to patriarchy is to beg the question of the behavior's origin. If men control society, why don't they just shut up and enjoy their supposed prerogatives? The answer is obvious when you consider sexual competition: men can't be quiet because that would give other men a chance to show off verbally. Men often bully women into silence, but this is usually to make room for their own verbal display. If men were dominating public language just to maintain patriarchy, that would qualify as a puzzling example of evolutionary altruism—a costly, risky individual act that helps all of one's sexual competitors (other males) as much as oneself. The ocean of male language that confronts modern women in bookstores, television, newspapers, classrooms, parliaments, and businesses does not necessarily come from a male conspiracy to deny women their voice. It may come from an evolutionary history of sexual selection in which the male motivation to talk was vital to their reproduction. The fact that men often do not know what they are talking about only shows that the reach of their displays often exceeds their grasp.

Cyrano's Panache

The verbal fireworks of male courtship are personified in the title character from Edmond Rostand's 1897 play *Cyrano de Bergerac*. Cyrano had a big nose, a big sword, and a big vocabulary. One might say they are all phallic symbols, but, given what we learned earlier about the penis, that would just identify them as sexually selected ornaments.

Much of the play concerns Cyrano's mission to convince his bookish, beautiful cousin Roxane to commit herself to the inarticulate but handsome baron Christian de Neuvillette. In preparing a translation of *Cyrano* for the New York stage in 1971, novelist Anthony Burgess noted of Roxane that "She loves Christian, and yet she rebuffs him because he cannot woo her in witty and poetic language. This must seem very improbable in an age that finds a virtue in sincere inarticulacy, and I was told to find an excuse for this near-pathological dismissal of a good wordless soldier whose beauty, on her own admission, fills Roxane's heart with ravishment." Our modern verbal displays remain a pale imitation of classic French wit—Cyrano's quatrains have given way to our anodyne psychobabble, self-help platitudes, and management buzzwords. We can be linguistically lazy now because we are surrounded by professional wordsmiths who entertain our sexual partners on our behalf: television, movie, comedy, and novel writers. We may never know whether our Pleistocene ancestors favored French-style wit, English-style irony, or German-style engineering. But they apparently favored some verbal fluency beyond the demands of flint-knapping and berry-picking.

The Cyrano story really illustrates verbal display by five males. First, the historical Cyrano de Bergerac: large-nosed 17th-century political satirist, wounded veteran, dramatist, free-thinking materialist, ridiculer of religious authority, and master of baroque prose and bold metaphors, whose *A Voyage to the Moon* of 1754 was arguably the first science-fiction novel. Second, the 19th-century playwright Edmond Rostand, whose dazzling versification throughout five acts of rhymed alexandrines secured his literary status. Third, Rostand's fictional character Cyrano, whose astonishing poetic fluency won Roxane's heart. Fourth, the play's translator, Anthony Burgess. Perhaps their lovers were equally fluent in private conversation, but we do not know, for they were not so motivated to broadcast their verbal genius to such wide audiences. The fifth male displayer is, of course, me, since I'm writing about Cyrano here. These endless chains of male verbal display constitute most of human literature and science.

Facing death at the end of the play, Cyrano's final words emphasized the similarities between ornamental bird plumage in nature, the white feather in his hat, and the style of his language:

> There is one thing goes with me when tonight
> I enter my last lodging, sweeping the bright
> Stars from the blue threshold of my salute.
> A thing unstained, unsullied by the brute
> Broken nails of the world, by death, by doom
> Unfingered—See it there, a white plume

Over the battle—A diamond in the ash
Of the ultimate combustion—My panache.

His reputation for wit and valor will outlast his death—as would his genes for those virtues, if Roxane had not secluded herself in that convent. His death-speech is a rather moving evolutionary metaphor, with the white plume of sexual selection flying high above the battleground of natural selection. This is not to suggest that Rostand of 1897 had read Darwin of 1871, only that both recognized that there is more to life than swords and noses, and more to female choice than lust for good wordless soldiers.

Poetic Handicaps

Cyrano's panache was manifest in his poetry. Literary souls sometimes praise poetry as a zone of linguistic freedom where words can swirl in dazzling flocks above the gray cityscape of pragmatic communication. A sexual selection viewpoint suggests a different interpretation. Poetry, in my view, is a system of handicaps.

Meter, rhythm, and rhyme make communication harder, not easier. They impose additional constraints on speakers. One must not only find the words to express meaning, but, to appropriate Coleridge, the right words with the right sounds in the right order and the right rhythm. These constraints make poetry more impressive than prose as a display of verbal intelligence and creativity. For example, literary scholar John Constable has noted that poetic meter is a kind of handicap in Zahavi's sense. A metric line must have a regular number of syllables. Across different poetic styles, languages, and cultures, this number is usually between six and twelve syllables. Constable showed that even successful writers such as George Eliot have trouble composing metric poetry. His evidence shows that on average they use shorter words when writing metric poetry than when writing prose, because shorter words are easier to fit together into regular line lengths. Meter imposes a measurable cost on the writer's verbal efforts, which makes it a good verbal handicap. Only those with verbal capacity to spare can write good metric lines.

Often, poetry demands a regular rhythm of stressed and unstressed syllables. This requires selecting words not only for their meaning and syllable number, but also for their stress pattern. Meter and rhythm are usually combined to form a double handicap. In iambic pentameter, for example, each line must be of exactly ten syllables, with alternating stresses on successive syllables. Moreover, poetry in many languages is expected to rhyme. Words must be selected so the last few phonemes (sound units) match across different lines. Rap musicians develop reputations largely for the ingenuity of their rhymes, especially the rhyming of rare, multi-syllabic words. Some poetic forms such as haiku, limericks, and sonnets also have constraints for the total number of lines (three, five, and fourteen, respectively). The most highly respected poetic forms such as the sonnet are the most difficult, because they combine all four rules, creating a quadruple handicap under which the poet must labor. Some poetic handicaps such as meter, rhythm, and rhyme are fairly universal across cultures, suggesting that our minds may have evolved some verbal adaptations for dealing with them. Specific forms of poetry are, of course, cultural inventions.

Good prose enhances the speaker's status. Good poetry is an even better indicator of verbal intelligence. This is why Cyrano was so impressive: we are clever enough to comprehend his wit, while acknowledging that we would have extraordinary difficulty matching it. If I had written this book in sonnets at Shakespeare's standard, you would not have understood human mental evolution any better, but you might have a higher opinion of my verbal ability.

In most cultures a substantial proportion of poetry is love poetry, closely associated with courtship effort. Poetry often overlaps with musical display, as in folk music with rhyming lyrics. Sung poetry demands the additional skill of holding a melody while maintaining meter, rhythm, rhyme, and line-number norms. In modern societies, poets who publish their work are little read, but poets who sing their work, backed up by guitars and sequencers, sell millions of albums and attract thousands of groupies. In considering whether ancestral poetry would have been considered sexually attractive, do not visualize Wallace Stevens, my favorite modernist poet, a drab New Haven insurance executive who wrote in the evenings after work. Instead, visualize Frank Sinatra, Jim Morrison, Courtney Love, or whichever songwriter/vocalist happens to be fashionable when you are reading this.

Our capacity for poetic language probably evolved after our capacity for prose. If the ability to produce good love poetry had been strongly selected during courtship ever since modern *Homo sapiens* originated a hundred thousand years ago, we would be much better at it. We would speak effortlessly in rhyming couplets, and find that quatrains of trochaic septameter take only a little effort. But we have not yet evolved the ability to handle multiple poetic handicaps very easily. Indeed, some among us may still believe that Keats rhymes with Yeats. Of course, if we had all evolved to the standard of Cyrano, then sexual selection would raise its standard again, perhaps favoring only those whose trochaic septameter quatrains were composed of alliterative word-triplets. The exact nature and number of poetic handicaps do not matter. What counts is that they function as proper biological handicaps, discriminating between those whose verbal displays can follow the rules, and those without sufficient verbal intelligence to play these bizarre word-games. At the moment, the meter, rhythm, and rhyme handicaps are sufficient hurdles that few of us can clear them.

Clearly, this analysis of poetry as a system of sexually selected handicaps aims to explain why poetry originated; it does not claim to account for poetry's content or contemporary human significance. Good poetry offers emotionally moving insights into the human condition, the natural world, and the transience of life. These psychologically appealing aspects may make it a more effective courtship display than if it droned on about nothing more than sex. (Indeed, because courtship is a way to arouse sexual interest in someone who is not already interested, courtship displays that make explicit reference to sex may be particularly unappealing.) Because humans are fascinated by many things, courtship displays can successfully appeal to human interests by talking about almost anything under the sun. This Darwinian account of poetry does not drain poetry of its meaning—on the contrary, it shows why its meaning is free to range over the entirety of human experience.

So Why Can't My Boyfriend Communicate?

For every word written in scientific journals about the evolution of our astonishing language ability, at least a hundred words have been written in women's magazines about men's apparent inability to articulate even the simplest thought or feeling. Women commonly complain that their sexual partners do not talk enough to them. If language evolved through sexual selection, and if sexual selection operates more powerfully on males than on females, you may legitimately wonder why your boyfriend or husband cannot share his feelings with you. Is it possible that, his early courtship efforts having brought success, he no longer feels driven to be as verbally energetic, interesting, and self-disclosing as he was before? The man who used to talk like Cyrano now talks like a cave-man. Once he was a poet, now he is prosaic. His verbal courtship effort has decreased.

I have already argued that effective verbal courtship is a reliable fitness indicator precisely because it is costly and difficult. Animals evolve to allocate their energies efficiently. If it took a million words to establish a sexual relationship with you, your boyfriend was apparently willing to absorb those costs, just as his male ancestors were. But if it takes only twenty words a day to maintain exclusive sexual access to you, why should he bother uttering more? His motivational system has evolved to deploy his courtship effort where it makes a difference to his reproductive success—mainly by focusing it where it improves his rate of sexual intercourse. Men apparently did not evolve from male ancestors who squandered high levels of verbal courtship effort on already-established relationships. Of course, if an established partner suspends sexual relations, or threatens to have an affair, evolution would favor motivations that produce a temporary resurgence of verbal courtship until the danger has passed. Frustratingly, a woman may find that the greater the sexual commitment she displays the less her man speaks.

This analysis may sound heartlessly unromantic, but evolution is heartlessly unromantic. It is stingy with courtship effort, stacking it heavily where it does the most good, and sprinkling it very lightly elsewhere. Human courtship, like courtship in other animals, has a typical time-course. Courtship effort is low when first assessing a sexual prospect, increases rapidly if the prospect reciprocates one's interest, peaks when the prospect is deciding whether to copulate, and declines once a long-term relationship is established. We all enjoy a desired partner besieging us with ardent, witty, energetic courtship. That enjoyment is the subjective manifestation of the mate preferences that shaped human language in the first place. As with any evolved preference, we may desire more than we can realistically get. Evolution's job is to motivate us, not to satisfy us.

So, when women universally complain about their slothfully mute boyfriends, we learn two things. First, women have a universal desire to enjoy receiving high levels of verbal courtship effort. Second, high levels of verbal courtship effort are so costly that men have evolved to produce them only when they are necessary for initiating or reviving sexual relationships. Far from undermining the courtship hypothesis for language evolution, this phenomenon provides two key pieces of evidence that support it.

The Scheherazade Strategy

Because verbal courtship is mutual, we might expect men to feel equally frustrated by women lapsing into habitual silence as a relationship ages. This seems less often lamented, either because men develop less hunger for conversation, or because women maintain their verbal courtship effort at a higher pitch for longer.

Earlier we saw that male mate choice grows stronger later in courtship, as men may be tempted to abandon a woman after she has become pregnant, and search for a new woman. In the Pleistocene age, females who could keep a useful male around for longer would have enjoyed more comfortable lives, and their children would have prospered. Through their courtship efforts, ancestral females could maintain male sexual commitment and paternal investment in their offspring. Sexual selection through male mate choice created modern women's drive to keep men sexually attracted to them over the long term. They do this, in part, by continuing to use verbal courtship long after men might prefer to read the newspaper.

The female incentives for sustained verbal courtship are illustrated by the classic Arabian folk tale of a thousand and one nights. The story goes like this. Shahriyar was a powerful Sassanid king who discovered his wife having sex with a slave. Mad with rage, he killed them both. To avoid further problems of female infidelity, he swore to sleep with a new virgin every night and to kill her in the morning. That way, no other man would have slept with her before him, and no other man could sleep with her after him. He did this for three years, until few young women were left in the city, except for the Grand Vizier's two daughters, Scheherazade and Dunyazad.

Scheherazade swore to save the women of the city from further danger, and offered herself next to Shahriyar. After Shahriyar deflowered her, Scheherazade begged him to let her say goodbye to her sister Dunyazad. Dunyazad, as previously arranged, asked Scheherazade to invent a story to help them pass their last night together in sisterly solidarity. The sultan, overcome with insomnia, agreed to hear her out. Scheherazade began a story that grew so complex and entertaining that she had still not finished it when dawn broke. Shahriyar was so enthralled by the story that he could not bear to kill the storyteller, so he agreed to spare Scheherazade 's life for one more day. The next night, the same thing happened: Scheherazade wove one story into the next, and was in the middle of a complicated plot as dawn broke. Again Shahriyar agreed to spare her life for one more day. This pattern continued for many months of storytelling and lovemaking.

After a thousand and one nights, Scheherazade had borne Shahriyar three sons, and she begged the king to allow her sons to be brought before him. Displaying the boys—a toddler, an infant, and a newborn—she asked for their sake to spare her life, observing that no other woman would love his sons as she would. The king embraced his sons and exclaimed that even before their arrival, he had fallen in love with Scheherazade for her creativity, eloquence, intelligence, wisdom, and beauty. The next morning he publicly spared Scheherazade 's life, and they lived happily together until death delivered them both to Paradise.

This story presents an uncannily accurate picture of the male mate choice pressures on ancestral human females, and the solution they apparently evolved. Shahriyar's fear of being cuckolded reflects what biologists call "paternity uncertainty": the male never knows for sure whether a female is being sexually faithful, and therefore whether his alleged children actually carry his genes. To guard against this paternity uncertainty, Shahriyar adopted an absurdly short-term mating strategy. By bedding a virgin every night, he knew she was not already pregnant with another man's child; by killing her the next morning, he knew that she would not be unfaithful in the future. This proved to be counterproductive: no heirs were produced to carry his selfish genes, and he had killed off most of the fertile women.

The pressures on Scheherazade were intense. Given a sexually jaded despot obsessed with his paternity uncertainty and caught in a pathologically short-term mating strategy, how could she elicit his long-term investment in herself and her offspring? Her verbal courtship ability proved her salvation. She invented stories that kept him entertained, and which persuaded him of her intelligence, creativity, and fitness. The thousand and one nights constitute a massive, long-term verbal courtship display. Shahriyar realized that Scheherazade's mind was an oasis of narrative fascination in his desert of sexual novelty-seeking. She made monogamy fun. She also made it pay genetically for both of them: Shahriyar's genes prospered jointly with Scheherazade's.

Evolution has extended human verbal display from the early stages of courtship through the entirety of sexual relationships. Talking keeps relationships interesting. Women use the Scheherazade strategy, but so do men. Long after partners grow overfamiliar with each other's bodies, the Scheherazade strategy—trying to keep conversations interesting throughout a relationship—keeps them from growing bored with each other's company. This probably brought mutual benefits to our ancestors. It allowed our female ancestors to keep useful males around, and it may have helped those males to overcome their sexual novelty-seeking when it became counterproductive.

As brain size increased over the last two million years, infants had to be born relatively earlier in their development so their heads could fit through the birth canal. All human babies are born prematurely relative to other primate babies. Human babies are less competent and more vulnerable at birth than almost any other mammal. This may have tipped the balance for men, making assistance to their own offspring more beneficial to their genes than seeking new mates. The sexual novelty-seeking characteristic of all male mammals was an ancient instinct, not easy to overcome. By evolving an appreciation of the cognitive novelties offered by good conversation with an established partner, men may have muted their obsession with the physical novelties of other women. This is why Shahriyar learned to listen, once Scheherazade started talking.

Language Outside Courtship

Human language did not evolve just for courtship, so that we could all talk like Cyrano and Scheherazade. It was shaped by many other selection pressures: for communication between relatives, social display to non-mates, coordination of group activities, and teaching things to children. Even if it originated as pure verbal courtship, like bird song, without any survival payoffs at all, it would soon have proved its other virtues. As Terence Deacon and others have observed, it is hard to imagine any social activity that would not benefit from language. The frustrations of visiting places where people speak foreign languages reveal the survival and social benefits of effective communication.

But a frustration is not the same as a selection pressure. We must remember that any theory of language's other social benefits must explain its apparent altruism with some hidden genetic benefit. If those hidden benefits turn out to be sexual, then we are back where we started. Much of the effort invested in apparently non-sexual uses of language may work as indirect courtship. Social display to non-sexual partners can improve one's mating prospects. Opposite-sex friends may become lovers, same-sex friends may have eligible sisters or brothers, and high-status tribe members impressed with your charms may gossip about you to others. Having a good reputation gives one a huge advantage before courting someone, and the two things that contribute to a good reputation are good words and good actions.

Language is useful in coordinating group activities, but here again we have an altruism problem. In the chapter on morality we saw that group benefits like big-game hunting and moral leadership could be favored by sexual selection. If an individual's ability to improve group success through verbal leadership is judged by potential mates, then apparently cooperative uses of language may conceal courtship functions.

Even when non-sexual pressures started to shape human language, sexual selection would have subverted those pressures. This is because sexual choice tries to preempt the effects of natural selection as much as possible. For example, consider language as a way to teach children about plants and animals. Survival selection might favor such pedagogy—one's children would be less likely to die of poisons and bites. Yet individuals might vary in teaching ability. If their differences remain genetically heritable (as they probably would, given the pressures of mutation on complex traits), and if teaching ability was reasonably important, sexual preferences would evolve to favor that ability. Individuals who mated with good teachers would produce children who taught their grandchildren more efficiently, allowing more grandchildren to carry one's genes forward. The ancestral versions of

David Attenborough would have been perceived as sexually charismatic, not just as good parents. At that point, teaching ability would have been favored by both survival selection and sexual selection.

Fact and Fantasy

Scheherazade attracted her sultan with fantasies. If sexual choice shaped language as an entertaining ornament and a fitness indicator, why does language have any factual content at all? Other sexually selected signals such as the songs of birds and whales do not say anything other than "I am fit—mate with me." We saw earlier that life stories, social gossip, and large vocabularies can work as good fitness indicators. They all demand content. But they do not seem to demand enough factual content to explain our interest in the truth, or the efficiency of language as a communication medium.

I think that, as with human morality, there was an equilibrium selection process at work. Every possible sexual signaling system can be viewed as an equilibrium in the grand game of courtship. There are more than a million sexually reproducing species on Earth, each with their own sexual signals. That means there are more than a million possible equilibria in the courtship game. At each equilibrium, individuals are displaying the best signals they can, and choosing the best mates they can, and nobody has any incentives to deviate from what they are already doing. In the vast majority of equilibria—(i.e. species)—apparently more than 99.9 percent of them—sexual signals convey no information other than fitness information. They are pure fitness indicators. Human language is the only signaling system that conveys any other sort of information in courtship. It is still a fitness indicator, but it is much more as well.

The Scheherazade problem is this: there could be "fantasy" equilibria where people impress mates by making up stories about fictional worlds, and "fact" equilibria where people impress mates by displaying real knowledge of the real world. As long as both displays are good fitness indicators, sexual selection should not favor fact over fantasy. Was it just blind luck that we ended up on a relatively factual equilibrium, where people care about truth and knowledge?

Imagine a fantasy equilibrium where verbal courtship display consists exclusively of spinning wild stories about battles waged with magic spells between wizards from alien civilizations. Individuals talk about nothing else. If the ability to invent wizard stories was a good fitness indicator, sexual selection would be perfectly happy with this equilibrium. The pointless waste of breath talking about wizards would not worry sexual selection any more than the peacock's tail does.

The trouble with a purely fantasy equilibrium is that the individuals would literally not know what they are talking about. How would they learn what any of their words mean? Their words refer only to fictional magic spells from alien civilizations. Their parents could not take them a hundred light-years away, point to a magic spell that creates a lethal hail of neutron stars, and say, "Look, that's a xoplix!" Words must be grounded in the real world in order to have any meaning. Humpback whale songs might accidentally be referring to actual events on alien worlds, but we wouldn't know, and neither would they. No animal playing a purely fantasy equilibrium could tell it from an ordinary fitness-indicator equilibrium.

The only way a signal can activate a concept in another individual's head is for the signal to be grounded, directly or indirectly, in some real-world meaning. This excludes all purely fantasy equilibria. Scheherazade's stories recombined real-world ideas in fantastic ways. She did not refer exclusively to fictional ideas. I suspect that there are only two kinds of sexual-signaling equilibria that are evolutionarily stable, in any naturally evolved species anywhere in the universe: pure fitness indicators, and language systems that make reference to objects and events in an organism's perceivable environment.

Scheherazade Versus Science

Language must be grounded in reality, but how tightly grounded? Sexual selection still has elbow room to favor Scheherazade equilibria (fantastic stories based on recognizable objects) or science equilibria (useful, true descriptions of the world). Now I am no longer sure which equilibrium our species is playing. Most people in most cultures throughout most of history have talked reasonably accurately about ordinary objects, people, and events, but they have talked absolute fantasy about astronomy, cosmology, theology, and any other phenomena that could not be directly observed.

One might think that a group of individuals playing a science equilibrium would out-compete a group playing a Scheherazade equilibrium, because science brings survival benefits. Wouldn't group competition favor sexual displays concerning falsifiable hypotheses and empirical facts, rather than sexual displays concerning Aladdin and his genie? The science-displayers would develop useful theories about the world as a side-effect of their sexual status games. The Scheherazades would not. The science-displaying groups should have had competitive advantages.

Indeed they have, but only in the last five hundred years. For all of human evolution we muddled along playing half-fact, half-fantasy games with our language. We learned useful words, certainly, but then immediately invented as many useless synonyms as possible so we could display our vocabulary sizes. We learned useful facts about other individuals through gossip, and then immediately embellished and distorted those facts to make more entertaining stories. We revealed our life stories, but only the good bits, and only as if we were always the protagonist, and never the chorus.

Language evolved as much to display our fitness as to communicate useful information. To many language researchers and philosophers, this is a scandalous idea. They regard altruistic communication as the norm, from which our self-serving fantasies might sometimes deviate. But to biologists, fitness advertisement is the norm, and language is an exceptional form of it. We are the only species in the evolutionary history of our planet to have discovered a system of fitness indicators and sexual ornaments that also happens to transmit ideas from one head to another with telepathy's efficiency, Cyrano's panache, and Scheherazade's delight.

11

The Wit to Woo

To many people, "evolutionary psychology" implies "genetic determinism." This common error makes it hard to understand how there could be an evolutionary account of human creativity. Darwin proposed his theory of natural selection to account for the existence of complex order, such as the structure of the eye. Yet creativity implies the generation of novel, unpredictable, non-deterministic behavior—apparently the opposite of order. Whereas the eye's structure makes parallel light-rays converge to a point, creativity makes ideas diverge in all directions. Creativity seems too chaotic, both in its mental processes and its cultural products, to count as a biological adaptation in the traditional sense. So how could it have evolved?

This chapter reviews how evolution favors unpredictable behavior in many animals, and suggests that these capacities for randomness may have been amplified into human creativity through sexual and social selection. We shall see that behaviors are often randomized by evolutionary design, not by accident. Creativity is not just a side-effect of chaotic neural activity in large brains: it evolved for a reason, partly as an indicator of intelligence and youthfulness, and partly as a way of playing upon our attraction to novelty. By understanding how natural selection can favor unpredictable strategies in competitive situations, we may better understand how sexual selection could favor the benign unpredictability of creativity and humor in courtship.

Evolution Against Genetic Determinism

Ever since the first nervous systems evolved, evolution has been striving to overcome "genetic determinism"—the direct coding of behaviors in genes. No scientist believes that genes preprogram every single behavior demonstrated by an organism during its lifetime. Evolution avoids such preprogramming by endowing animals with senses for registering what is going on in the environment, and reflexes for letting those senses influence movements. These senses and reflexes allow behavior to track environmental variables faster than genetic evolution can. One key variable is the location of food. A flatworm's eyes can notice that food is available in a certain location, without having to wait for the flatworm species to evolve the belief that food is there. If you believe in the existence of senses and spinal cords, you are not a genetic determinist in the strict sense.

Evolution did not stop with eyes and spines. It took perfectly good spinal cords and expanded their first several segments into great bastions of antideterminism called brains, then added layer upon layer of thinking and feeling between sensory input and motor output. The job of evolutionary psychology is to analyze how evolution constructs these mental adaptations that turn environmental cues into fitness-promoting behaviors. The larger the brain, the more sophisticated the environmental cues it can use to guide behavior, and the more sophisticated that behavior. Into the grand, generation-long cycle of genetic evolution, brains insert millions of faster feedback loops. On a second-by-second basis, senses and brains track new opportunities to promote survival and reproduction. Their whole reason for existence is to keep genes from having to change every time the environment does.

Genes rarely determine specific behaviors, but they often determine the ways in which environmental cues activate behaviors. Many behaviors are fairly predictable if you know what an organism is perceiving at the moment. This predictability comes from the demands of optimality: for any given environmental situation, there is often one best thing to do. Animals that do the right thing survive and reproduce better; animals that deviate from optimal behavior tend to die. This pressure for optimal behavior makes many behaviors predictable.

However, there are situations in which it is a very bad idea to be predictable. If another organism is trying to predict what you will do in order to catch you and eat you, you had better behave a bit more randomly. Selection may favor brain circuits that randomize responses, to produce adaptively unpredictable behavior. The benefits of randomization were first understood in a deep way by game theorists. What they said about randomization will help us understand human creativity later.

Matching Pennies

John von Neumann had an astonishingly creative mind, even compared with other Hungarian mathematicians. By the age of 30 in 1933, he had developed the modern definition of ordinal numbers, specified an axiomatic foundation of set theory, and written a standard textbook on quantum physics. When he worked on the Manhattan Project, he had a key insight about how to make the atomic bomb work, and he also originated a fundamental concept of computer science, the "Von Neumann architecture." But these were just warm-up exercises for his work on the theory of games, which became the foundation of both modern economics and modern evolutionary biology.

Von Neumann realized that many games are best played by randomizing what you do at each step. Consider a game called "Matching Pennies." In this game, there are two players, and they each have a penny. At each turn, each player secretly picks heads or tails: they turn their pennies heads-up or tails-up under their hands. Then the coins are revealed. If the first player, in the role of "matcher," has turned up the same side as the opponent (e.g. if both coins are heads), then the matcher wins the opponent's penny. If the coins don't match (e.g. if one is heads, the other tails), then the matcher must give a penny to the opponent. The first play is not so interesting, but as the game is repeated, one can form predictions about the opponent's behavior. The possibility of prediction makes Matching Pennies a strategically intricate game.

The roles of "matcher" and "non-matcher" seem different, but their goals are fundamentally the same: predict what the opponent will do, and then do whatever is appropriate (matching or not matching) to win the turn. All that matters is to find out the opponent's intentions. The ideal offensive strategy is to be the perfect predictor: figure out what the opponent is doing based on his or her past behavior, extrapolate that strategy to the next move, make the prediction, and win the money. But there is an easy way to defeat this prediction strategy: play unpredictably. Von Neumann remarked, "In playing Matching Pennies against an at least moderately intelligent opponent, the player will not attempt to find out the opponent's intentions, but will concentrate on avoiding having his own intentions found out, by playing irregularly heads and tails in successive games."

In particular, if a player picks heads half the time and tails half the time, then no opponent, no matter how good a predictor he or she is, can do better than break even in this game. This half-heads, half-tails strategy is an example of what game theorists call a "mixed strategy," because it mixes moves unpredictably. In their seminal 1944 book *The Theory of Games and Economic Behavior*, John von Neumann and Oskar Morgenstern proved an important theorem. Roughly speaking, they showed that in every competitive game between two players that has more than one equilibrium, the best strategy is mixed. We have already seen in the chapters on morality and language that many important games have more than one equilibrium. We know from evolution how important competition is. The theorem implies that when any two animals are interacting and they have a conflict of interest, they would often do well to randomize their behaviors at some level. When being predictable can make you lose a penny, unpredictability is recommended. When being predictable can make you lose your life to a predator, unpredictability is highly recommended.

The importance of randomness has long been appreciated in military strategy, competitive sports, and poker. In World War II, submarine captains sometimes threw dice to determine their patrol routes, generating a zigzagging course that would not be predictable to enemy ships. Some modern fighter aircraft are equipped with "electronic jinking" systems that can automatically randomize their evasion maneuvers when guided missiles try to intercept them ("jinking" means zigzagging very abruptly and randomly). Professional tennis players are coached to "mix it up" when they serve and return shots. Plays in American football are carefully randomized to be unpredictable. Random drug tests make it harder for Olympic athletes to predict when they can abuse steroids. These are all "mixed strategies" that work by being unpredictable. Game theory showed the common rationale for randomness in many situations like these, where players have conflicts of interest and benefit from predicting each other's behavior.

Strategic Randomness in Biology

In 1930, Sir Ronald Fisher showed that animals play a game similar to Matching Pennies. They must evolve a strategy to determine whether to produce male or female offspring. If an animal could predict which sex will be in higher demand in the next generation, it could gain an advantage by producing the rarer, more sought-after sex. In an all-female population, a single male could do very well, spreading his genes through the entire gene pool in one generation. Likewise for a female in an all-male population. So, should animals try to out-predict their evolutionary opponents? Fisher said no. As in Matching Pennies, the best they can do is to randomize, by producing half males and half females. The sex ratio is balanced strategically, not because there is some biological law that says it has to be a 50/50 split. (As W. D. Hamilton showed, in some parasites with unusual mating systems, the optimal strategy is some other ratio, such as 3 males to 11 females, and such species duly evolve that biased sex ratio.)

At the level of behavior, biologists were slower to recognize the uses of randomness. In 1957 Michael Chance published a minor classic titled "The role of convulsions in behavior." Researchers had long been puzzled by the fact that laboratory rats sometimes go into strange convulsions when lab technicians accidentally jangle their keys. Why should certain sounds induce seizures that look so maladaptive, resulting in rats injuring themselves against the cage walls? Chance found that the rats were responding to the key-jangles as if they indicated the approach of a dangerous predator. If provided with hiding places (little rat-huts) in their cages, they simply ran and hid when keys were jangled. Only if they had nowhere to hide did they go into convulsions. The convulsions may therefore have evolved as last-ditch defensive behaviors rather than pathologies. Wild convulsions, including "death throes," would make it harder for predators to catch and hold the convulser. The aptly named Dr. Chance argued that rats evolved defensive strategies that exploit randomness.

Shortly after Chance's work on rats, Kenneth Roeder found that bat sounds can induce similarly randomized behavior in moths. Bats eat moths, locating them at night by chirping and listening for ultrasonic echoes. If you're a moth, and you suddenly get hit by a blast of ultrasound, you can be pretty sure a gaping bat-mouth is close behind. Roeder found that moths in this situation produce an extraordinarily unpredictable range of evasive movements, including tumbling, looping, and power dives. Moth genes for predictable behavior usually got digested in bat stomachs rather than passed on to baby moths.

Protean Behavior

In 1970, British ethologists P. M. Driver and D. A. Humphries suggested that these rat and moth behaviors were examples of "protean behavior." They named this kind of adaptive unpredictable behavior after the mythical Greek river-god Proteus. Many enemies tried to capture Proteus, but he eluded capture by continually, unpredictably changing from one form into another—animal to plant to cloud to tree. Driver and Humphries's 1988 book *Protean Behavior: The Biology of Unpredictability* presented a detailed theory of randomized behavior, supported by a wide range of field observations. Unfortunately they did not make the connection to mixed strategies in game theory, so these prophets of genetic indeterminism did not have the influence they deserved in evolutionary theory.

The logic of proteanism is simple. If a rabbit fleeing from a fox always chose the single apparently shortest escape route, the consistency of its behavior would make its escape route more predictable to the fox, its body more likely to be eaten, and its genes less likely to replicate. Predictability is punished by hostile animals capable of prediction. Instead of fleeing in a straight line, rabbits tend to zigzag erratically—a protean escape behavior that makes rabbits much harder to catch. Like the moth, the rabbit probably evolved special brain mechanisms to randomize its escape path.

Protean escape is probably the most widespread and successful adaptation against being eaten by predators, and is used by virtually all mobile animals on land, under water, and in the air. Proteanism explains why it is harder to predict the movements of a common housefly for the next ten seconds than the orbit of Saturn for the next ten million years. Yet there is more to proteanism than escape behavior. The effectiveness of almost any behavior can be enhanced by making its details unpredictable to evolutionary opponents. For example, predators also use proteanism to confuse prey. When a weasel is stalking a vole, it may do a "crazy dance." The weasel jumps about like a mad thing, chases its tail, shakes its head, licks its feet, all the while positioning itself closer and closer to its bemused prey. The seemingly pointless series of weird actions baffles the vole. The vole is caught in a web of confusion. Australian aborigine hunters did similar wild dances to mesmerize the kangaroos they hunted. Perhaps our hominid ancestors did too.

Animal play behavior also reveals the importance of proteanism. Most animal play is play-chasing and play-fighting. At the level of movement patterns, play is a way of practicing pursuit and evasion. But at the psychological level, it is a way of practicing prediction and proteanism.

Unpredictability can be useful at many levels. When threatened, octopuses and cuttlefish use "color convulsions."

Their pigmented skin cells, which are under direct control of the nervous system, display an unpredictable series of color patterns to confound the perceptual expectations of predators. One moment the cuttlefish has black stripes, the next it has red spots, which makes it hard for predators to keep in mind what they're supposed to be chasing. The lesson of proteanism is very general: whenever one animal benefits from being able to predict something about another animal's behavior or appearance, the second animal might benefit from making its behavior or appearance unpredictable.

Proteanism Versus Science

Proteanism may be one reason why the behavioral sciences are so much better at description than prediction. We can sometimes explain behavior after the fact, and can often make statistical predictions about average future behavior. But it is almost impossible to predict whether a particular rabbit in a particular situation will hop left or right.

The physical sciences offer many examples of unpredictability, but it is usually there by accident, not design. Quantum theory accepted the "noisiness" of elementary particles. But it did not assume that the randomness was put there just to frustrate physicists. Chaos theory showed that the behavior of many systems is very sensitive to the starting conditions. Many systems that unfold deterministically over the short term become unpredictable over the long term. But chaos theory does not attribute any strategic intention to chaotic systems. The behavioral sciences have tried to follow the physical sciences in this regard, viewing unpredictability as noise. If the same animal in the same situation does different things on different occasions, this is usually considered to be behavioral "noise." Yet that is exactly what moths and rabbits evolved to do—avoiding predators through unpredictability. Psychology's favorite brand of statistics, called the analysis of variance, assumes that all behavior can be explained as the interaction of environmental determinants and random, nonadaptive noise. There is no place for proteanism in the analysis of variance, because analysis of variance does not distinguish between random errors and adaptive unpredictability.

Proteanism does not fit into this framework of scientific explanation. It is both adaptive and noisy, both functional and unpredictable—like human creativity. The difficulty of predicting animal behavior may be much more than a side-effect of the complexity of animal brains. Rather, the unpredictability may result from those brains having been selected over evolutionary history to baffle and surprise all of the would-be psychologists who preceded us. To appreciate why psychology is hard, we have to stop thinking of brains as physical systems full of quantum noise and chaos, or as computational systems full of informational noise and software bugs. We have to start thinking of brains as biological systems that evolved to generate certain kinds of adaptive unpredictability under certain conditions of competition and courtship. If you're not looking for proteanism, you won't find it.

How Proteanism Works

Proteanism does not imply that all of your brain cells are firing randomly in total cortical anarchy. The randomness is injected into your behavior at a particular level appropriate to the situation. If you are fleeing "randomly," your trajectory through the environment may be unpredictable. But you are still maintaining order at many other levels: coordinated nerve firings to activate muscles, coordinated muscle movements to power limbs, coordinated limb movements to maintain an efficient gait, and eye–foot coordination to avoid obstacles. Proteanism implies the strategic ability to use randomness just when it is needed to make yourself unpredictable. It does not imply a masochistic enslavement to Fortuna, the pagan goddess of chance. Here proteanism foreshadows human creativity, since creativity implies the strategic use of novelty to achieve a social effect, not the random combination of random ideas in a chaotic style.

A capacity for proteanism in one situation does not imply an ability to act like a random number generator in all situations.

Psychologists have tested human capacities for "randomness" since the 1950s, but they have usually given paper-and-pencil tests that do not tap into natural proteanism abilities. For example, when people are asked to write down a random series of "heads" or "tails" on paper, they fail statistical tests of randomness: they alternate too much (heads, tails, heads, tails) and do not produce enough long runs (heads, heads, heads, heads). By the mid-1970s, after dozens of experiments on the generation of random series, psychologists came to believe that people are hopelessly bad at randomizing their responses.

However, these tests did not usually provide any incentives to behave randomly. When incentives are provided, people do rather better. In the 1980s, psychologist Alan Neuringer found that rats and people can produce almost perfectly random sequences when given good feedback and good incentives for performance. Also, the social situation matters. Amnon Rapoport and David Budescu found that when people play Matching Pennies for real money, they get very good at randomizing very quickly. You do not even have to tell them to randomize. They just do it naturally, to be unpredictable.

When I give talks on protean behavior, I usually ask two members of the audience to play Matching Pounds. This is like Matching Pennies, but played for higher stakes: British £1 coins. I give the players ten pounds each, and they can walk away with whatever they win after they have played ten rounds. The chance to win as much as £10 in five minutes concentrates the minds of British academics wonderfully. The resulting drama of prediction, counterprediction, greed, fear, frustration, and incredulity is something to behold. I do not instruct the players to behave randomly; they just figure out that they had better do so. Those who alternate too predictably between heads and tails quickly lose £3 or £4 to their opponents. Most players learn that it is much easier just to randomize than to try to out-predict one's opponent. Our innate capacities for proteanism reveal themselves only in strategic situations where unpredictability becomes important to behavior.

Normal people can randomize pretty well, but autistic people cannot. Psychologist Simon Baron-Cohen found that people with autism are very poor at randomizing their strategies in games like Matching Pennies. He suggested this was because they lack the "theory of mind" that ordinary people use to understand the beliefs and desires of other people. Autistic people seem unable to realize that other people can form predictions about what they will do next, so they usually alternate heads and tails in a totally predictable way. Randomizing your strategies in new situations seems to require the ability to understand that opponents are trying to predict your moves. Of course, rabbits don't need to understand fox minds in order to zigzag unpredictably, because the rabbits evolved brain circuits dedicated to playing the evolutionarily ancient game of pursuit and evasion. They do not need a theory of mind in order to zigzag when frightened. A theory of mind may be required for proteanism only when we are playing evolutionarily novel games such as Matching Pennies.

Protean Primates Versus Machiavellian Mind-Readers

In the 1990s primate researchers became enthusiastic about the idea of Machiavellian intelligence—the ability that apes and humans have evolved for predicting and manipulating the behavior of other individuals. Apes and humans live in social groups where one's survival and reproduction prospects depend on one's social relationships. Once primatologists understood evolution from the selfish-gene viewpoint, they saw social interaction in a new light. Before, social behavior was thought to be for "pair-bonding" and "group cohesion." Now, it became viewed as a strategic game of politics, alliances, reciprocity, kinship, aggression, and peacemaking. A key to success in these strategic games is the ability to predict the behavior of other individuals. The Machiavellian intelligence theory suggests that great apes evolved larger brains and higher intelligence to better predict one another's behavior.

Suppose this view is right. Would evolution stop there, with everyone able to predict and manipulate everyone else's behavior? Or would counter-strategies evolve? In a society of Machiavellian psychoanalysts, individuals that are harder to predict and manipulate would have the usual protean advantages.

In their important 1984 paper on "mind-reading and manipulation," John Krebs and Richard Dawkins identified only two defenses an animal might use against having its actions predicted by an opponent: concealment and deception. You can try to hide your intentions (the poker-face strategy), or you can create a false impression about your intentions (the bluffing strategy). However, they overlooked the classic third option: randomness. The protean strategy. Doubtless each of these strategies is useful under particular conditions, and in a species with high Machiavellian intelligence, all of them would evolve. However, the protean strategy has one big advantage: it stops prediction dead in its tracks. The poker-face and bluffing strategies remain vulnerable to the evolution of better intention-sensing and deception-foiling abilities. But there is no way to improve prediction when you meet genuine randomness.

The Mad Dog Strategy

Despots throughout history have often used a form of social proteanism to maintain power. They have unpredictable rages that terrify subordinates. Caligula, Hitler, and Joan Crawford were all alleged to have increased their power over underlings through this "mad dog strategy," which keeps subordinates in line by imposing stressful levels of uncertainty on them.

Imagine a despot who had a fixed threshold for getting angry. Subordinates could quickly learn that threshold and do anything just below the anger threshold with impunity. If King Arthur only got upset by knights actually having sex with Queen Guinevere, the knights could still court her, kiss her, and plot with her. But if Arthur's anger-threshold was a random variable that changed every day, subordinates could never be sure what they could get away with. Maybe he was happy for them to carry her flag at the joust yesterday, but maybe he will chop off their heads for even looking at her today.

Against the mad dog strategy, any insult, however slight, risks retaliation. But mad dog despots don't incur the time and energy costs of having a fixed low anger threshold—the uncertainty does most of the work of intimidating subordinates. Despotism is the power of *arbitrary* life and death over subordinates. If a despot can't kill people at random, he isn't a real despot. And if he doesn't kill people at random, he probably can't retain his despotic status. Social proteanism lies at the root of despotic power.

The mad dog strategy is just the most dramatic example of how unpredictability can bring social benefits. The advantages of an unpredictable punishment threshold also apply to sexual jealousy, group warfare, and moralistic aggression to punish antisocial behavior. Fickleness, moodiness, inconstancy, and whimsy may be other manifestations of social proteanism. However, we need more research on human and ape capacities for adaptively unpredictable social behavior. Given the importance of mixed strategies in game theory, and the fact that many social interactions can be interpreted as games, it would be surprising if randomized behaviors did not play a large role in human social interaction.

If great apes differ from monkeys in having better social prediction abilities, it seems likely that they would also have evolved better social proteanism abilities to avoid being predictable. How does this relate to human creativity? The mad dog strategy sounds sexually repulsive, not the sort of behavior that sexual choice might favor. Yet I shall argue that the same capacities for strategic randomization that underlie the mad dog strategy were transformed, through sexual selection, into our human capacities for creativity, wit, and humor. There are at least three ways that social proteanism may have smoothed the way for human creativity to evolve. One has to do with the brain mechanisms underlying creativity, the second with sexually selected indicators of proteanism ability, and the third with playfulness as an indicator of youthfulness.

Random Brains

Social proteanism may have provided a set of brain mechanisms for randomizing that could have been modified to play an important role in human creativity. Proteanism depends on the capacity for the rapid, unpredictable generation of highly variable alternatives. Creativity researchers agree that creativity depends on exactly this sort of mechanism, though they disagree about whether to call it "divergent thinking," "remote association," or something else. As far back as 1960, psychologist Donald Campbell insisted on the importance of randomness in creativity. He saw an analogy between creative thought and genetic evolution: both work through an interplay between "blind variation" and "selective retention." It is fairly clear how the brain might do the "selective retention" using well-studied aspects of judgment, evaluation, and memory. But how could the brain produce large numbers of "mutant" ideas when creativity is demanded?

Perhaps brain areas that originally evolved for proteanism were modified in the service of creativity. Instead of randomizing escape plans and social strategies, these brain areas might have been re-engineered to randomly activate and recombine ideas. As with all forms of proteanism, this random activation would happen at the appropriate level of behavior. If one is improvising jazz music, one might activate random melody fragments and very quickly sort through them using various unconscious filters. One would not activate random memories of life events, random limb movements, or random moral ideals.

It is hard to test such a theory at the moment, but it will become easier with advances in neuroscience and behavior genetics. The theory that creativity derives from proteanism suggests that some of the same brain systems should be active in playing Matching Pennies and in doing various creative tasks. It also suggests that some of the same genes associated with high randomization abilities in strategic games should also predict high creativity (after controlling for general intelligence, of course). However, this random-brain theory is not very satisfying, because it does not identify what selection pressures favored creativity. To do that, we have to ask why evolution would favor amplified displays of the brain systems used in proteanism.

Creativity as a Display of Proteanism

A second way to connect proteanism to creativity is through indicator theory. If proteanism was important to survival and reproduction among our group-living ancestors, then mate choice would have created the usual incentives to pay attention to it. In particular, individuals who showed better social proteanism abilities should have been favored as sexual partners, because their offspring would inherit these abilities, which would confer social benefits. Once sexual selection started focusing on proteanism as a criterion for mate choice, reliable indicators of proteanism might evolve. Any social behavior that clearly demonstrated randomization ability would tend to be included in courtship.

Some forms of everyday creativity, especially humor, could be viewed as proteanism displays. They harness randomization abilities in the service of courtship, not competition. When your train of thought proves fascinatingly unpredictable to a potential mate, perhaps you are also showing that your social strategies can be devastatingly unpredictable to your social competitors. Creativity displays make unpredictability attractive, not intimidating. Perhaps creativity evolved through sexual selection as a reliable indicator of social proteanism ability.

This idea makes some of the same predictions as the first hypothesis about brain systems for randomization. It suggests that individuals who are poor at social proteanism should be poor at creativity. However, I don't find this idea completely satisfactory, because social proteanism abilities may have been less important than other abilities. Given a choice between an individual good at randomizing when he attacks and a very strong individual capable of winning any attack, mate choice may favor the strong over the random. The pressures of social competition may have been strong enough to favor good proteanism abilities, but it is not clear that they were so strong that sexual selection would have favored specific indicators of proteanism. Let's consider a third possible way to connect proteanism to creativity.

Playfulness as a Youth Indicator

Most mammals start out cute, playful, and innovative, and gradually become grim, pragmatic, and habit-ridden. Ashley Montagu and many others have observed that humans retain some aspects of juvenile playfulness longer into adulthood. This has been considered one of the prime symptoms of human "neoteny," the slowing-down of behavioral maturation relative to physical maturation. The traditional explanation for human neoteny is that slower cognitive development might permit a longer period of useful learning. Certainly there may have been good reasons for specific kinds of social learning to persist longer into adulthood over hominid evolution. But I see no reason why this would generalize into the sort of playfulness that we see in adult humans but not in adult chimpanzees.

Playfulness has large time and energy costs. Indeed, biologists struggled for a long time to identify what possible benefits could offset the costs of play behavior, even for young animals. A consensus has emerged that most animal play is practice. Play-fighting, play-chasing, and play-fleeing are ways of practicing some of the most important skills that adult animals need for competing, eating, and avoiding being eaten. But once these basic skills are mastered, what possible selection pressure could favor the retention of playfulness into adulthood?

One clue is that adult human playfulness is not uniform across all situations. When human hunter-gatherers are foraging, they do not walk playfully like John Cleese in the Monty Python "Ministry of Silly Walks" sketch. They walk along with the silent, steady efficiency of any other adult mammal making its living. But when they are socializing in a group—especially a mixed-sex group—they may very well hop, skip, jump, and do the Chicken Walk.

Playful, creative behaviors could function as indicators of

youthfulness. Their persistence into human adulthood may be not a side-effect of neoteny, but a result of direct sexual selection for youth indicators. We have already seen how large human breasts may have evolved as youth indicators. The same reasoning would work here for playfulness and creativity: if playfulness usually decreases from juveniles to older adulthood for all mammals, then playfulness may be a reliable cue of youthfulness, health, and fertility.

Playfulness is also a general fitness indicator. The energy and time costs of play were sufficient to make biologists wonder why play could ever have evolved even in young animals. These costs do not go away for adults—if anything, they increase. Juveniles have to compete only for survival, but sexually mature adults also have to compete sexually and take care of offspring. The costs of playfulness for adults with so many demands on their time and energy may be higher than the costs for juveniles. And as adults grow older, the relative energy costs of playfulness must keep increasing. Middle-aged and older adults often revert to the playfulness of youth if they fall in love again with someone new, though their playfulness does not usually show the same incandescent physical energy as that of young adults. Thus, the costs of playfulness generally increase as age increases, and this makes playfulness a potentially reliable indicator of youth, fertility, energy, and fitness.

Still, creativity is a mental capacity, whereas play is a physical manifestation of creativity. It is easy to see how running around and acting playful for several hours could be favored by sexual selection as a fitness indicator. It is less clear how the quieter forms of creativity could be favored. They are not necessarily manifest in whole-body movements. They may be displayed mostly in verbal courtship, which has low energy costs. Creativity may also be displayed in art or music, which only have moderate performance costs.

However, there is good evidence that even less physical forms of creativity can work as energy indicators. Psychologist Dean Keith Simonton found a strong relationship between creative achievement and productive energy. Among competent professionals in any field, there appears to be a fairly constant probability of success in any given endeavor. Simonton's data show that excellent composers do not produce a higher proportion of excellent music than good composers—they simply produce a higher total number of works. People who achieve extreme success in any creative field are almost always extremely prolific. Hans Eysenck became a famous psychologist not because all of his papers were excellent, but because he wrote over a hundred books and a thousand papers, and some of them happened to be excellent. Those who write only ten papers are much less likely to strike gold with any of them. Likewise with Picasso: if you paint 14,000 paintings in your lifetime, some of them are likely to be pretty good, even if most are mediocre. Simonton's results are surprising. The constant probability-of-success idea sounds very counterintuitive, and of course there are exceptions to this generalization. Yet Simonton's data on creative achievement are the most comprehensive ever collected, and in every domain that he studied, creative achievement was a good indicator of the energy, time, and motivation invested in creative activity.

Creativity and Intelligence

People's scores on psychological tests of creativity are correlated with their scores on standard intelligence tests. The correlation is moderate, but not perfect. In particular, high intelligence appears to be a necessary but not sufficient condition for high creativity. Many creativity researchers believe that people who become famous for their "creativity" usually have an IQ of at least 120. The evidence from psychological testing implies that creativity is a rather good indicator of general intelligence, not just an indicator of youthfulness and proteanism ability.

A similar story comes from behavior genetics. The heritability of creativity is fairly modest, much lower than that of general intelligence. In studies that look at both creativity and intelligence together, the heritability of creativity appears to be carried almost entirely by the heritability of intelligence. In this respect, creativity is like vocabulary size: it looks heritable in its own right, but it is probably heritable only because it depends so strongly on general intelligence, which is highly heritable.

So, what is this "general intelligence"? I have mentioned intelligence repeatedly throughout this book as an important criterion of mate choice, but I have not discussed it explicitly in much detail. There are two reasons for this. First, intelligence research remains controversial. A few vocal critics who do not understand modern intelligence research have had an undue influence on public opinion. Despite the fact that more is known about the nature, importance, and genetics of intelligence than about almost anything else in psychology, I did not want to get side-tracked into such debates. Perhaps my ideas are already controversial enough. Second, I am still thinking about the relationships between intelligence, fitness, genes, and sexual selection. I can make some plausible guesses about how they may have interacted during human evolution, but these guesses should be taken as provisional speculations.

Perhaps what psychologists call "general intelligence" or "the *g* factor" will turn out to be a major component of biological fitness. If so, the high heritability of general intelligence may reflect, in part, the heritability of fitness itself. There are a few pieces of evidence that support a link between general intelligence and biological fitness. A recent study at the University of New Mexico found a 20 percent correlation between performance on an intelligence test and a compound measure of body symmetry. Body symmetry is often used as a proxy for heritable fitness, so this result suggests that there is a relationship between general intelligence and heritable fitness. Intelligence is also known to correlate positively with body height, physical health, longevity, and social status. These intercorrelations may arise because all these traits tap into biological fitness to some extent. Much more research needs to be done to address this question, however.

If the correlation between intelligence and fitness holds up, then any intelligence indicator may work as a reliable fitness indicator. If that is true, then any creative behavior that depends on intelligence can work as a fitness indicator too. Cyrano's creativity may have evolved for the same reason as his vocabulary size: to advertise his fitness to potential mates.

Neophilia

Creativity may have evolved as a sexually selected indicator of proteanism ability, youthful energy, and intelligence, but that still does not explain what is distinctive about creativity. Creative people are delightful because they are full of surprises. They produce novelty. They are unpredictable, but in a good way. To account for creativity's psychological appeal, perhaps we should consider the charms of novelty.

Neophilia, an attraction to novelty, runs deep in animal brains. Brains are prediction machines. They run an internal model of what is happening in the world, and pay attention when the world deviates from their model. Violations of expectation attract attention. Attention guides behavior to adjust the world to one's desires, or guides learning to adjust one's world-model to reality. Both functions of attention are crucial to the effectiveness of nervous systems as behavior-control systems, and both depend on registering violations of expectation. Sensitivity to violations of expectation can be shown even in very small, primitive nervous systems.

The attention-attracting power of novelty is one of the most fundamental psychological biases that could have influenced the evolution of courtship displays. In *The Descent of Man*, Darwin observed that "It would even appear that mere novelty, or change for the sake of change, has sometimes acted like a charm on female birds, in the same manner as changes of fashion with us." In Darwin's view, novelty-seeking was an irrepressible force in sexual selection that could account for the rapid evolution of sexual ornaments. In recent years more direct evidence has emerged for neophilia in mate choice. Females of several bird species have been found to prefer males who display larger song repertoires, with greater diversity and novelty. Such neophilic mate choice may account for the creativity of male blackbirds, nightingales, sedge warblers, mockingbirds, parrots, and mynahs.

Primates are especially neophilic, as illustrated by the fictional chimpanzee "Curious George." They are playful, exploratory, and inventive. Apes in zoos are easily bored, and must be given especially rich environments and plenty of other apes to socialize with. It is not yet clear whether this neophilia affects their sexual choice, but primatologist Meredith Small has claimed that "The only consistent interest seen among the general primate population is an interest in novelty and variety." Chimpanzee females sometimes take considerable risks to mate with novel mates from outside their own groups.

In modern human societies, neophilia is the foundation of the art, music, television, film, publishing, drug, travel, pornography, fashion, and research industries, which account for a significant proportion of the global economy. Before such entertainment industries amused us, we had to amuse one another on the African savanna. Our neophilia may have demanded ever more creative displays from our mates. If other apes are neophilic and modern humans are extremely neophilic, perhaps our ancestors were too.

In this view, human creativity evolved through sexual selection as an anti-boredom device. Perhaps as our ancestors evolved larger and cleverer brains, their neophilia increased as well. Boredom became more frustrating. Sexual partners who were regarded as tedious after a few days or weeks could not have established the longer-term relationships that yielded large reproductive payoffs. Less creative brains that offered less ongoing novelty to sexual partners did not leave as many offspring. This would have been sufficient for creativity to evolve.

Potentially, the cognitive variety offered by one creative individual can compensate for the physical variety offered by a string of short-term sexual partners. Scheherazade retained her sultan's interest by producing a stream of novel stories, to compensate him for giving up the stream of novel women he had previously enjoyed. This does not imply that creativity evolved as a "pair-bonding mechanism." Rather, individuals who wish to retain a sexual partner's interest over the long term found it strategically effective to act more creative, playful, and innovative in their relationship. Basically, this kept boredom from driving their lovers into the arms of another.

Sexual choice did not favor unpredictability at all levels of behavior. Predictable kindness and predictable sexual fidelity were probably valued. For couples to successfully cooperate, they must be able to anticipate each other's needs and plans. At the other extreme, superficial or dangerous forms of unpredictability were unlikely to have proven attractive. Epileptic fits may be protean in form, but are not considered creative. The mad dog strategy may be effective in terrifying subordinates (and lovers who wish to leave), but it is not sexually attractive.

The attractive forms of novelty tend to rely on a uniquely human trick: the creative recombination of learned symbolic elements (e.g. words, notes, movements, visual symbols) to produce novel arrangements with new emergent meanings (e.g. stories, melodies, dances, paintings). This trick allows human courtship displays not just to tickle another's senses, but to create new ideas and emotions right inside their minds, where they will most influence mate choice. Scheherazade did not produce a random series of nonsense words to play upon the sultan's neophilia. She took existing words that already had a meaning, and put them together in new combinations that evoked new characters, plots, and images. To produce novelty at the cognitive level, one must use standardized signals at the perceptual level.

Creativity is not just a production line for churning out random ideas. It depends on both selective retention and blind variation. A capacity for novelty production will yield interesting entertainment only if it is combined with a huge knowledge base, virtuoso expression, and good critical judgment. It also demands the social intelligence necessary to figure out how to express a novel idea in a comprehensible way. As all writers know, it is one thing to have an idea in one's head, and quite another to put it on paper in a way that will evoke it in someone else's head. In his classic 1950 book *The Creative Process*, Brewster Ghiselin noted that "Even the most energetic and original mind, in order to reorganize or extend human insight in any valuable way, must have attained more than ordinary mastery of the field in which it is to act, a strong sense of what needs to be done, and skill in the appropriate means of expression." A creative display demands skill and motivation, not just inspiration.

Creative Problem-Solving Versus Creative Display

Creativity research has focused much more on creative problem-solving than on creative courtship display. It is easy to envision natural selection favoring animals who solve their survival problems more creatively. Psychologists tend to think of Wolfgang Köhler's experiments from the 1920s, in which chimpanzees figured out how to stand on a box and use a stick to knock some bananas down from a height so they could eat them. Such examples lead us to think of creativity being favored for its survival payoffs, and this focus is reinforced by research funding priorities. Creativity research justifies its costs as a way to discover how people might improve their ability to solve technical problems. Corporations want more creative thinkers so they can patent more innovations, not so their workers can attract better mates. The problem-solving viewpoint has been reinforced by the mass of biographical research on the creativity of great scientists and inventors.

Many creativity researchers suggest that an idea's creativity should be measured by two criteria: novelty and utility. Utility concerns the idea's appropriateness for solving a well-defined problem. Novelty is somewhat incidental, reflecting the difficulty of solving that problem and thus how rarely people have solved it in the past. In this problem-solving perspective, human creativity is subject to the same bottom line as R & D divisions in a corporation. The blue-sky dreaming has to yield dividends sooner or later: novelty cannot be justified as an end in itself, only as a means of finding otherwise elusive solutions. Cognitive psychology is especially concerned with problem-solving. Since Herbert Simon's work on artificial intelligence and problem-solving in the 1950s, cognitive psychology has gradually taken over creativity research. Creativity is sometimes seen as little more than a way to solve slightly harder-than-average problems.

It is possible, but rather dreary, to see the world as a mixture of problems and solutions. One could even speak of courtship as a problem and displays as a solution. But this problem-oriented viewpoint rather misses the point of human creativity, and indeed of courtship display in general.

Consider the creativity demanded by slapstick comedy. The great physical comedians of the silent-film era, Buster Keaton and Harold Lloyd, were not in the business of solving problems. On the contrary. Their genius lay in taking unproblematic everyday acts, and turning them into elaborately inventive displays of clumsiness. The climbing of a ladder became an opportunity for exploring the dozens of inappropriate ways in which a human body can interact with a ladder and a floor. Comedy depends on showing how many ways something can go wrong—on violating expectations, not solving problems.

Perhaps in considering the evolution of creativity, we should focus more on humor and less on technical invention. I think that neophilic laughter rather than technophilic profit was the fitness payoff that mattered in the evolution of creativity. Laughter may seem a rather weak thread from which to hang such a grand ornament as human creativity, yet laughter is an important part of human nature. It is universal within our species, manifest in distinct facial and vocal expressions. It emerges spontaneously during childhood, and is deeply pleasurable. It shows all the hallmarks of a psychological adaptation.

An appreciation of humor is an important part of mate choice too. One of the strongest and most puzzling findings from evolutionary psychology research has been the value that people around the world place on a good sense of humor. Indeed, this is one of the few human traits important enough to have its own abbreviation (GSOH) in personal ads. Perhaps we can finally understand why a GSOH is so frequently requested and so frequently advertised by singles seeking mates. A capacity for comedy reveals a capacity for creativity. It plays upon our intense neophilia. It circumvents our tendencies towards boredom. Creativity is a reliable indicator of intelligence, energy, youth, and proteanism. Humor is attractive, and that is why it evolved.

In his 1964 book *The Act of Creation*, Arthur Koestler struggled in vain to find a survival function for creative wit, humor, and laughter. He wrote:

> What is the survival value of the involuntary, simultaneous contraction of fifteen facial muscles associated with certain noises which are often irrepressible? Laughter is a reflex, but unique in that it serves no apparent biological purpose; one might call it a luxury reflex. Its only utilitarian function, as far as one can see, is to provide temporary relief from utilitarian pressures. On the evolutionary level where laughter arises, an element of frivolity seems to creep into a humorless universe governed by the laws of thermodynamics and the survival of the fittest.

Looking for survival value in a sexually attractive biological "luxury" is arguably the most typical mistake of 20th-century theorizing about human evolution. This book has repeatedly celebrated this "element of frivolity" that sexual selection introduces into the cosmos. Humor—the wit to woo—is one of its most delightful products.

Where Partnerships Can Be Joined or Loosened in an Instant

Our creative capacities remain hard to fathom at the psychological level, despite the emergence of some reasonable evolutionary theories about their origins. When caught in creativity's flow, the mind seems to let itself go more liquid than solid. The best description of this state was written by William James in an 1880 article for *The Atlantic Monthly*:

> Instead of thoughts of concrete things patiently following one another in a beaten track of habitual suggestion, we have the most abrupt cross-cuts and transitions from one idea to

> another, the most rarefied abstractions and discriminations, the most unheard of combinations of elements, the subtlest associations of analogy; in a word, we seem suddenly introduced into a seething cauldron of ideas, where everything is fizzling and bobbling about in a state of bewildered activity, where partnerships can be joined or loosened in an instant, treadmill routine is unknown, and the unexpected seems the only law.

One of William James's best friends was the philosopher Charles Sanders Peirce, who saw himself as a spokesman for the indeterminate, the chaotic, and the random. Peirce had little patience for those who viewed the human mind as a deterministic system running on the fixed rails of heredity and environment. The human mind, in his view, was an arena of refined chaos, where description is difficult and prediction is impossible. Yet Peirce, like James, was sympathetic to Darwinism, and viewed the mind as a natural evolutionary outcome.

Perhaps science will one day regain the sophistication about human creativity that it attained in 1880s Harvard, when James and Peirce saw no conflict between a Darwinian theory of mental evolution and an indeterminist theory of mental processes. They would have viewed our current debates about "genetic determinism" with amusement. They understood that inherited mental capacities could produce unpredictable behaviors, not just by accident, but by design. This chapter has been a sort of footnote to Peirce's joyful indeterminism. We have seen that many games demand mixed strategies, and many evolutionary situations demand unpredictable behavior. Human creativity may be the culmination of a long trend toward ever more sophisticated brain mechanisms that produce ever less predictable behaviors. These capacities may make psychology maddeningly difficult as a predictive science, but they also make life worth living outside the lab.

Human Evolution as Romantic Comedy

We learn something important about human creativity, I think, from the observation that romantic comedy is a rather more successful film genre than documentaries on the lives of great inventors. This is not just because romantic comedy depicts attractive people progressing through a successful courtship by exploiting each other's neophilia. It is also because romantic comedies form part of our own courtship efforts. We can (indirectly) pay Hollywood scriptwriters to make our intended romantic partners laugh. But our ancestors could not do this, and even now it does not suffice. If we prove boring during the conversation after the film, our dates may say they had a lovely time, but let's be just friends. You can't buy love. You have to inspire it, partly through humor, the premier arena for advertising your creativity.

Theories of human evolution are scientific hypotheses, but they are also stories. To develop a good new hypothesis, it can help to choose a story from an overlooked genre. The traditional evolution stories could be filmed largely as action adventures, war stories, or political intrigues. In casting one would automatically visualize Mel Gibson in a fur loincloth with a steely gaze, glistening pectorals, and hearty clansmen, battling for independence from Neanderthal oppressors. Or Sigourney Weaver fighting Pleistocene monsters in dark tunnels to protect endangered children, after her less intelligent male comrades have been disemboweled.

I am making a different pitch, for romantic comedy as the genre least likely to mislead us, if we think of human evolution as a narrative. My rationale is that in action, war, and intrigue, people mostly just die. But in romantic comedy, people sometimes get pregnant. Evolution is a multi-generation epic that depends on some couples courting and having children. Although action adventures better fulfill Aristotle's insistence on the dramatic unities of time and place, maybe we should pay more attention to Darwin's insistence on our unbroken chain of descent. Human evolution could be imagined as a million-year-long version of *Bringing Up Baby*, in which ancestral Katharine Hepburns and Cary Grants fell in love through a combination of slapstick, verbal repartee, and amusing adventures with wild animals. Evolution may be heartless, but it is not humorless.

Sexual Personae

By viewing human evolution as a romantic comedy, we might understand not only our creative capacities for producing witty novelties, but also our ability to reinvent ourselves with each new sexual relationship. People act differently when they're in love with different people. We tend to match our expressed interests and preferences to those of a desired individual. One develops a crush on a mountain-climber, and suddenly feels drawn to the sublime solitude of the Alps. One dates a jazz musician, and feels prone to sell one's now puerile-seeming heavy metal albums. Should an otherwise perfect lover confide her secret belief in the healing power of crystals, one may find yesterday's sneering skepticism about such nonsense replaced by a sudden open-mindedness, a certain generosity of faith that must have lain dormant all these years. In courtship, we work our way into roles that we think will prove attractive.

Chimpanzees have some capacities for "tactical deception," for pretending to do something other than what they are really doing. But they cannot pretend to be someone other than who they are. Sexual courtship may have been the arena in which we evolved the capacity for dramatic role-playing. With each new lover, we experience a shift in image and identity. These shifts are rarely as dramatic as the changes of sexual personae adopted by David Bowie or Madonna with each new album. But they are more profound. Often, we may find it difficult to relate to our former selves from previous romances. Events experienced by that former self, which seemed so vivid at the time, become locked away in a separate quadrant of memory's labyrinth, accessible only if we happen to run into the former lover. Our minds undergo these sexual revolutions, reshaping themselves to each new lover like an advertising company dreaming up new campaigns for capturing new market niches.

Acting is not the prerogative of a few highly strung

professionals, but a human birthright, automatically activated whenever we fall in love. In courtship, all the world became a stage, and all the proto-humans merely players. Perhaps we evolved the ability to creatively role-play because sexual choice favored those who were better at adopting an attractive series of sexual personae. Our identity shifts operate not only at the level of consciousness and identity, but at all observable levels: ornamentation, clothing, posture, gesture, accent, facial expression, attitude, opinion, and ideology.

Creative Ideologies Versus Reliable Knowledge

Sexual selection for creativity raises some worries about the reliability of human knowledge. According to traditional views, animals with delusions should be eliminated by natural selection. Evolution should produce species with brains that interpret the world more and more accurately, enabling behavior to be guided more adaptively. Such reasoning is central to the field of "evolutionary epistemology," which studies how evolutionary processes can generate reliable knowledge. Evolutionary epistemologists such as Karl Popper, Donald Campbell, and John Ziman have credited evolution with a tendency to endow animals with reasonably accurate models of the world. This idea seems to solve many of the traditional philosophical worries about the validity of human perception and belief.

For most kinds of knowledge embodied in most of our psychological adaptations, I think that their argument is correct. Natural selection has endowed us with an intuitive physics that allows us to understand mass, momentum, and movement well enough to deal with the material world. We also have an intuitive biology that allows us to understand plants and animals well enough to survive, and an intuitive psychology that lets us understand people. Especially since the 1980s, psychologists have been busy investigating these intuitive forms of knowledge in children and adults. Our hundreds of adaptations for sensation, perception, categorization, inference, and behavior embody thousands of important truths about the world.

However, when we come to verbally expressed beliefs, sexual selection undermines these reliability arguments. While natural selection for survival may have endowed us with pragmatically accurate perceptual systems, mate choice may not have cared about the accuracy of our more complex belief systems. Sexual selection could have favored ideologies that were entertaining, exaggerated, exciting, dramatic, pleasant, comforting, narratively coherent, aesthetically balanced, wittily comic, or nobly tragic. It could have shaped our minds to be amusing and attractive, but deeply fallible. As long as our ideologies do not undermine our more pragmatic adaptations, their epistemological frailty does not matter to evolution.

Imagine some young hominids huddling around a Pleistocene campfire, enjoying their newly evolved language ability. Two males get into an argument about the nature of the world, and start holding forth, displaying their ideologies.

The hominid named Carl proposes: "We are mortal, fallible primates who survive on this fickle savanna only because we cluster in these jealousy-ridden groups. Everywhere we have ever traveled is just a tiny, random corner of a vast continent on an unimaginably huge sphere spinning in a vacuum. The sphere has traveled billions and billions of times around a flaming ball of gas, which will eventually blow up to incinerate our empty, fossilized skulls. I have discovered several compelling lines of evidence in support of these hypotheses. . . ."

The hominid named Candide interrupts: "No, I believe we are immortal spirits gifted with these beautiful bodies because the great god Wug chose us as his favorite creatures. Wug blessed us with this fertile paradise that provides just enough challenges to keep things interesting. Behind the moon, mystic nightingales sing our praises, some of us more than others. Above the azure dome of the sky the smiling sun warms our hearts. After we grow old and enjoy the babbling of our grandchildren, Wug will lift us from these bodies to join our friends to eat roasted gazelle and dance eternally. I know these things because Wug picked me to receive this special wisdom in a dream last night."

Which ideology do you suppose would prove more sexually attractive? Will Carl's truth-seeking genes—which may discover some rather ugly truths—out-compete Candide's wonderful-story genes? The evidence of human history suggests that our ancestors were more like Candide than Carl. Most modern humans are naturally Candides. It usually takes years of watching BBC or PBS science documentaries to become as objective as Carl.

Runaway sexual selection for ideological entertainment would not have produced accurate belief-systems, except by accident. If ideological displays were favored as fitness indicators, the only truth they had to convey was truth about fitness. They need not be accurate world-models any more than the eyes of a peacock's tail need to represent real eyes. *Das Kapital* demonstrated Karl Marx's intelligence, imagination, and energy, but its reliability as a fitness indicator does not guarantee the truth of dialectical materialism. The majesty of Brigham Young's religious visions were sufficient to attract 27 wives (who averaged 24.5 years old at marriage—with wives number 12 through 21 marrying him when he was in his mid-40s), but that does not guarantee the veracity of his belief that dead ancestors can be retroactively converted to the Mormon faith.

When we considered the evolution of language, we saw that sexual selection rarely favors displays that include accurate conceptual representations of the world. Across millions of species throughout the Earth's history, there have been only two good examples of sexual selection for world-representing truth: human language and human representational art. Even so, human language's ability to refer to real objects and events does not guarantee the reliability of human ideologies expressed through language.

Sexual selection usually behaves like an insanely greedy tabloid newspaper editor who deletes all news and leaves only advertisements. In human evolution, it is as if the editor suddenly recognized a niche market for news in a few big-brained readers. She told all her reporters she wanted wall-to-wall news, but she never bothered to set up a fact-checking department. Human ideology is the result: a tabloid concoction of religious conviction, political idealism, urban myth, tribal myth, wishful thinking, memorable anecdote, and pseudo-science.

Richard Dawkins has suggested that these ideological phenomena all result from "memes"–virus-like ideas that evolved at the cultural level to propagate themselves by grabbing our attention, remaining memorable, and being easy to transmit to others. The meme idea offers a novel perspective on human culture, but it begs several questions. Why do people display such ideas so fervently in young adulthood, especially during courtship? Why do people compete to invent new memes that will make them famous? Why were most memes invented by men? Why did natural selection leave us so vulnerable to ideological nonsense? Perhaps by viewing ideological displays as part of courtship, we can answer such questions. Mostly, we use our memes to improve our sexual and social status; they do not just use us.

This sexual selection theory of ideology poses a serious challenge to evolutionary epistemology. Natural selection can favor accurate intuitive models of the world, but it seems incapable of producing communication systems that allow those models to be shared. Sexual selection can favor rich communication systems such as language, but it tends to distort verbally expressible world-models, making them more entertaining than accurate. There seems to be a trade-off between reliable individual cognition and social communication—we can be mute realists or chatty fabulists, but not both. This is far from the evolutionary epistemology view, in which truth-seeking cognition evolved with truth-sharing language to give us a double-barreled defense against falsehood.

Our ideologies are a thin layer of marzipan on the fruitcake of the mind. Most of our mental adaptations that patiently guide our behavior remain intuitively accurate. They are our humble servants, toiling away at ground level, unaffected by the strange signals and mixed metaphors flying overhead from one consciousness to another during the mental fireworks show of courtship. Sexual selection has not impaired our depth perception, voice recognition, sense of balance, or ability to throw rocks accurately. But it may have profoundly undermined the reliability of our conscious beliefs. This is the level of epistemology that people care about when they challenge other people's claims to "knowledge" in the domains of religion, politics, medicine, psychotherapy, social policy, the humanities, and the philosophy of science. It is in these domains that sexual selection undermines the evolutionary epistemology argument, by turning our cognitive faculties into ornamental fitness-advertisements rather than disciples of truth.

Creative Science

Given minds shaped by sexual selection for ideological entertainment rather than epistemic accuracy, what hope do we have of discovering truths about the world? History suggests that we had very little hope until the social institutions of science arose. Before science, there was no apparent cumulative progress in the accuracy of human belief systems. After science, everything changed.

From a sexual selection perspective, science is a set of social institutions for channeling our sexually selected instincts for ideological display in certain directions according to strict rules. These rules award social status to individuals for proposing good theories and gathering good data, not for physical attractiveness, health, kindness, or other fitness indicators. Scientists learn to derogate the normal human forms of ideological display: armchair speculation, entertaining narratives, comforting ideas, and memorable anecdotes. (Of course, this spills over into derogation of popular science books that try to present serious ideas in attractive form.) Science separates the arenas of intellectual display (conferences, classrooms, journals) from other styles of courtship display (art, music, drama, comedy, sports, charity). Science writing is standardized to channel creativity into inventing new ideas and arguments instead of witty phrases and colorful metaphors. Scientists are required to provide intellectual displays to young single people (through undergraduate teaching, graduate advising, and colloquium-giving), but are discouraged from enjoying any sexual benefits from these displays, so are kept in a state of perpetual quasi-courtship until retirement.

These scientific traditions are ingenious ways of harnessing human courtship effort to produce cumulative progress towards world-models that are abstract, communicable, and true. It is surprising that science works so well, given the absence of referential content in the sexual signals of all other species, and our Scheherazade-style genius for fictional entertainment. Science is not asexual or passionless. But neither is it a result of some crudely sublimated sex drive. Rather, it is one of our most sophisticated arenas for human courtship, which is the most complex and conscious form of mating that has ever evolved on our planet.

Epilogue

This book has explored only a few of the human mind's unusual abilities, and only a few of the possible ways of applying sexual selection theory to account for them. I have not pretended to offer a complete account of human evolution, the human mind, or human sexual choice. My theory is quite limited in scope, and my presentation of it even more so. Like art, music may be an evolutionary product of sexual choice, but analyzing it would have required repeating too many arguments and analogies from the chapter on art, and introducing too many new ideas—it is a scientifically challenging and emotionally charged topic, and one I hope to address elsewhere. Likewise for the relationship between sexual selection, human intelligence, learning, and cultural dynamics. I have hardly mentioned some of the central topics in cognitive science, such as perception, categorization, attention, memory, reasoning, and the control of bodily movement, which may have evolved under some influence from sexual choice—or they may not have. My sexual choice theory also hopped over that treacherous patch of philosophical quicksand known as "consciousness." I have stressed repeatedly that the sexual choice theory aims to account for just some of the distinctly human aspects of our minds, not the huge number of psychological adaptations that we share with other animals, including, for example, all the intricacies of great ape social intelligence, primate vision, and mammalian spatial memory. Finally, the sexual choice theory is descriptive, not prescriptive—it is a partial theory of human origins and a partial description of human nature, not a theory of human potential or a description of human limits.

Setting boundaries on human behavior is the job of law, custom, and etiquette, not evolutionary psychology.

Despite these limitations, the sexual choice theory is ambitious in trying to offer some new theoretical foundations for understanding human culture. I agree with E. O. Wilson's book *Consilience* that all areas of human knowledge should strive for mutual consistency through a biologically grounded view of human nature. The social sciences and humanities would benefit, I think, from turning to evolutionary psychology as their conceptual basis, rather than Marxism, psychoanalysis, and French philosophy. However, evolutionary psychology will not replace art history or linguistics, any more than physics could replace organic chemistry or paleontology. These sciences all describe phenomena at different levels, demanding the use of different concepts, models, and research methods. To argue that sexual choice has powerfully shaped human nature is not, for example, to suggest that economics should focus on human sexual behavior instead of markets, prices, and strategies. However, it may suggest that more attention to unconscious sexual strategies might help economists understand patterns of earning and spending.

Understanding the origins of human morality, art, and language is unlikely to diminish our appreciation of ethical leadership, aesthetic beauty, or witty conversation. On the contrary, if these human capacities evolved through sexual choice, then our appreciation of them, depending on a relatively hardwired set of sexual preferences, should be immune to any of the alleged wonder-reducing effects of scientific explanation. In any case, I trust that the enjoyment of worldly delights is better accompanied by true understanding than by romantic obscurantism.

One's understanding of human sexuality and human behavior depends, to some extent, on one's sex. Throughout this book, I have tried to write first as a scientist, second as a human, and only third as a male. Yet some of my ideas have probably been too influenced by my sex, my experiences, and my intuitions. The trouble is, I don't know which ideas are the biased ones, or I would have fixed them already. Perhaps others will be kind enough to identify them. A woman might have written a book about mental evolution through sexual selection with different emphases and insights. Indeed, I hope that women will write such books, so we can triangulate on the truth about human evolution from our distinctive viewpoints. Evolutionary psychology has made rapid progress in part because it includes a nearly equal sex ratio of researchers, with both men and women drawing upon their experiences to develop new ideas and experiments. Personal experience is not very useful in testing scientific theories, but it can be invaluable in formulating and refining them. I hope that each sex will continue to correct the other's biases and oversights within the scientific arena, without any pretense that either knows everything about a two-sexed species.

Scientific theories never dictate human values, but they can often cast new light on ethical issues. From a sexual selection viewpoint, moral philosophy and political theory have mostly been attempts to shift male human sexual competitiveness from physical violence to the peaceful accumulation of wealth and status. The rights to life, liberty, and property are cultural inventions that function, in part, to keep males from killing and stealing from one another while they compete to attract sexual partners. Feminist legal scholars have been right to point out this male bias in moral and political theory. The bias has been exacerbated by trying to ground ethical debates in survival rights rather than reproduction rights. Since most homicides and wars are perpetrated by adult males, and males kill mostly other males, a survival rights viewpoint tends to marginalize women and children.

Sexual selection offers a different perspective, in which human rights to mate choice and courtship can be better appreciated. For rape to be viewed as a serious crime from a survival rights viewpoint, for example, it must be characterized as "a crime of violence, not sex"—a description that raises many difficulties in cases of date rape. By contrast, a sexual choice viewpoint leads naturally to the view that even non-violent rape is a serious crime, because it violates human rights to exercise sexual choice. A sexual selection framework might also clarify the ethical arguments against sexual harassment, sexual stalking, incest, pedophilia, and female genital mutilation. Such a framework might also lead some to question the medical prioritization of "essential" therapies (e.g. expensive treatments that prolong the survival of the very old by a couple of years) over "cosmetic" therapies (e.g. cheap treatments that dramatically improve the courtship prospects of the young). It might also lead some to challenge educational policies that prioritize "academic fundamentals" (e.g. skills that increase worker productivity on behalf of corporate shareholders and tax-collectors) over "extracurricular activities" (e.g. sports, drama, dance, music, and art skills that increase individual sexual attractiveness).

This book has stressed that there are many possible ways for individuals to advertise their fitness when trying to attract a mate. Each animal species has evolved its own set of fitness indicators. Likewise, each human culture has developed its own set of learned fitness indicators, such as distinct ways of acquiring and displaying social status. Humans are in the unique position of being able to argue about what kinds of indicators we should encourage in our societies. Evolutionary psychology should not pretend that the male display of monetary wealth and the female display of physical beauty are the only fitness indicators available to our species. This book has argued that both human sexes have evolved many ways of displaying creative intelligence and other aspects of fitness through storytelling, poetry, art, music, sports, dance, humor, kindness, leadership, philosophical theorizing, and so forth. Marxists, feminists, artists, and saints have long understood that human intelligence, creativity, kindness, and leadership can be displayed in many ways other than by climbing economic status hierarchies to acquire material luxuries. I agree, and this book has focused on the traditional hominid and hippie modes of display: body ornamentation, rhythmic dance, irreverent humor, protean creativity, generosity, ideological ardor, good sex, memorable storytelling, and shared consciousness. I hope that the sexual choice theory increases your confidence that people can appreciate your mind's charms directly, in ordinary conversation, unmediated by your ability to work, save, shop, and spend.

Our modern quality of life depends on our ability to benefit from millions of acts of courtship, in which we are neither the producer nor the intended receiver. One's life may be saved by a side-impact airbag designed by an engineer in Stockholm, striving for local status in a Volvo design team. Or one may be uplifted by a novel written by the long-dead Balzac trying to impress his aristocratic Russian mistresses. The signal difference between modern life and Pleistocene life is that we have the social institutions and technologies for benefiting from the courtship efforts of distant strangers.

It is our responsibility to design social institutions that reap maximum social benefits from individual instincts for sexual competitiveness. In the terminology of game theory, we may not be able to keep individuals from playing as selfish competitors in the mating game, but we may choose, to some extent, which mating game our society plays. We cannot keep people from playing equilibrium strategies, but we can recognize that there are many possible equilibria available, and debates over social values can be viewed as equilibrium selection methods. One society, for example, may organize human sexual competition so that individuals become alienated workaholics competing to acquire consumerist indicators of their spending ability. In another possible society, individuals could compete to display their effectiveness in saving poor villages from economic stagnation and saving endangered habitats from destruction. In my view, conspicuous charity is at least as natural as conspicuous consumption, and we are free to decide which should be more respected in our society. In other words, discovering better ways of managing human sexual competitiveness should be the explicit core of social policy.

Existing political philosophies all developed before evolutionary game theory, so they do not take equilibrium selection into account. Socialism pretends that individuals are not selfish sexual competitors, so it ignores equilibria altogether. Conservatism pretends that there is only one possible equilibrium—a nostalgic version of the status quo—that society could play. Libertarianism ignores the possibility of equilibrium selection at the level of rational social discourse, and assumes that decentralized market dynamics will magically lead to equilibria that yield the highest aggregate social benefits. Far from being a scientific front for a particular set of political views, modern evolutionary psychology makes most standard views look simplistic and unimaginative.

Likewise for standard views on "bioethics." The possibilities of genetic screening and genetic engineering seem to raise new ethical challenges for our species. Some bioethicists warn that parents should have no right to "play God" by giving their offspring unfair genetic advantages over others. They worry that new reproductive technologies may lead to runaway fashions for certain physical or mental traits. They even imagine that capricious divergence in such fashions may lead our species to split apart into distinct subspecies with different bodies, minds, and lifestyles. However, sexual selection theory suggests that such warnings have come about 500 million years too late. Animals have been playing God ever since they first evolved powers of sexual choice. Finding mates with good genes is one of the major functions of mate choice. Every female insect, bird, or mammal that selects a male based on fitness indicators is engaging in a form of genetic screening. Sometimes their choices are based on sensory appeal or novelty, leading to runaway fashions for bodily ornaments and courtship behaviors. Divergence in sexual preferences has been splitting species apart for millions of years, generating most of the biodiversity on our planet. We could outlaw genetic screening for heritable traits, but I imagine that our jails would have difficulty housing all of the sexually reproducing animals in the world that exercise mate choice—the female humpback whales alone would require prohibitively costly, high-security aquariums. Our current debates about reproductive technologies might benefit from recognizing the antiquity of sexual choice mechanisms that evolved specifically to give one's offspring unfair genetic advantages over others.

A sexual selection framework suggests one final point about human values. Mate choice is intrinsically discriminatory and judgmental, built to rank potential mates by reducing their rich subjectivity to a crass list of physical, mental, and social features. It scrutinizes individuals for infinitesimally harmful mutations and trivial biological errors, anxiously anticipating any heritable weakness that natural selection would have spurned in the Pleistocene. It discounts everything that humans have in common, focusing only on differences. And it pays the most attention to the fitness indicators that amplify those differences to the greatest extent. When we are actually choosing long-term sexual partners, there may be good reasons to listen to our mate choice circuits. But for the rest of the time we do not have to view people through the lens of mate choice. The better we understand our mate choice instincts, the easier they may be to override when they are socially inappropriate. There is much more to modern human social life than courtship, and much more to people than their fitness indicators.

When our automatic sexual judgments assert themselves, tempting us to discriminate and objectify when we should be sympathizing, we might try remembering the following. First, all living humans are evolutionary success stories whose 80,000 or so genes have already managed to prosper through thousands or millions of generations. Second, all normal humans are incredibly intelligent, creative, articulate, artistic, and kind, compared with other apes and with our hominid ancestors. Third, through the contingencies of human romance and genetic inheritance, almost everyone you meet will produce at least one great-grandchild who will be brighter, kinder, and more beautiful than most of your great-grandchildren. Such lessons in humility, transience, and empathy come naturally from an evolutionary perspective on human nature.

Over the long term, our species, like every other, has just two possible evolutionary fates over the long term: extinction, or further splitting apart into a number of daughter species—each of which will either go extinct or split again. If we avoid extinction, each of our daughter species will probably develop distinctive styles of courtship display, and different ways to channel their sexual competitiveness into various forms of physical, artistic, linguistic, intellectual, moral, and economic display. Some may continue to live on our home planet, and some may move elsewhere. Some may shape their own evolution naturally through sexual selection, while others may shape their evolution consciously through genetic technologies. We cannot imagine the minds that our far-future descendants might evolve, any more than our ape-like ancestors could have imagined ours. That does not matter. Our responsibility is not to speculate endlessly about the possible futures of our daughter species, but to become, with as much panache as we can afford, their ancestors.

Acknowledgments

For useful feedback on various chapters, thanks to Rosalind Arden, David Buss, John Constable, Leda Cosmides, Helena Cronin, James Crow, Oliver Curry, Dan Dennett, John Endler, Dylan Evans, Jennifer Freyd, Kristen Hawkes, Nicholas Humphrey, James Hurford, Marek Kohn, Robert Kruszynski, Henry Plotkin, David Shanks, Peter Singer, Randy Thornhill, and Peter Todd.

For publishing expertise, thanks to my agent John Brockman, my editors Ravi Mirchandani and Roger Scholl, and my copyeditor John Woodruff.

For supporting my research, thanks to the National Science Foundation (USA), the Max Planck Society (Germany), and the Economic and Social Research Council (UK).

For sharing ideas and inspiration related to the book, thanks to the following colleagues: At Stanford University: my Ph.D. thesis advisor Roger Shepard; also, during their 1989–1990 visits, David Buss, Leda Cosmides, Martin Daly, Jennifer Freyd, John Tooby, and Margo Wilson. At the University of Sussex: Margaret Boden, Dave Cliff, Inman Harvey, Phil Husbands, John Maynard Smith, and Michael Wheeler. At the Max Planck Institute's Center for Adaptive Behavior and Cognition: Gerd Gigerenzer, Dan Goldstein, Ralph Hertwig, Ulrich Hoffrage, Tim Ketelaar, Alejandro Lopez, Laura Martignon, and Peter Todd. At the London School of Economics: Helena Cronin, Oliver Curry, Dylan Evans, Nicholas Humphrey, Colin Tudge,

Richard Webb, and Andy Wells. At University College London: Ken Binmore, Chris McManus, Amy Parish, Henry Plotkin, Andrew Pomiankowski, Camilla Power, David Shanks, and Volker Sommer.

Others who contributed important ideas, inspiration, support, and/or feedback related to the book have included Laura Betzig, Robert Boyd, Ellen Dissanayake, Dean Falk, Robert Frank, Steve Gangestad, Arthur Jensen, Chris Knight, Bjorn Merker, Steven Mithen, Randy Nesse, Brad Payne, Robert Plomin, Don Symons, Andy Whiten, George Williams, David Sloan Wilson, and John Ziman.

For their unwavering support and generosity over the years, special thanks to my parents Frank and Carolyn Miller, and to my friends Helena Cronin and Peter Todd.

For choosing me, thanks to my Scheherazade, my partner Rosalind Arden.

Glossary

adaptation A biological trait that evolved through natural selection or sexual selection to promote survival or reproduction in a particular way.

adaptive radiation The branching out of a number of species from a common ancestor, as a result of that ancestor having evolved a useful new adaptation that allows it to spread into new ecological niches.

altruism Helping others without direct benefit to oneself. Apparent altruism can evolve only through indirect or hidden benefits to one's genes.

anthropology The study of human evolution (physical anthropology) and human cultures (cultural anthropology).

archaic *Homo sapiens* Ancestral hominids that lived in Africa, Europe, and Asia from about 400,000 to about 100,000 years ago, fairly similar to modern humans, with large brains.

archeology The study of prehistoric artifacts and human remains.

artificial selection The selective breeding and domestication by humans of other species, e.g. breeding dairy cattle for maximum milk yield.

assortative mating Sexual choice for traits similar to one's own, e.g. tall women favoring tall men.

Australopithecine One of a set of hominids that lived about 4 to 1 million years ago; bipedal, with strong jaws and small, ape-sized brains. The earlier ones were probably ancestral to humans.

band A social group of hunter-gatherers, usually around 20 in number, occupying a local territory.

behavior genetics The study of the inheritance of human and animal behavior, often using twin and adoption studies (to separate genetic from environmental effects) and molecular genetics methods to identify specific genes.

behaviorism A school of psychology, flourishing from about 1920 to 1970, that tried to explain behavior through learned associations between stimuli and responses, without reference to minds, intentions, behavior genetics, or evolutionary functions.

bonobo A species of great ape previously known as the "pygmy chimpanzee." Very sexual and very clever, bonobos are found in Zaire and are closely related to the common chimpanzee.

bowerbird One of the 18-odd species of birds in New Guinea and Australia in which males attract females by building ornamental nests called bowers.

brain size A convenient indicator of the number and complexity of psychological adaptations that have evolved in a species. Brain size can be estimated from fossil skulls, and correlates 40 percent with intelligence in modern humans.

cognitive psychology An area of psychology that studies the mental processes that underlie perception, categorization, judgment, decision-making, memory, learning, and language.

cognitive science The interdisciplinary study of intelligence based on the computer metaphor for the mind, excluding research on individual differences in intelligence, its heritability, or its evolution.

condition-dependence A trait's sensitivity to an animal's health and energy level. For example, dance ability is condition-dependent because tired, sick animals can't dance very well.

consortship Exclusive association between a male and a female in estrus, during which the male tries to keep the female sexually separated from other males.

conspicuous consumption Costly indicators of wealth displayed to achieve social status—the human cultural analog of sexually selected handicaps.

convergent evolution The independent evolution in separate lineages of adaptations that serve the same function.

copulatory courtship Energetic, prolonged copulation that provides mutual evidence of fitness through mutual pleasure.

courtship effort The time, energy, skill, and resources spent trying to impress potential sexual partners.

Darwinian aesthetics The evolutionary analysis of what people find beautiful, by viewing human aesthetic preferences as adaptations for favoring habitats, foods, tools, and sexual partners that promote one's reproductive success.

death A misfortune that precludes further courtship or reproduction.

developmental stability An organism's ability to grow a complex body part in its normal form, despite various environmental and genetic stresses. For body parts that are normally symmetrical, symmetry is an indicator of developmental stability.

dimorphism bodily differences between males and females.

discriminative parental solicitude The tendency of parents to direct their care and attention to offspring that are more likely to survive and reproduce.

display A conspicuous behavior shaped by evolution to advertise fitness, condition, motivation, or desperation.

dominance The ability to intimidate other individuals into giving up food, territory, or sexual partners.

ecological niche The position of a species within an ecology, including its habitat, food supply, and relations to predators and parasites.

equilibrium In game theory, any situation in which no player can do better by changing their strategy, given what other players are already doing.

equilibrium selection Any process that leads a population to play one equilibrium rather than another in a strategic game that has more than one possible equilibrium. It can occur through genetic evolution, cultural history, or individual learning.

estrus Signs of ovulation manifest in a female's body or behavior, evolved to attract males and incite male–male competition.

ethology The study of the mechanisms and functions of animal behavior in the wild.

evolution Descent with cumulative genetic modification, due to natural selection, sexual selection, and various random effects.

evolutionary psychology The study of human psychological adaptations, including their evolutionary origins, adaptive functions, brain mechanisms, genetic inheritance, and social effects.

extended phenotype An organism considered as a set of adaptive effects that reach out into the environment to promote its survival and reproduction. It can include evolved traits like beaver dams, spider webs, bowerbird bowers, and hominid handaxes.

female The sex that produces larger gametes called eggs.

fitness (1) The relative reproductive success (including survival ability) of one set of genes relative to others. (2) Good physical or mental condition that might prove genetically heritable.

fitness indicator An adaptation that evolved to advertise an individual's fitness during courtship and mating, typically by growing an ornament or performing a behavior that a lower-fitness individual would find too costly to produce.

fitness matching The assortative mating for fitness that happens in a competitive mating market when individuals mate with the highest-fitness sexual partner who is willing to mate with them.

foraging Finding wild plant and animal foods to eat.

function How an adaptation evolved to promote survival or reproduction under ancestral conditions.

***g* factor** The basic dimension of general intelligence and brain efficiency that accounts for the positive correlations between scores on mental tests. Basically, it is what IQ tests try to measure.

game theory The study of interdependent decision-making in situations where each player's payoffs depend on how their own strategies interact with the strategies of other players. Game theory is studied mostly by economists.

gamete A reproductive cell such as a sperm or egg.

gene A piece of DNA long enough to code for some biological information but short enough to survive many generations of sexual recombination. The gene is the basic unit of replication and selection in evolution.

gene–culture co-evolution The hypothesis that the human brain enlarged to learn more culture, which allowed cultures to become more complex, which in turn selected for larger brains, and so forth.

gene pool The total set of genes in a population.

genetic algorithm A computer program that evolves solutions to specified problems by applying selection, mutation, and genetic recombination to populations of simulated individuals that represent possible solutions.

genome The complete set of genetic information in an organism. The human genome contains over 60,000 genes and 3 billion DNA base pairs.

group selection Competition between groups that favors group-benefiting adaptations such as altruism or equilibria with high mutual payoffs.

handaxe A stone artifact with a roughly triangular outline, two symmetric faces, and a sharp edge around the circumference. Handaxes were made from about 1.6 million years ago until 50,000 years ago by various hominids.

handicap A costly, reliable indicator of fitness, often a result of sexual selection.

handicap principle The idea that fitness indicators can be reliable only if they impose such costs that low-fitness pretenders cannot afford them.

heritability For traits that vary between individuals, the proportion of that variation that is explained by genetic differences between the individuals. Heritability can range from 0 to 100 percent.

Holocene The geological era from 10,000 years ago to the present.

hominid Any of the bipedal apes of the last few million years, whether our direct ancestors or not.

Homo erectus A medium-brained hominid that flourished from about 1.8 million years ago to about 400,000 years ago (in Africa) and 50,000 years ago (in parts of east Asia).

human nature The complete set of psychological adaptations that has evolved in our species.

hunter-gatherers Humans or proto-humans living in small bands without farming or animal herding. Females typically raise the children and gather water, firewood, fruits, tubers, vegetables, berries, and nuts. Males sexually compete by playing status games such as warfare, hunting, and pretending to have spiritual powers. Before 10,000 years ago all humans were hunter-gatherers.

ideology A system of beliefs that has become sufficiently popular in a culture that believers stand a decent chance of finding a like-minded mate.

indicator A trait that evolved to advertise a particular aspect of an individual's fitness, condition, or motivation.

intelligence Mental fitness, as measurable by intelligence tests and displayed in verbal courtship. In this book, intelligence means the highly heritable "*g* factor" that underlies individual differences in a vast array of behavioral and cognitive abilities.

kin selection An evolutionary process that tends to favor generosity to blood relatives, in proportion to their genetic relatedness.

lek A place where males congregate to attract females with songs, dances, and visual ornaments.

lineage A line of common descent; a succession of organisms linked by genetic inheritance.

love An emotional adaptation for focusing courtship effort on a particular individual.

Machiavellian intelligence theory The idea that the large brains and high social intelligence of apes and humans evolved to deceive and manipulate others within a social group.

male The sex that produces smaller gametes called sperm.

marketing Designing, producing, advertising, delivering, pricing, and selling products that satisfy consumer preferences: the economic analog of sexual selection through mate choice.

marriage A socially legitimated sexual relationship in which sexual fidelity and parental responsibilities are maintained through the threat of social punishment.

mate choice Choice of sexual partners. This book prefers "sexual choice," which is less confusing for Anglo-Australian cultures, in which a "mate" is a non-sexual friend who reciprocates beer-buying behavior.

meme A unit of cultural information transmitted by imitation.

mixed strategy A strategy that randomizes behaviors in a certain proportion to keep an opponent guessing about a player's next move.

Modern Synthesis The integration of Darwinian evolutionary theory and Mendelian genetics that was achieved in the 1930s.

monogamy An exclusive sexual relationship of one male with one female.

morphology The physical structure of an organism; its body form.

mutation A spontaneous change in the structure or sequence of a DNA strand that changes how a gene works. Usually a bad idea.

mutation–selection balance An evolutionary equilibrium in which selection removes harmful mutations at the same average rate that harmful mutations arise.

mutual choice When both sexes are choosy about their sexual partners.

natural selection Changes in the gene pool of a species due to differences in the ability of individuals to survive and reproduce. Against current biological fashion, this book follows Darwin in using "natural selection" to cover differences in survival ability only, and "sexual selection" to cover differences in reproduction ability.

Neanderthal A species of hominid that flourished in Europe and western Asia from about 300,000 years ago until 50,000 years ago. Stocky, large-nosed, and large-brained, they were apparently not our direct ancestors.

neoteny The persistence of juvenile traits into adulthood, including bulbous heads, small jaws, and playful gregariousness.

ornament (1) In biology, a trait that evolved through sexual choice to appear sexually attractive. (2) In aesthetics, a hard-to-fake display of artistic skill and time, viewed as wasteful decadence by the Bauhaus and other 20th century modernist movements.

ornamental mind theory The idea that the human mind evolved through sexual choice as a set of entertainment systems used in courtship.

parasite load The number of parasites carried by an organism. High parasite loads impair condition, reducing health and sexual attractiveness.

parental investment Any care, protection, or effort given by parents that increases offspring fitness at a cost to the parent.

phenotype The observable traits of an organism, including body and behavior.

Pleistocene The geological epoch that began 1.64 million years ago and ended 10,000 years ago, during which almost all of human evolution happened.

polyandry ("many men") A sexual relationship in which one female copulates regularly with more than one male partner. Sometimes observed in Tibet.

polygamy ("many marriages") A legal system in which an individual can legitimately marry more than one spouse.

polygyny ("many women") A sexual relationship in which one male copulates regularly with more than one female partner, found in most human cultures throughout history.

population A group of individuals that tend to mate with each other.

population genetics The area of biology that models how evolution changes gene frequencies in populations.

primatology The scientific study of the 300-odd species of primates, including apes, monkeys, lemurs, lorises, tarsiers, marmosets, and tamarins.

promiscuity Mating by a female with many males to maximize sperm competition within her reproductive tract; favored by chimpanzees.

protean behavior Adaptively unpredictable behavior, as when prey zigzag randomly to escape from predators. Named for the mythical Greek shape-shifter Proteus.

psychological adaptation An inherited behavioral capacity that evolved to promote survival or reproduction in a particular way under ancestral conditions.

reciprocal altruism The theory that mutual generosity can evolve if individuals take turns giving and receiving benefits across many encounters.

reportability The ability to talk about one's subjective experiences.

reproductive success The number of viable offspring produced by an individual. Reproductive success is the basic currency of evolutionary success.

ritualization Evolutionary modification of a behavior for greater effectiveness as a display, through standardization, repetition, and amplification.

runaway brain theory The idea that the human brain evolved through runaway sexual selection.

runaway sexual selection A positive-feedback process that amplifies the size and complexity of sexual ornaments.

savanna Open grassland with scattered shrubs and trees, and alternating dry and wet seasons (rather than winters and summers), typical of East Africa where humans evolved.

Scheherazade strategy Keeping a sexual partner interested in oneself by telling good stories and being a good conversationalist.

selection pressure Any feature of the physical, biological, social, or sexual environment that causes some individuals to survive or reproduce better than others.

selfish gene A gene that acts as if it is trying to replicate itself; the gene considered as the unit of evolutionary selection.

sensory bias theory The idea that animal senses are more responsive to some stimuli than others, and that this can influence sexual selection to produce ornaments with sensory appeal.

serial monogamy A mating strategy in which individuals go through a series of monogamous sexual relationships (lasting a few weeks to several years) over the course of their lives. Serial monogamy has probably been the norm in human evolution.

sexual choice Choice of some sexual partners in preference to others. It has been a driving force behind sexual selection and evolution.

sexual preferences The criteria for sexual choice, whether perceptual, cognitive, emotional, or social.

sexual reproduction The production of offspring by combining an egg from a mother with sperm from a father; the prerequisite for sexual selection.

sexual selection Evolutionary change due to heritable differences in the ability to attract sexual partners, repel sexual rivals, or do anything else that promotes reproduction.

signal Any behavior that evolved to convey information from one animal (the signaler) to another (the receiver). Most signals convey information about a signaler's fitness, condition, motivation, or location.

social selection Selection for the ability to promote one's survival and reproduction by attaining social status and managing social relationships, including sexual relationships.

sociolinguistics The study of the social variations and uses of human language, especially as a function of age, sex, class, and ethnicity.

speciation The splitting apart of one population to form two species that no longer interbreed with each other.

species A group of organisms willing to breed with one another. The species is the basic unit of biological classification.

sperm competition Competition between sperm to fertilize an egg, which occurs when a female has mated with two or more males.

status Socially recognized merit, often used as a fitness indicator in sexual choice.

strategic handicap A costly behavioral display, such as dance or speech, that can easily be turned off if an animal is in poor condition. A strategic handicap is very efficient as a fitness indicator.

survival of the fittest A catchy but misleading phrase invented by Herbert Spencer to describe natural selection, which led biologists to neglect sexual selection.

Theory of Mind The ability to attribute beliefs and desires to other individuals, in order to better understand their behavior. The theory of mind is a key component of Machiavellian intelligence theory.

tribe A small society with a distinctive language and culture, typically a cluster of kin groups that interbreed.

universal Typical of all normal humans across cultures and history, suggesting an evolutionary rather than a cultural origin.

verbal courtship Talking to attract a sexual partner.

virtues Behavioral abilities and motivations that are socially approved and sexually desired.

waste The apparently pointless costs (in time, energy, resources, and risk) of sexual display that keep the displays reliable as indicators of fitness (in biology) or wealth (in modern culture).

Notes

The number preceding each note is the page number to which it refers.

CHAPTER 1

CENTRAL PARK

General references

Sexual selection: popular overview: Gould & Gould (1997); textbook: Andersson (1994); history: Cronin (1991); evolution of sex: Michod (1995).

Popular works on sexual selection in humans: Buss (1994), Etcoff (1999), Hersey (1996), Ridley (1993).

Academic works on sexual selection in humans: Abramson & Pinkerton (1995), Baker & Bellis (1995), Barkow (1989), Betzig (1986, 1997), Betzig *et al.* (1988), Buss (1999), Daly & Wilson (1988), Knight (1995), Symons (1979).

Previous sexual selection models of human evolution: Darwin (1871), Ellis (1905, 1934), Parker (1987), Sloman & Sloman (1988), Baum (1996).

Popular introductions to evolutionary psychology: Buss (1994), Pinker (1994, 1997), Matt Ridley (1993), Wright (1994).

New evolutionary psychology textbook: Buss (1999).

Academic evolutionary psychology: Barkow (1989), Barkow *et al.* (1992), Buss (1999), Cosmides & Tooby (1994), Crawford & Krebs (1998), Cummins & Allen (1998), Hirschfeld & Gelman (1994), Simpson & Kenrick (1997).

Notes

2. Survivalist theory that the mind evolved for toolmaking and other useful technologies: Byrne (1995), Gibson & Ingold (1993), Kingdon (1993).
3. Charles Darwin on sexual selection through mate choice: Darwin (1859, 1871).
3. Peacock tails: Petrie *et al.* (1991).
5. Steven Pinker on art, music, etc. as side-effects: Pinker (1997, chapter 8).
5. Pinker quote "The mind is a neural computer . . ." Pinker (1997, p. 534).
5. Pinker quote "if music confers no survival advantage . . .": Pinker (1997, p. 544).
5. Pinker quote that the arts are "biologically frivolous": Pinker (1997, p. 531).
6. Creative intelligence for Machiavellian manipulation: Byrne & Whiten (1988), Whiten & Byrne (1997).
6. Morality as a tit-for-tat accountant: Ridley (1996), Cosmides & Tooby (1992, 1997).
6. Theories of language evolution that neglect storytelling, poetry, etc.: Pinker (1994); see also Hurford *et al.* (1998).

6. David Buss and Randy Thornhill's research on human sexual preferences: Buss (1994, 1999), Thornhill (1997, 1998).
7. My research on sexual selection and human evolution: Miller (1993, 1994a, 1996, 1997a,b, 1998a,b, 1999a,b,c, 2000a,b, Miller & Todd (1995, 1998), Todd & Miller (1997a,b, 1999).
7. Essential role of adaptations in evolution: Dawkins (1976, 1982, 1986), Williams (1996); see also Bell (1997), Ridley (1997); conceptual issues: Cronin (1991), Rose & Lauder (1996), Sober & Wilson (1998), Sterelny & Griffiths (1999); basics of behavioral adaptation: Alcock (1998), Krebs & Davies (1997).
8. What makes sexual selection so special: Darwin (1871); Gould & Gould (1997), Ridley (1993); see also Miller (1994b, 1998a, 1999a,b), Miller & Todd (1995, 1998).
8. Definitions of natural selection versus sexual selection: Keller & Lloyd (1992).
9. Sexual selection as more accurate and efficient than natural selection: Miller & Todd (1995).
11. Sexual selection as a positive-feedback system: Fisher (1930), Iwasa & Pomiankowski (1995).
12. Machiavellian intelligence: Byrne & Whiten (1988), Humphrey (1976), Whiten & Byrne (1997), Wilson *et al.* (1996).
12. Robin Dunbar on minds for tracking social relationships: Dunbar (1996).
12. Theory of Mind: Baron-Cohen (1995), Whiten (1991).
14. Steven Pinker and John Tooby arguing for a focus on uniform traits rather than on traits with heritable individual differences: Pinker (1997), Tooby & Cosmides (1990a); for a different view on uniform traits versus sexually selected traits, see Miller (2000a,b).
15. Historical neglect of sexual selection: Andersson (1994), Cronin (1991).
15. Ideological biases, sexual prudery, sex differences, and sexual selection: Bender (1996), Buss & Malamuth (1996), Dahlberg (1981), Ellis (1905, 1934), Gilligan (1982), Goldberg (1993), Haraway (1989), Hrdy (1981, 1999), Illouz (1997), Knight (1995), Landau (1991), Lovejoy (1981), Marcuse (1956), Margulis & Sagan (1991), Miller (1993), Nietzsche (1968), Paglia (1990), Small (1993), Sommers (1994), Wrangham & Peterson (1996).
17. Evolution of brains and intelligence: Allman (1999), Foley & Lee (1991), Jerison (1973), Jerison & Jerison (1988), Mackintosh (1994).
17. Time lag between big brains and technological progress, and basics of human evolution generally: Boyd & Silk (1997), Conroy (1997), Foley (1997), Jones *et al.* (1992), Kingdon (1993), Landau (1991), Pitts & Roberts (1997), Stringer & McKie (1996).
18. Clear survival payoffs for human creative intelligence only apparent in last few tens of thousands of years: Kingdon (1993), Stringer & McKie (1996).
19. Steven Pinker on elephant trunks and language evolution: Pinker (1994, pp. 332–333).
20. More than three hundred species of primates, each with different facial appearance: see photographs in Rowe (1996).
21. Evolution of language: Aitchison (1996), Bickerton (1995), Deacon (1998), Hurford *et al.* (1998), Pinker (1994).
22. Fossil-oriented approaches to understanding human mental evolution: Aiello & Dean (1990), Gould (1977), Mithen (1996), Pitts & Roberts (1997).

22. John Pfeiffer on the "Upper Paleolithic symbolic revolution": Pfeiffer (1982).
24. Leslie Aiello on costs of large guts: Aiello & Wheeler (1995).
24. Steven Mithen on fossil evidence: Mithen (1996).
24. Understanding human mental evolution by analyzing present human adaptations and comparisons with other primates: Buss (1999), Goodall (1986), Peterson & Goodall (1993), Pinker (1997), Thornhill (1997), de Waal (1989, 1996); see also Rose & Lauder (1996).
25. The DNA revolution: Hamer & Copeland (1998); Plomin *et al.* (1997).
26. Robert Plomin and colleagues finding a gene associated with high intelligence: Chorney *et al.* (1997).
26. 98 percent similarity between chimp and human DNA according to DNA hybridization method: Gibbons (1998).
26. Analysis of Neanderthal DNA: Krings *et al.* (1997).
29. Jean-Paul Sartre on existential rootlessness: see his classic *Being and nothingness.*
30. Cognitive psychology approach to mental evolution: Cosmides & Tooby (1992, 1994, 1997), Donald (1991), Pinker (1997).
31. Nietzsche and Veblen as quasi-Darwinian intellectual heroes: see Nietzsche's *Genealogy of morals, Beyond good and evil,* and *Will to power*; and Veblen's *The theory of the leisure class.*

CHAPTER 2

DARWIN'S PRODIGY

General references

History of sexual selection ideas: Cronin (1991), Andersson (1994, pp. 3–31), Moller (1994, pp. 3–7).

Darwin biographies: Browne (1995), Desmond & Moore (1994), Mayr (1991).

All quotes from Darwin's The Descent of man (1871) are from the 1981 facsimile edition published by Princeton University Press; the quoted page numbers are the same as those of the original.

Notes

34. Biographical details on Darwin: Browne (1995), Desmond & Moore (1994).
35. Details of Darwin's development of sexual selection theory: Cronin (1991).
36. Darwin quote "The sole object . . .": Darwin (1871, Vol. 1, pp. 2–3).
36. Darwin quote "the second part . . .": Darwin (1871, Vol. 1, p. 5).
37. Darwin quote "not on a struggle . . .": Darwin (1859, p. 88).
38. Darwin quote "if man can in a short time . . .": Darwin (1859, p. 89).
41. Darwin quote "All animals present . . .": Darwin (1871, Vol. 2, p. 124).
42. Darwin quote "The case of the male Argus . . .": Darwin (1871, Vol. 2, pp. 92–93).
43. Wilhelm Wundt: Richards (1987).
45. Pre-Darwinian evolutionary theory: Richards (1987).
47. Darwin quote "He who admits . . .": Darwin (1871, Vol. 2, p. 402).
48. Alfred Wallace on natural and sexual selection: see his *Contributions to the theory of natural selection* (1870) and *Darwinism: An exposition of the theory of natural selection, with some of its applications* (1889); see also Cronin (1991).

49. Wallace quote "The enormously lengthened plumes . . .": Wallace (1889, p. 293).
50. Stephen Jay Gould's *Ontogeny and phylogeny*: Gould (1977).
51. Victorian sexual attitudes: for an amusing look at 19th-century sexual anxieties and sexual art, see Dijkstra (1986); for sexual selection ideas in 19th-century American novels, see Bender (1996).
53. August Weismann quote "sexual selection . . .": from Weismann's *The evolution theory* (1904, p. 237).
53. Weismann quote "Darwin has shown . . .": Weismann (1904, p. 238); another rare supporter of sexual selection theory around 1900 was Havelock Ellis (1905, 1934).
53. The eclipse of Darwinian theory around 1900: Bowler (1983).
53. Thomas Hunt Morgan: see his book *Evolution and adaptation* (1903).
53. Sir Ronald Fisher: Fisher (1915, 1930); biography of Fisher: Box (1978).
54. Fisher as inventor of statistical tests: Gigerenzer & Murray (1987).
54. Fisher quotes from 1915: Cronin (1991).
56. Morgan quote on runaway: Andersson (1994, p. 24).
56. Fisher quote "the potentiality of a runaway process . . .": Fisher (1930, p. 137).
57. Fisher quote "both the feature preferred . . .": Fisher (1930, p. 145).
57. Julian Huxley's 1938 papers opposed to sexual selection: Huxley (1938a,b).
58. Huxley quote "It was rather the opposite . . .": Huxley (1942, p. 35).
58. Huxley quotes "improvement in efficiency . . .": Huxley (1942, p. 562); "increased control . . .": Huxley (1942, pp. 564–565).
59. Freud's Lamarckian ideas: Sulloway (1979).
59. Behaviorism's history: Hilgard (1987).
59. Ernst Mayr's focus on speciation: Mayr (1991).
60. B. F. Skinner's *Science and human behavior*: Skinner (1953).
61. For a humorous look at the Bauhaus and modernist aesthetics, see Wolfe (1982).
61. Haldane quote "the results may be biologically advantageous . . .": Haldane (1932).
61. Huxley quote "favour the evolution . . .": Cronin (1991).
62. Maynard Smith on fruit fly courtship: Maynard Smith (1956).
62. Peter O'Donald on sexual selection: O'Donald (1980).
62. George Williams's *Adaptation and natural selection*: Williams (1966).
63. Influence of female primatologists: Haraway (1989), Small (1993).
63. The sociobiology of the 1970s: Wilson (1975).
63. Zahavi's handicap principle: Zahavi (1975); see also Zahavi & Zahavi (1997).
64. Dawkins's balanced appraisal of the handicap principle: Dawkins (1976).
64. Maynard Smith skeptical about handicaps: Maynard Smith (1978; but see 1985).
64. O'Donald's *Genetic models of sexual selection*: O'Donald (1980).
64. Runaway process works: Lande (1981), Kirkpatrick (1982).
65. Other key works in the revival of sexual selection: Trivers (1972), Williams (1975), Maynard Smith (1978), Symons (1979), Daly & Wilson (1983), Bradbury & Andersson (1987).
65. Flurry of experimental work on flies, widowbirds, frogs, scorpionflies: Andersson (1994); on humans: see Buss (1989).
65. Reviews of sexual selection: Andersson (1994), Harvey & Bradbury (1993), Miller (1998a), Ryan (1997).

CHAPTER 3

THE RUNAWAY BRAIN

General references

Runaway sexual selection: Ridley (1993).

Logic of sex differences in human mating strategies: Buss (1994).

Evolution of intelligence and big brains: Jerison (1973), Jerison & Jerison (1988), Mackintosh (1994).

Notes

70. Sexual selection's greatest hits: Gould & Gould (1997).
70. Fisher's runaway process: Fisher (1930); for an accessible description of runaway, see Ridley (1993).
72. Genetic correlations and recent technical work on runaway: Andersson (1994), Bakker & Pomiankowski (1995), Balmford (1991), Boyce (1990), Bradbury & Andersson (1987), Hasson (1990), Iwasa & Pomiankowski (1995), Miller & Todd (1993), Pomiankowski et al. (1991), Todd & Miller (1993), Zahavi (1991).
73. My Ph.D. thesis: Miller (1993).
73. Positive-feedback processes in evolution: Dawkins & Krebs (1979); Cliff & Miller (1995), Ridley (1993).
73. E. O. Wilson on gene–culture co-evolution: Lumsden & Wilson (1981); other models of gene–culture co-evolution: Boyd & Richerson (1985), Donald (1991).
73. Richard Dawkins on memes: Dawkins (1976); see also Blackmore (1999).
73. Nicholas Humphrey on social intelligence: Humphrey (1976).
73. Machiavellian intelligence: Whiten & Byrne (1988).
73. Richard Alexander on group competition: Alexander (1989); see also Chagnon (1983).
74. Matt Ridley on my theory: Ridley (1993).
75. Our ancestors as moderately polygynous: see Baker & Bellis (1995), Barkow (1989), Betzig (1986, 1997), Buss (1994, 1999), Daly & Wilson (1983, 1988), Ellis & Symons (1990), Hrdy (1999), Symonds (1979), Wrangham & Peterson (1996).
75. Sex differences in body size: Martin *et al.* (1994), Rogers & Mukherjee (1992).
75. Polygyny in hunter-gatherers: Betzig *et al.* (1988), Brown (1991), Chagnon (1983), Gregor (1985), Hill & Hurtado (1996), Shostak (1981).
76. Harems and Moulay Ismail: Betzig (1986); on Roman polygyny: Betzig (1992); for a critique of such examples, see Einon (1998).
76. Rise of monogamy in European Christian cultures: MacDonald (1995).
76. Polygyny among American presidents: Betzig & Weber (1993).
76. Unpredictability of sexual selection: Iwasa & Pomiankowski (1995).
77. Diversity of bowerbird nests: Borgia (1986); diversity of primates: Rowe (1996).
77. Our computer simulations on runaway: Miller (1994b), Miller & Todd (1993, 1995), Todd & Miller (1991, 1993, 1997b); for a general introduction to genetic algorithms: Mitchell (1998), see also Ziman (2000).
78. Speed and pattern of human brain evolution: Aiello & Dean (1990), Allman (1999), Jerison (1973); sexual selection and brain evolution: Falk (1997), Jacobs (1996).
79. Brain evolution happened in several bursts: Stanyon *et al.* (1993). Brain evolution in bursts is consistent with the "punctuated equilibrium" model proposed by Gould and Eldredge (1977), but their model assumed that the

bursts would be driven by external environmental changes rather than by chance shifts in the endogenous mating preferences of a population.

81. The intensity of sexual competition appears to predict brain size in primates: Toshiyuki Sawaguchi found evidence of a 65 percent correlation between relative neocortex size and a sexual dimorphism index across 19 genera of monkeys and apes: Sawaguchi (1997).
81. Male brains are about 8 percent larger than female; the difference is not explained by differences in stature or body mass: Ankney (1992), Falk *et al.* (1999), Pakkenberg & Gundersen (1997).
82. Some sex differences in particular psychological adaptations: Halpern (1992).
82. No sex difference in general intelligence: Jensen (1998, pp. 536–542).
82. Arthur Jensen quote "The sex difference in psychometric *g* . . .": Jensen (1998, p. 540).
82. No sex difference in scores on Raven's Standard Progressive Matrices: Court (1983).
82. More male geniuses and idiots: Feingold (1992), Lubinski & Benbow (1992); on reasons for male risk-seeking in behavior and brain development see Daly & Wilson (1983).
82. Greater male variability in IQ not a reflection of greater variability in general intelligence: Feingold (1994), Jensen (1998, pp. 537, 541).
82. Evidence of greater male cultural output: Miller (1999b).
82. More data on age profiles of culture production: Simonton (1988).
83. Evolutionary views on sex differences in culture production and social status: Browne (1998), Goldberg (1993), feminist views on sex differences in culture production: Battersby (1989), Haraway (1989), Paglia (1990), Sommers (1994); on the possibility of an evolutionary feminism: Buss & Malamuth (1996), Fisher (1982, 1992, 1999), Gowaty (1997), Hrdy (1997, 1999), Lancaster (1991), Parish (1993), Smuts (1995).
84. Brain size correlates 40 percent with general intelligence within each sex: Jensen (1998, pp. 146–149), Wicket *et al.* (1994) – but brain cells may be more tightly packed in female brains: see Wittelson *et al.* (1995).
85. Trivers on parental investment: Trivers (1972).
85. The logic of eggs, sperm, and sex differences: Ridley (1993); see also Baker (1996), Buss (1994, 1999), Daly & Wilson (1983), Short & Balaban (1994).
88. Sex differences in willingness to have short-term sexual relationships: Clark & Hatfield (1989); see also Baker & Bellis (1995), Betzig (1997), Buss (1994, 1999), Ellis & Symons (1990), Fisher (1992), Grammer (1991), Paul & Hirsh (1996), Perusse (1993), Scheib (1994), Symons (1979), Thornhill & Gangestad (1996), Thornhill & Thornhill (1992), Wrangham & Peterson (1996).
90. Darwin quote "It is, indeed, fortunate . . .": Darwin (1871, Vol. 2, pp. 328–329).
90. Genetic correlation between heights of the sexes: Rogers & Mukherjee (1992); see also Lande (1987).
94. Mutual mate choice: Johnstone (1997), Johnstone *et al.* (1996), Kirkpatrick *et al.* (1990), Roth & Sotomayor (1990); evidence of mutual choice in monogamous species: Jones & Hunter (1993), Trail (1990).
95. David Buss on short- versus long-term mating: Buss & Schmitt (1993); Buss (1994).
95. Doug Kenrick on increased choosiness in long-term relationships: Kenrick *et al.* (1990).

96. Concealed ovulation favoring longer-term relationships: Alexander & Noonan (1979), Fisher (1992), Margulis & Sagan (1991); critical view: Pawlowski (1999).

CHAPTER 4

A MIND FIT FOR MATING

General references

Evolution of sex: Maynard Smith & Szathmary (1995), Michod (1995), Ridley (1993).

Fitness indicators: Johnstone (1995), Rowe & Houle (1996).

Notes

99. Good introduction to sex, mutations, and fitness: Ridley (1993).
99. Evolution of sexual reproduction: Michod (1995), Michod & Levin (1988); important early works on the evolution of sex: Maynard Smith (1978), Williams (1975).
100. Number of species on Earth: May (1992).
100. Mutations in evolution: Kondrashov (1988a,b, 1995), Michod (1995).
100. Adam Eyre-Walker and Peter Keightley on harmful mutation rate in humans: Eyre-Walker & Keightley (1999); for commentary see Crow (1999); other comments on the alarmingly high mutation rate in humans: Crow (1997), Kondrashov (1995).
101. Theory that sex evolved to limit mutational damage: Bernstein *et al.* (1985), Kondrashov (1988a,b, 1995), Atmar (1991).
101. For general overviews of current genetics: Coen (1999), Lynch & Walsh (1998), Maynard Smith (1998).
101. Genetic dominance and mutation repair: Maynard Smith (1998).
103. Mutations, fitness, and development: Houle *et al.* (1992, 1994), Iwasa *et al.* (1991), Moller & Swaddle (1998).
104. Fitness indicators—especially in males—as a genetic sieve for the species: Atmar (1991).
104. Sexual selection for fitness indicators and good genes: textbook: Andersson (1994); recent review: Moller & Alatalo (1999); see also Grafen (1990), Hamilton & Zuk (1982), Iwasa *et al.* (1991), Maynard Smith (1985), Michod & Hasson (1990), Moller (1994), Pomiankowski & Moller (1995).
104. The human mind as a set of fitness indicators: Miller (1997b, 1998a,b, 1999a,b,c, in press a, in press b).
106. On the human body as a set of sexually selected fitness indicators: Gangestad (1993), Gangestad & Thornhill (1997), Gangestad *et al.* (1994), Grammer & Thornhill (1994), Jensen-Campbell *et al.* (1995), Miller & Todd (1998), Moller *et al.* (1995), Shackelford & Larsen (1999), Thornhill (1998), Thornhill & Gangestad (1994, 1996), Thornhill & Grammer (1999).
107. Definitions of fitness: Keller & Lloyd (1992).
107. Evolutionary fitness: Dawkins (1976, 1986), Williams (1996).
110. W. D. Hamilton on fitness: Hamilton & Zuk (1982).
112. David Buss and others on the attractiveness of strong males: Buss (1994, 1999), Ridley (1993).
113. Sir Ronald Fisher on fitness: Fisher (1915, 1930).

114. George Williams quote "It is to the female's advantage . . .": Williams (1966, p. 184).
115. Heritable versus inherited: Tooby & Cosmides (1990a), Plomin *et al.* (1997).
115. Leks: Balmford (1991); sage grouse: Boyce (1990).
116. Lek paradox: Kirkpatrick & Ryan (1991), Pomiankowski & Moller (1995), Reynolds & Gross (1990), Rice (1988).
117. The hunt for evolutionary forces that could maintain variation in fitness—the heritability of fitness debate: Gangestad (1997), Kondrashov (1988a,b), Maynard Smith (1985), Michod & Levin (1988), Moller & Swaddle (1998), Rice (1988).
117. Fitness varies across space and time: Andersson (1994), Ridley (1993).
117. William Hamilton on sexual selection against parasites: Hamilton *et al.* (1990), Hamilton & Zuk (1982); see also Tooby (1982).
119. Mutations maintain heritable fitness; mutation selection balance: Crow (1997, 1999), Houle (1998), Houle *et al.* (1992, 1994), Kondrashov (1988a,b, 1995), Pomiankowski *et al.* (1991), Rice (1988).
124. Peacocks: peahens prefer large tails: Petrie *et al.* (1991); large-tailed peacocks produce healthier offspring: Petrie (1994).
124. Zahavi's handicap model: Zahavi (1975, 1991), Zahavi & Zahavi (1997), Grafen (1990); see also Andersson (1994), Johnstone (1995); critics of handicap model: see Andersson (1994, chapter 3) for history.
125. Economists on cheap talk: Crawford (1998), Farrell & Rabin (1996).
127. Alan Grafen on handicaps: Grafen (1990); on fitness indicators as amplifiers of variation see Hasson (1990).
127. Condition-dependence: Andersson (1986), Griffith *et al.* (1999), Rowe & Houle (1996).
128. Analogies between biological handicaps and conspicuous consumption: Miller (1999a).
129. Rowe and Houle's model: Rowe & Houle (1996).
131. Stephen Jay Gould on unique human capacities as evolutionary side-effects: Gould (1977).
131. Criteria for mental adaptations used by evolutionary psychologists: Buss (1999), Hirschfeld & Gelman (1994), Pinker (1994, 1997), Tooby & Cosmides (1990a,b, 1992); need to go beyond these traditional criteria to deal with fitness indicators: Miller (1997b, 2000a,b).
132. Human mental traits supposed to show low heritability: Tooby & Cosmides (1990a); but on reasons why mental traits should remain heritable see Bailey (1998), Buss (1991), Gangestad (1997), Wilson (1994); evidence that many human traits are heritable: Jensen (1998), Neisser *et al.* (1996), Plomin *et al.* (1997), Wright (1998).
134. Costs of large brains: Aiello & Wheeler (1995), Foley & Lee (1991).
137. Richard Alexander's group warfare theory: Alexander (1989).

Chapter 5

ORNAMENTAL GENIUS

General references

Sensory bias theory: Bradbury & Vehrencamp (1998), Endler & Basolo (1998).

Notes

140. Animal signals as manipulation: Dawkins & Krebs (1978), Krebs & Dawkins (1984); see also Bradbury & Vehrencamp (1998).
142. Sensory biases in the evolution of signals: Eberhard (1985, 1996), Endler (1992), Guilford & Dawkins (1991), Ryan (1990), Ryan & Keddy-Hector (1992), Zahavi (1991); see also Burley (1988).
143. Michael Ryan on frog calls: Ryan (1990, 1997).
144. Simulations by Magnus Enquist: Enquist & Arak (1993); for a critique of these simulations see Bullock & Cliff (1997).
145. John Endler on lighting and animal visual sensitivities: Endler (1992), Endler & Basolo (1998).
147. Michael Ryan's list of sensory biases: Ryan (1990), Ryan & Keddy-Hector (1992).
148. Tim Guilford and Marion Stamp Dawkins on psychology and signal reception: Guilford & Dawkins (1991).
148. Evolution of pleasure and emotions: Tiger (1992), Tooby & Cosmides (1990b).
150. Pattie Maes on emotions for selecting behavior: Maes (1991).
153. Mind-as-computer metaphor: Gigerenzer & Murray (1987), Miller & Todd (1998).
156. Mutual choice and assortative mating: Johnstone (1997), Keller *et al.* (1996), Mascie-Taylor (1988), Sloman & Sloman (1988).
162. Why animals of different species develop different tastes: Howard & Berlocher (1998), Lambert & Spencer (1995), Lande (1981), Todd & Miller (1991, 1997b).
163. Evolutionary innovations: Nitecki (1990).
165. Threshold effect may have been overstated: Dawkins (1986), Eigen (1992).
166. On the threshold problem, and why natural selection gets stuck: Eldredge (1989).
169. William Eberhard on male genitals: Eberhard (1985, 1996).
169. Hugh Paterson on species as consensual systems of mate choice: Lambert & Spencer (1995).
170. Innovation through sexual choice: Miller (1994b), Miller & Todd (1995), Todd & Miller (1997b).
170. Evolution of bird wings and John Ostrom's theory: Shipman (1998).
172. Technological innovation similar to evolutionary innovation: Ziman (2000).
173. Marketing as a way to connect consumer preferences to products: Schmitt & Simonson (1997).

CHAPTER 6

COURTSHIP IN THE PLEISTOCENE

General references

Popular introductions to human evolution: Stringer & McKie (1996); Diamond (1992), Johanson & Edgar (1996), Pitts & Roberts (1997).

Technical introductions to human evolution: Boyd & Silk (1997), Conroy (1997), Foley (1997).

Pleistocene life: Dunbar (1996), Eaton *et al.* (1988), Foley (1996, 1997), Hrdy (1999), Tooby & Cosmides (1992).

A child's introduction to Pleistocene life: Johanson & O'Farrell (1990).
Hunter-gatherers: Barnard (1999), Hill & Hurtado (1996), Lee (1979), Shostak (1981).
Popular introductions to primates: Byrne (1995), Goodall (1986), Peterson & Goodall (1993), Rowe (1996), de Waal (1989, 1996), de Waal & Lanting (1997).
Prehistory very different from history: Betzig (1986, 1992, 1997), Diamond (1992), MacDonald (1995), Marcuse (1956), Perusse (1992), Vining (1986).

Notes

179. Importance of the Pleistocene era in human evolution: Boyd & Silk (1997), Buss (1999), Pinker (1997), Tooby & Cosmides (1990b).
180. Life in the Pleistocene: Dunbar (1996), Eaton, Shostak, & Konner (1988), Foley (1996, 1997), Hrdy (1999), Tooby & Cosmides (1992).
183. Sexual selection in primates: Dixson (1998), Dunbar (1988), Small (1993).
185. "Capricious" female primate preferences for male personalities?: Small (1993), Smuts (1985).
185. Relentless sociosexual strategizing in primates: Byrne & Whiten (1988), Whiten & Byrne (1997).
186. Pleistocene mating and sexual selection in human evolution: Buss (1994), Cashdan (1997), Daly & Wilson (1983, 1988), Ellis & Symons (1990), Ellis (1905, 1934), Fisher (1982, 1992), Hrdy (1981), Parker (1987), Symons (1979), Thornhill & Gangestad (1996), Wrangham & Peterson (1996).
186. Short- versus long-term mating: Buss (1994), Buss & Schmitt (1993), Fisher (1992), Scheib (1994).
187. Rape in human evolution: Thornhill & Thornhill (1992); on sexual harassment versus sexual selection generally: Clutton-Brock & Parker (1995a), Studd (1996).
188. Human sexual behavior and courtship: Brown (1991), Eibl-Eibesfelt (1989), Grammer (1989, 1991).
188. Role of consumerism in modern dating: Illouz (1997).
189. Sarah Blaffer Hrdy on single mothers as the norm: Hrdy (1999).
190. Males may have given some food to females: Lovejoy (1981).
190. Male mammals not usually very involved parents: Clutton-Brock (1991); however, sexual selection could in principle favor good parents: Hoelzer (1989), Wolf *et al.* (1997).
191. Males often more trouble than they're worth: Shostak (1981), Miller (1998a).
193. Mothers juggling courtship with mothering: Knight (1995), Power (1999).
194. Step-fathers more likely to kill children: Daly & Wilson (1988, 1995, 1998).
196. Fitness matching and assortative mating by quality: Buss (1994), Mascie-Taylor (1988), Noë & Hammerstein (1995), Real (1991), Sloman & Sloman (1988).
198. Fitness spreading: Buss (1985), Sloman & Sloman (1988).
201. Mark Kirkpatrick on Darwin's model of sexual selection for monogamous birds: Kirkpatrick *et al.* (1990).
202. Critique of the "species recognition markers" hypothesis: Zahavi & Zahavi (1997).
204. Consortships: Goodall (1986).
205. Simple rules for mate search: Miller & Todd (1998); see also Frey & Eichenberger (1996), Real (1991), Roth & Sotomayor (1990), Todd & Miller (1999), Wiegmann *et al.* (1996).

205. Picking sexual prospects by ranking cues: Gigerenzer & Goldstein (1996), Gigerenzer & Todd (1999), Miller (1997b).
206. 37 percent rule: Miller & Todd (1998).
207. "Try a Dozen" rule: Todd & Miller (1999).
209. Evolutionary psychology's emphasis on high-status men and fertile women: Buss (1994, 1999), Buss & Schmitt (1993), Cashdan (1997), Pinker (1997), Symons (1979), Thornhill & Gangestad (1996).
209. Men giving meat to women: Fisher (1982), Hawkes (1991, 1993), Power (1999).
211. David Buss's evidence on female preference for high-status males: Buss (1989, 1994).
212. Evolutionary logic of male preference for young women: Symons (1979), Thornhill (1998), Thornhill & Gangestad (1996).
212. Helen Fisher on relationships normally lasting no more than five years: Fisher (1992).
213. Doug Kenrick on age preferences: Kenrick & Keefe (1992).
214. Gould on neoteny: Gould (1977).
216. Diversity of fitness indicators in social versus sexual attractiveness: Cunningham *et al.* (1997).
216. Brief popular introduction to parental solicitude: Daly & Wilson (1998); more on discriminative parental solicitude: Daly & Wilson (1988, 1995), Daly *et al.* (1997), Trivers (1974).
217. Homosexual behavior in animals: Bagemihl (1998); homosexual behavior in bonobos: de Waal & Lanting (1997); see also Parish (1993).
221. Chris Knight and Camilla Power on collective female rituals: Knight (1995), Knight *et al.* (1995), Power (1999); see also Dunbar *et al.* (1999), Hurford *et al.* (1998).
222. Evolution of human skin color: Darwin (1871), Kingdon (1993).
223. Afrocentric approach to human evolution: Darwin (1871), Stringer & McKie (1996).

CHAPTER 7

BODIES OF EVIDENCE

General references

Popular overview of physical attractiveness: Etcoff (1999).

Evolution of the human body through sexual selection: Barber (1995), Ellis (1934), Hersey (1996), Morris (1967, 1985), Thornhill (1998).

Good reviews of the psychology of physical attractiveness: Berscheid & Walster (1978), Buss (1999), Cunningham *et al.* (1997), Gangestad (1993), Ridley (1993), Symons (1979, 1995), Thornhill (1998).

Notes

224. Physical sex differences: Martin *et al.* (1994), Morris (1985).
224. Riskiness of male development: Daly & Wilson (1983, 1988), Ellis (1905, 1934).
225. Criteria for identifying sexually selected traits: Darwin (1871), Andersson (1994).

226. Sex differences highly diagnostic of sexual selection: Martin *et al.* (1994).
228. Universal preferences for smooth skin, taller men, and symmetric faces: Buss (1994), Etcoff (1999), Thornhill (1998).
229. Darwin on population and racial differences in sexually selected traits: Darwin (1871).
229. Developmental stability, symmetry, and fitness indicators: Moller & Swaddle (1998); see also Gangestad & Thornhill (1997), Gangestad *et al.* (1994), Moller & Alatalo (1999).
230. Evolution of the penis: Baker & Bellis (1995), Barber (1995), Dixson (1998), Eberhard (1985), Margulis & Sagan (1991), Morris (1985).
230. Ape penis sizes: Dixson (1998), de Waal & Lanting (1997).
231. Amazing animal penises: Wallace (1980).
231. Popular introduction to sperm competition: Baker (1996); technical reviews of sperm competition: Baker & Bellis (1995); Birkhead & Moller (1998).
231. Men produce larger-than-normal ejaculate when female partners return from long trip: Baker & Bellis (1995).
231. Sperm competition better predicted by testicle size: Harvey & Harcourt (1984); see also Birkhead & Moller (1998).
233. Margulis quote "penis dimension . . .": Margulis & Sagan (1991, p. 161).
233. Penis display promoting the shift to bipedal walking: Sheets-Johnstone (1990).
234. Universality of male open-legged sitting posture: Eibl-Eibesfelt (1989).
234. William Eberhard on penis evolution and copulatory courtship: Eberhard (1985, 1996).
238. Anatomy of clitoris: Angier (1999, chapter 4).
238. Male scientists' views of the clitoris: Gould (1987), Symons (1979), Eibl-Eibesfelt (1989), Morris (1967, 1985).
239. Female scientists on the clitoris: Fisher (1992), Hrdy (1981), Margulis & Sagan (1991), Small (1993); the clitoris debate: Angier (1999, chapter 4), Sherman (1989).
239. Natalie Angier quote "She is likely . . .": Angier (1999, pp. 72–73).
241. Female genital mutilation: Angier (1999).
241. Evolution of breasts: Cant (1991), Caro & Sellen (1990), Gallup (1982), Low *et al.* (1987), Morris (1985).
244. Breast symmetry as a fitness indicator: Manning *et al.* (1997), Moller *et al.* (1995), Singh (1995).
246. Evolution of buttocks: Cant (1991), Caro & Sellen (1990), Low *et al.* (1987), Morris (1985).
247. Waists and hips: Singh (1993, 1995), Tassinary & Hansen (1998).
249. Books on body and face aesthetics: Etcoff (1999); Morris (1985).
249. Face aesthetics: Grammer & Thornhill (1994), Langlois *et al.* (1994), Perrett *et al.* (1998), Thornhill (1998), Thornhill & Gangestad (1994, 1996).
249. Face is rich in fitness information: Shackelford & Larsen (1999), Thornhill (1998); evidence against this view: Kalick *et al.* (1998).
250. Ashley Montagu on neoteny: Montagu (1989).
251. Healthy Pleistocene diet: Eaton *et al.* (1988).
253. Basic game theory: Kreps (1990), Sigmund (1993), Skyrms (1996).
254. Greater male interest in ritualized rule-following than relationships: Gilligan (1982).
255. Male competition, violence, and sports: Chagnon (1983), Daly & Wilson (1988), Wilson & Daly (1985), Wrangham & Peterson (1996).

CHAPTER 8

ARTS OF SEDUCTION

General references

The evolution of art: Dissanayake (1988, 1992), Turner (1991); see also Aiken (1998), Cooke & Turner (1999), Eibl-Eibesfelt (1989).

Notes

259. Ellen Dissanayake's books on art as an adaptation: Dissanayake (1988, 1992).
260. John Pfeiffer on the "creative explosion": Pfeiffer (1982); on prehistoric art see Sandars (1985), Taylor (1996).
260. Red ochre for body ornamentation: Knight *et al.* (1995), Power (1999); see also Low (1979).
260. Ernst Grosse quote "long ago . . .": Grosse (1897, p. 312).
261. Images of artist as genius besieged by wicked temptresses: Dijkstra (1986).
261. Evolution of pleasure: Tiger (1992), Tooby & Cosmides (1990b).
262. Social functionalism in anthropology: Haviland (1996), Knight (1995).
262. Robin Dunbar on language for managing social relations: Dunbar (1996).
263. Social functionalism requires pro-social altruism, which does not usually evolve: Ridley (1996); for a contrary view, see Sober & Wilson (1998).
263. Paleolithic "pornography": Hersey (1996), Taylor (1996).
263. Robert Layton quote "a pragmatic one . . .": Layton (1991, p. 7).
265. Steven Pinker quote "Many writers . . .": Pinker (1997, pp. 535–536).
267. Bowerbirds: for popular overview see Borgia (1986); see also Andersson (1994, pp. 172–174), Borgia (1995), Diamond (1988).
270. Richard Dawkins on extended phenotypes: Dawkins (1982).
271. Darwin quote "self-adornment, vanity, . . .": Darwin (1871, Vol. 2, p. 342).
272. Modern research supporting Darwin's claims about the costs of scarification: Ludvico & Kurland (1995), Singh & Bronstad (1997).
272. Darwin quote "It is extremely improbable . . .": Darwin (1871, Vol. 2, p. 343).
272. Freud on art as sublimated sexuality: Freud (1910).
272. Felix Clay quote "How the pleasure . . .": Clay (1908, pp. 162–163).
273. Sexual content of prehistoric art: Hersey (1996), Sandars (1985), Taylor (1996).
274. Tantric Buddhist art: Rawson (1978).
275. Darwinian aesthetics: Thornhill (1998); general aesthetics: Feagin & Maynard (1997), Higgins (1996).
275. Higher artistic output by males: Chadwick (1990), Paglia (1990), Miller (1999b).
276. Instincts don't proclaim their adaptive functions: Pinker (1997), Tooby & Cosmides (1990b, 1992).
278. Grant Allen's books: *Physiological aesthetics* (1877); *The color sense* (1879).
278. Felix Clay's *The origin of the sense of beauty*: Clay (1908).
278. Heyday of physiological/evolutionary aesthetics in 1870s through 1890s: Clay (1908).
278. Nancy Aiken's *The biological origins of art*: Aiken (1998).
279. Sensory biases in human aesthetics: Gombrich (1984), Rentschler *et al.* (1988).
279. Nicholas Humphrey on monkey aesthetics: personal communication, April 1999.

280. Desmond Morris on *The biology of art*: Morris (1962).
280. Review of research on monkey painting and Dali quote "The hand of the chimpanzee . . .": Lenain (1997).
281. Works of art as displays of virtuosity: Gombrich (1984), Zahavi (1978); on art as a costly, sexually selected behavior, see Power (1999).
281. Thorstein Veblen quote "The marks of expensiveness . . .": Veblen (1899, p. 80).
281. Ellen Dissanayake on "making things special": Dissanayake (1992).
282. Franz Boas quote "The enjoyment of form . . .": Boas (1955, p. 349).
282. Franz Boas quote "Among primitive peoples . . .": Boas (1955, p. 356).
282. Enrst Gombrich on virtuosity in decorative arts: Gombrich (1984).
283. Friedrich Nietzsche on the functions of beauty: Nietzsche (1968).
285. Arthur Danto quote "We have entered a period . . .": quoted in Dissanayake (1992, p. xiv).
286. Thorstein Veblen on spoons: Veblen (1899).
286. Franz Boas quote "The appreciation of the esthetic value . . .": Boas (1955, p. 19).
286. Walter Benjamin on the age of mechanical reproduction: Benjamin (1970).
288. Handaxes as ornaments: Kohn (1999), Kohn & Mithen (1999).
290. H. G. Wells on handaxes as projectiles: Kohn (1999, p. 59).
290. Spears from 400,000 years ago in Germany: Thieme (1997).
290. Marek Kohn quote "is a highly visible indicator . . .": Kohn (1999, p. 137).
291. Marek Kohn quote "A handaxe is a measure . . .": Kohn (1999, p. 122).

CHAPTER 9

VIRTUES OF GOOD BREEDING

General references

Popular introductions to the evolution of morality: Ridley (1996), Wright (1994).

Basic evolution of morality: Alexander (1987), Boone (1998), Petrinovich (1998), Singer (1994), Sober & Wilson (1998).

History of evolutionary thinking about altruism: Cronin (1991), Richards (1987).

Important works on sexual selection in the evolution of morality: Boone (1998), Hawkes (1991), Smith & Bird (in press), Tessman (1995).

Notes

292. David Buss's finding that kindness is the top-ranked trait in 37 cultures: Buss (1989), Buss *et al.* (1990).
292. Oscar Wilde's *An ideal husband*: the 1998 film version starring Cate Blanchett left out some of the most insightful passages on sexual choice and morality.
294. Darwin on violent competition within species: Darwin (1859).
295. Immanuel Kant on people as ends versus means: Singer (1994).
295. Evolutionary definitions of altruism: Keller & Lloyd (1992), Sober & Wilson (1998).
296. Good explanation of how economists view morality: Binmore (1994, 1998).
297. W. D. Hamilton's 1964 paper on kin selection is in his collected papers: Hamilton (1996).
298. Clear, accessible explanations of kin selection: Cronin (1991), Dawkins (1976).

298. The "genetic similarity fallacy" in kinship: Daly *et al.* (1997).
299. Human morality derived from kinship: Hamilton (1996), Wilson (1975).
300. Reciprocal altruism theory: Trivers (1971); see also Axelrod (1984), Cosmides (1989), Poundstone (1992), Ridley (1996).
301. Importance of punishment routines in maintaining reciprocity: Clutton-Brock & Parker (1995b); see also Boyd & Richerson (1992); cf. Skinner (1988).
301. The contract view of morality based on reciprocity considerations: Kahane (1995), Skyrms (1996).
301. Gerald Wilkinson on reciprocity among vampire bats: Wilkinson (1984); Frans de Waal on reciprocity among chimpanzees: de Waal (1996); review of cooperation among animals: Dugatkin (1997).
302. Leda Cosmides and John Tooby experiments on cheater detection: Cosmides (1989), Cosmides & Tooby (1992, 1994, 1997); more on detecting cheats and liars: Ekman (1992); replications and extensions of Cosmides's results: Gigerenzer & Hug (1992).
304. Matt Ridley's *Origins of virtue*: Ridley (1996).
304. T. H. Huxley on evolution and ethics from 1896: Huxley (1989).
305. Sigmund Freud on innate depravity: Freud (1930).
305. Innate depravity view among evolutionists: Sober & Wilson (1998), Wilson (1975).
305. Richard Dawkins quote "Be warned . . .": Dawkins (1976, p. 3).
305. Frans de Waal quote "Humans and other animals . . .": de Waal (1996, pp. 16–17).
306. Robert Frank's *Passions within reason*: Frank (1988), on the reality of kindness see also Tooby & Cosmides (1996).
306. Elliot Sober and David Sloan Wilson: Sober & Wilson (1998).
306. Adam Smith's divergent views of morality: Binmore (1984, 1988).
306. Innate goodness views of Jean-Jacques Rousseau: Singer (1994).
307. Irwin Tessman on sexual selection for morality: Tessman (1995).
307. Amotz Zahavi on altruism: Zahavi (1991), Zahavi & Zahavi (1997).
307. James Boone on magnanimity: Boone (1998); see also Smith & Bird (2000).
308. Views of hunting as useful provisioning: Fisher (1982), Hill & Hurtado (1996), Knight (1995), Lovejoy (1981), Power (1999), Stanford (1999).
308. Man the hunter: Lee & DeVore (1968); woman the gatherer: Dahlberg (1981).
309. Kristen Hawkes on the inefficiency of hunting: Hawkes (1991, 1993).
309. Chimps hunting: Goodall (1986), Stanford (1999).
310. Helen Fisher's *The sex contract*: Fisher (1982).
310. Owen Lovejoy on male hunting as good fathering: Lovejoy (1981).
310. Women refusing sex to poor hunters: Knight (1995).
310. Hunted meat as a public good: Hawkes (1993).
312. Yanomamo club-fights: Chagnon (1983).
313. Arabian babblers: Zahavi & Zahavi (1997).
314. Popular history of game theory: Poundstone (1992); accessible introductions to game theory: Kreps (1990), Sigmund (1993), Skyrms (1996); more technical introductions to game theory and evolution: Maynard Smith (1982); Binmore (1994, 1998), Dugatkin & Reeve (1998).
315. Biography of John Nash: Nasar (1998).
316. Equilibrium selection: Binmore (1994, 1998), Boyd & Richerson (1990), Roth & Sotomayor (1990), Skyrms (1996); experiments on equilibrium selection: Kagel & Roth (1995).

316. Brian Skyrms's *Evolution of the social contract*: Skyrms (1996).
317. James Boone quote "Now imagine that . . .": Boone (1998, p. 16).
318. Robert Boyd & Peter Richerson on equilibrium selection and group selection: Boyd & Richerson (1990); cf. models of group selection for morality that don't consider equilibrium selection explicitly: Alexander (1987), Sober & Wilson (1998).
319. Frans de Waal on moral leadership in chimps: de Waal (1989, 1996).
319. Status and reproductive success: Barkow (1989), Betzig (1986, 1992, 1997), Betzig & Weber (1993), Betzig *et al.* (1988), Buss (1999), Ellis (1995), Perusse (1993), Vining (1986); the male drive for status: Barkow (1989), Daly & Wilson (1988), Fisher (1999).
325. Robert Frank's *Luxury fever*: Frank (1999).
325. Tipping in restaurants as an example of altruism: Frank (1988).
326. John D. Rockefeller, Sr.: Chernow (1998); Rockefeller's early charity: Chernow (1998, p. 50).
327. Quote on Rockefeller "He was thought much of . . .": Chernow (1998, p. 53).
327. Helen Fisher and Camilla Power on courtship as sex-for-meat exchange: Fisher (1982), Power (1999).
328. Gary Becker's economic analysis of marriage: Becker (1991).
329. Commodification of human courtship: Illouz (1997).
330. Women attracted to "agreeableness" in men: Botwin *et al.* (1997), Buss *et al.* (1990), Graziano *et al.* (1997), Jensen-Campbell *et al.* (1995); note that kindness is also apparently favored by female baboons: Smuts (1985).
330. The 5-factor personality model: Matthews & Deary (1998); see also Botwin *et al.* (1997), Buss (1991); cross-cultural validity of 5-factor model: Williams *et al.* (1998).
330. Agreeableness appears to be moderately heritable: for popular overviews of genetics and personality see Hamer & Copeland (1998), Harris (1998); on the heritability of personality traits, including agreeableness: Bouchard (1994), Loehlin (1992), Plomin *et al.* (1997), Rowe (1994).
330. Psychopathy: Hare (1999); Lykken (1995), Mealey (1995); relationship to Machiavellian intelligence: Wilson *et al.* (1996).
331. Hans Eysenck on psychoticism: Eysenck (1992).
331. Evolutionary perspectives on human sexual fidelity: Buss (1994), Fisher (1982), Lovejoy (1981).
332. The self-serving sexual morality of unfaithful men: Paul & Hirsch (1996).
332. Rarity of sexual fidelity in primates: Goodall (1986), Small (1993).
333. Psychological research on romantic love: Sternberg & Barnes (1988); cross-cultural universality of romantic love: Hatfield & Rapson (1996), Jankowiak (1995).
333. Evolutionary analysis of sexual fantasy: Baker & Bellis (1995), Ellis & Symons (1990).
334. Sexual selection for good, generous fathers: Hewlett (1992); see also Daly & Wilson (1998), Hawkes (1991, 1993), Hill & Hurtado (1996).
335. Sarah Blaffer Hrdy's *Mother nature*: Hrdy (1999).
336. Meritocratic instincts underlying morality: Binmore (1994, 1998).
337. Egalitarianism: Erdal & Whiten (1994), Rawls (1971).
337. The Nietzschean virtues: Nietzsche (1968).
338. Nietzsche quote "(1) virtue as force . . .": Nietzsche (1968, p. 172).
338. Nietzsche quote "a luxury of the first order": Nietzsche (1968, p. 179).

338. Nietzsche quote "the charm of rareness . . .": Nietzsche (1968, p. 175).
338. Nietzsche quote "processes of physiological prosperity or failure": Nietzsche (1968, p. 149).

CHAPTER 10

CYRANO AND SCHEHERAZADE

General references

Overviews of language: Crystal (1987), Pinker (1994).
Popular introductions to the evolution of language: Aitchison (1996), Bickerton (1995), Deacon (1998), Pinker (1994).
State of the art in scientific theories of language evolution: Hurford *et al.* (1998).
Other works on the evolution of language: Lieberman (1991), Maynard Smith & Szathmary (1995).
Language evolution through sexual selection: Burling (1986); see also Baum (1996), Crow (1998), Dessalles (1998), Locke (1998).

Notes

343. Getting past the ape language and innateness controversies: Hurford *et al.* (1998), Pinker (1994).
343. Trained use of visual symbols by Kanzi: Savage-Rumbaugh & Lewin (1994).
344. Controversy about innateness of language: Chomsky (1980), Pinker (1994).
344. Pinker quote "Language is a complex, specialized skill . . .": Pinker (1994, p. 18).
345. Need for an adaptationist analysis of language's costs and benefits: Hurford *et al.* (1998).
346. Richard Dawkins and John Krebs on signal evolution: Dawkins & Krebs (1978), Krebs & Dawkins (1984).
347. Modern theory of signal evolution: Hauser (1996), Bradbury & Vehrencamp (1998).
347. Handicap principle makes signals reliable: Bradbury & Vehrencamp (1998, chapter 20); see also Grafen (1990), Johnstone (1995), Zahavi & Zahavi (1997).
348. Chris Knight on the unreliability of displaced reference: Knight (1998).
348. Language used to convey propositional information: Pinker (1994).
350. Carl Rogers on "non-directive" therapy: Rogers (1995).
352. Ontogeny and phylogeny: Gould (1977).
352. Alan Turing on the imitation game: Turing (1950).
354. Language evolution through social status: Burling (1986), Locke (1998), Dessalles (1998).
354. Robbins Burling quotes "All that is needed . . ." and "We need our very best language . . .": Burling (1986).
354. John Locke on verbal plumage: Locke (1998, pp. 90–101).
354. Locke quote "Yo' rap is your thing . . .": Locke (1998, p. 91).
358. Male frogs with deep voices: Ryan (1985, 1990); on voice pitch see also Andersson (1994, chapter 14).
359. Cooperative model of language transmission: Crystal (1987).
361. Sociolinguistics: Talbot (1998), Tannen (1991).
365. Consciousness as a mental clearing-house allowing reportability: Mithen (1996).

366 Jennifer Freyd on shareability: Freyd (1983).
366. Robin Dunbar's "social grooming" view of language: Dunbar (1996).
366. Jean-Louis Dessalles on language relevance: Dessalles (1998).
368. Dunbar's analysis of human conversational behavior: Dunbar *et al.* (1997).
368. Dunbar quote "Female conversations . . .": Dunbar *et al.* (1997, pp. 242–243).
369. Vocabulary sizes: Aitchison (1994), G. A. Miller (1996).
369. Call repertoires of other species: primates: Hauser (1996), Savage-Rumbaugh & Lewin (1994); birds: Catchpole & Slater (1995).
369. 10 words a day learned: Pinker (1994).
370. Most vocabulary not used very often: Crystal (1987).
370. Universality of this power law: Constable (1997).
370. Richards on Basic English: Richards (1943).
371. Richards quote "it is possible to say . . .": Richards (1943, p. 20).
371. Richards quote "willing, serviceable little workers . . .": Richards (1943, p. 29).
371. Pidgins and creoles: Bickerton (1995), Crystal (1987), Pinker (1994).
372. Bird song repertoires: Catchpole & Slater (1995).
372. Birds with larger repertoires sire healthier offspring: Hasselquist *et al.* (1996).
372. Verbal intelligence tests and vocabulary size: Jensen (1998).
373. Heritability of vocabulary carried by heritability of intelligence: Plomin *et al.* (1997).
373. Intelligence correlates with body symmetry: Furlow *et al.* (1997).
373. Assortative mating for vocabulary size: Mascie-Taylor (1988).
375. Sex differences in language ability: Halpern (1992), Jensen (1998).
376. Sex differences in language use and conversational style: Talbot (1998), Tannen (1991); see also Fisher (1999).
376. Greater male language output in public speaking and writing: Burling (1986), Locke (1998), Miller (1998b), Tannen (1991).
377. Cyrano de Bergerac: Rostand (1993).
377. Anthony Burgess quote "She loves Christian . . .": Rostand (1993, p. ii).
378. Cyrano's death: Rostand (1993, pp. 174–175).
379. Poetry as a system of handicaps: Constable (1997).
382. Sexual conflict over desired levels of communication in relationships: Tannen (1991).
383. Scheherazade strategy: Miller (1993, 1997a, 1998a).
386. Terence Deacon on the many uses of language: Deacon (1998).
388. Equilibrium selection: Binmore (1994, 1998), Boyd & Richerson (1990), Samuelson (1997), Skyrms (1996).
388. Language as an outcome of equilibrium selection: Skyrms (1996).
388. In most species, sexual signals convey only fitness information: Andersson (1994), Bradbury & Vehrencamp (1998).

CHAPTER 11

THE WIT TO WOO

General references

Basics of creativity: Boden (1991, 1994); see also Ghiselin (1952), Hofstadter (1995), Koestler (1964), Simonton (1993), Smith *et al.* (1995), Sternberg (1999), Vernon (1970).

Evolution of creativity: Corballis (1991), Mithen (1998).

Notes

392. Genetic determinism and evolutionary psychology: Wright (1994).
393. Evolution of behavior, genetic determinism, and optimality: Krebs & Davies (1997).
394. John von Neumann biographical material: Poundstone (1992).
395. Game theory: Von Neumann & Morgenstern (1944); see also Kreps (1990), Skyrms (1996).
395. Von Neumann quote "In playing Matching Pennies . . .": Von Neumann & Morgenstern (1944, p. 144).
395. On mixed strategies see also Maynard Smith (1982), Bergstrom & Godfrey-Smith (1998).
395. Mixed strategies in human behavior: Driver & Humphries (1988).
396. Sex ratio as a strategic outcome of randomization: Fisher (1930).
396. Unusual sex ratios in parasites: Hamilton (1996, chapter 4).
396. Michael Chance on convulsions: Chance (1957).
397. Kenneth Roeder on moth evasion tactics: Roeder & Treat (1961).
397. Protean behavior: Chance & Russell (1959), Driver & Humphries (1988); our simulations of the evolution of protean behavior: Miller & Cliff (1994); our experiments on human pursuit/evasion abilities: Blythe *et al.* (1996, 1999).
398. Animal play behavior: Bekoff & Byers (1998).
399. Randomness versus science: Bergstrom & Godfrey-Smith (1998), Gigerenzer & Murray (1987), E. E. Miller (1997), G. F. Miller (1997a); worries that creativity makes a science of behavior impossible: Skinner (1988).
400. Human randomization abilities: Wagenaar (1972), Neuringer (1986), Neuringer & Voss (1993).
400. Randomization in competitive games: Rapoport & Budescu (1992), Rapoport *et al.* (1997).
402. Autistic children have trouble randomizing in Matching Pennies: Baron-Cohen (1992); randomization is also particularly impaired by damage to frontal lobes due to head injury (Spatt & Goldenberg, 1993) and alcoholism (Pollux *et al.*, 1995).
402. Machiavellian intelligence: Byrne & Whiten (1988), Whiten & Byrne (1997); proteanism versus Machiavellianism: Miller (1997a).
403. John Krebs and Richard Dawkins on mind-reading: Krebs & Dawkins (1984).
403. Mad dog strategy: Miller (1997a); despots: Betzig (1986).
405. Donald Campbell on random variation and selective retention: Campbell (1960, 1988).
407. Ashley Montagu on neoteny: Montagu (1989).
408. Animal play behavior and playfulness: Bekoff & Byers (1998).
408. Dean Keith Simonton on creativity and productive energy: Simonton (1988, 1993).
409. Creativity and intelligence: Sternberg (1999), Vernon (1970).
409. Heritability of creativity: Plomin *et al.* (1997).
410. General intelligence: Brody (1992), Jensen (1998).
410. New Mexico study on intelligence and body symmetry: Furlow *et al.* (1997).
410. Relationship of intelligence to general fitness: Houle (on press).
411. Darwin quote "It would even appear . . .": Darwin (1871, Vol. 2, p. 230).
411. Female birds favor large song repertoires: Catchpole & Slater (1995).

412. Meredith Small quote "The only consistent interest . . .": Small (1993, p. 153).
412. Neophilia and sensation seeking: Zuckerman (1994).
413. Brewster Ghiselin quote "Even the most energetic . . .": Ghiselin (1952, p. 29).
414. Creativity research focused on problem-solving: Smith *et al.* (1995), Sternberg (1999).
414. Wolfgang Köhler on chimpanzee problem-solving: Köhler (1925).
414. Herbert Simon on problem-solving: Simon (1957).
415. Sexual preferences for good sense of humor: Buss (1989, 1994).
416. Arthur Koestler quote "What is the survival value . . .": Koestler (1964, p. 31).
416. William James quote "Instead of thoughts . . .": quoted from Simonton (1993).
417. Charles Sanders Peirce: Hausman (1997), Kuklick (1979).
419. Sexual personae: Paglia (1990).
419. Tactical deception in chimpanzees: Whiten & Byrne (1997).
420. Evolutionary epistemology: Campbell (1960, 1988), Dennett (1995), Plotkin (1998), Popper (1972), Shepard (1997), Ziman (1978).
420. The mind as well adapted by natural selection to the physical aspects of the world: Shepard (1987, 1997).
421. On the evolution of Candide-like religious beliefs: Boyer (1994), Hinde (1999).
422. Human ideologies as courtship displays: G. F. Miller (1996).
423. Epistemology and memes: Dawkins (1993); see also Blackmore (1999), Dennett (1995), Sperber (1994).
424. How poorly adapted the human mind is for science: Wolpert (1992).
424. Issues in the social channeling of human sexuality: Freud (1930), Marcuse (1956), Posner (1992).

EPILOGUE

Notes

426. Evolution of music: Wallin *et al.* (1999); see also Catchpole & Slater (1995).
426. Evolution of intelligence: Jensen (1998), Jerison (1973), Jerison & Jerison (1988).
426. Evolution of consciousness: Bock & Marsh (1993), Chalmers (1996), Dennett (1991, 1995), Humphrey (1992), Searle (1992, 1997).
427. E. O. Wilson on consilience: Wilson (1998); other arguments for consistency between evolutionary psychology and the social sciences: Campbell (1988), Runciman (1998).
427. Male versus female perspectives in evolutionary psychology: Angier (1999), Buss & Malamuth (1996), Fisher (1992, 1999), Gowaty (1997), Hrdy (1981, 1997, 1999), Lancaster (1991), Parish (1993), Smuts (1995), Sommers (1994).
428. Evolution and ethics: Huxley (1989), Wright (1994).
430. Social debate as an equilibrium selection method: Binmore (1994, 1998).
431. Bioethics and issues in genetic screening, etc.: Petrinovich (1998), Silver (1998).

Bibliography

Literature cited from the endnotes

KEY SUGGESTIONS FOR FURTHER READING:

Matt Ridley (1993). *The red queen: Sex and the evolution of human nature.* New York: Viking.
Helena Cronin (1991). *The ant and the peacock: Altruism and sexual selection from Darwin to today.* Cambridge University Press.
James L. Gould and Carol G. Gould (1997). *Sexual selection.* New York: Scientific American Library.
Amotz Zahavi and Avishag Zahavi (1997). *The handicap principle: A missing piece of Darwin's puzzle.* Oxford University Press.
Robert Boyd and Joan Silk (1997). *How humans evolved.* New York: Norton.
David Buss (1999). *Evolutionary psychology: The new science of mind.* New York: Allyn & Bacon.
Steven Pinker (1997). *How the mind works.* New York: Norton.
Dean Hamer and Peter Copeland (1998). *Living with our genes.* New York: Doubleday.
Edward O. Wilson (1998). *Consilience: The unity of knowledge.* New York: Knopf.

The scientific literatures on sexual selection, evolutionary psychology, and human evolution have grown far too large for everything relevant to be included here. This selection tends to favor books over academic journal papers, as books are easier for most people to get hold of. I have also favored works that are popularly accessible, recent, comprehensive, and/or fun. Some classic older works cited in text or endnotes (e.g. Immanuel Kant's *Critique of judgment*) are not included here, since reprints can easily be tracked down.

Abramson, P. R., & Pinkerton, S. D. (eds.). (1995). *Sexual nature, sexual culture.* University of Chicago Press.
Aiello, L., & Dean, C. (1990). *Human evolutionary anatomy.* San Diego: Academic Press.
Aiello, L. C., & Wheeler, P. (1995). The expensive tissue hypothesis: The brain and the digestive system in human and primate evolution. *Current Anthropology*, *36*, 199–221.
Aiken, N. E. (1998). *The biological origins of art.* Westport, CT: Praeger Press.
Aitchison, J. (1994). *Words in the mind: An introduction to the mental lexicon* (2nd ed.). Oxford: Blackwell.

Aitchison, J. (1996). *The seeds of speech: Language origin and evolution.* Cambridge University Press.

Alcock, J. (1998). *Animal behavior: An evolutionary approach* (6th ed.). Sunderland, MA: Sinauer.

Alexander, R. D. (1987). *The biology of moral systems.* New York: Aldine de Gruyter.

Alexander, R. D. (1989). The evolution of the human psyche. In P. Mellars & C. Stringer (eds.), *The human revolution*, pp. 455–513. Edinburgh University Press.

Alexander, R. D., & Noonan, K. M. (1979). Concealment of ovulation, parental care, and human social evolution. In N. A. Chagnon & W. Irons (eds.), *Evolutionary biology and human social behavior*, pp. 402–435. North Scituate, MA: Duxbury Press.

Allman, J. M. (1999). *Evolving brains.* San Francisco: W. H. Freeman/Scientific American.

Andersson, M. (1986). Evolution of condition-dependent sex ornaments and mating preferences: Sexual selection based on viability differences. *Evolution, 40*, 804–816.

Andersson, M. (1994). *Sexual selection.* Princeton University Press.

Angier, N. (1999). *Woman: An intimate geography.* New York: Houghton Mifflin.

Ankney, C. D. (1992). Sex differences in relative brain size: The mismeasure of woman, too? *Intelligence, 16*, 329–336.

Atmar, W. (1991). On the role of males. *Animal Behavior, 41*, 195–205.

Axelrod, R. (1984). *The evolution of cooperation.* New York: Basic Books.

Bagemihl, B. (1998). *Biological exuberance: Animal homosexuality and natural diversity.* London: St. Martin's Press.

Bailey, J. M. (1998). Can behavior genetics contribute to evolutionary behavioral science? In C. Crawford & D. Krebs (eds.), *Handbook of evolutionary psychology*, pp. 211–233. Mahwah, NJ: Erlbaum.

Baker, R. R. (1996). *Sperm wars: The science of sex.* New York: Basic Books.

Baker, R. R., & Bellis, M. A. (1995). *Human sperm competition.* London: Chapman & Hall.

Bakker, T. C. M., & Pomiankowski, A. (1995). The genetic basis of female mate preferences. *Journal of Evolutionary Biology, 8*, 129–171.

Balmford, A. (1991). Mate choice on leks. *Trends in Ecology and Evolution, 6*, 87–92.

Barber, N. (1995). The evolutionary psychology of physical attractiveness, sexual selection, and human morphology. *Ethology and Sociobiology, 16*, 395–424.

Barkow, J. (1989). *Darwin, sex, and status.* University of Toronto Press.

Barkow, J., Cosmides, L., & Tooby, J. (eds.). (1992). *The adapted mind: Evolutionary psychology and the generation of culture.* Toronto: Oxford University Press.

Barnard, A. (1999). Modern hunter-gatherers and early symbolic culture. In R. Dunbar, C. Knight, & C. Power (eds.), *The evolution of culture*, pp. 50–68. Edinburgh University Press.

Baron-Cohen, S. (1992). Out of sight or out of mind? Another look at deception in autism. *Journal of Child Psychology and Psychiatry, 33*, 1141–1155.

Baron-Cohen, S. (1995). *Mindblindness: An essay on autism and theory of mind.* Cambridge, MA: MIT Press.

Battersby, C. (1989). *Gender and genius.* London: Women's Press.

Baum, E. B. (1996). Did courtship drive the evolution of mind? *Behavioral and Brain Sciences, 19*, 155–156.

Becker, G. (1991). *A treatise on the family.* Cambridge, MA: Harvard University Press.

Bekoff, M., & Byers, J. A. (eds.). (1998). *Animal play: Evolutionary, comparative, and ecological perspectives*. Cambridge University Press.

Bell, G. (1997). *Selection: The mechanism of evolution*. New York: Chapman & Hall.

Bender, B. (1996). *The Descent of love: Darwin and the theory of sexual selection in American fiction, 1871–1926*. Philadelphia: University of Pennsylvania Press.

Benjamin, W. (1970). *Illuminations*. London: Jonathan Cape.

Bergstrom, C., & Godfrey-Smith, P. (1998). On the evolution of behavioral heterogeneity in individuals and populations. *Biology and Philosophy*, *13*, 205–231.

Bernstein, H., Byerly, H. C., Hopf, F. A., and Michod, R. E. (1985). DNA damage, mutation, and the evolution of sex. *Science*, *229*, 1277–1281.

Berscheid, E., & Walster, E. (1978). *Interpersonal attraction* (2nd ed.). Reading, MA: Addison-Wesley.

Betzig, L. (1986). *Despotism and differential reproduction: A Darwinian view of history*. Hawthorne, NY: Aldine de Gruyter.

Betzig, L. (1992). Roman polygyny. *Ethology and Sociobiology*, *13*, 309–349.

Betzig, L. (ed.). (1997). *Human nature: A critical reader*. Oxford University Press.

Betzig, L., & Weber, S. (1993). Polygyny in American politics. *Politics and the Life Sciences*, *12*(1), 45–52.

Betzig, L., Borgerhoff-Mulder, M., & Turke, P. (eds.). (1988). *Human reproductive behavior: A Darwinian perspective*. Cambridge University Press.

Bickerton, D. (1995). *Language and human behavior*. Seattle, WA: University of Washington Press.

Binmore, K. (1994). *Game theory and the social contract*. Vol. 1. *Playing fair*. Cambridge, MA: MIT Press.

Binmore, K. (1998). *Game theory and the social contract*. Vol. 2. *Just Playing*. Cambridge, MA: MIT Press.

Birkhead, T. R., & Moller, A. P. (eds.). (1998). *Sperm competition and sexual selection*. London: Academic Press.

Blackmore, S. (1999). *The meme machine*. Oxford University Press.

Blythe, P., Miller, G. F., & Todd, P. M. (1996). Human simulation of adaptive behavior: Interactive studies of pursuit, evasion, courtship, fighting, and play. In P. Maes *et al.* (eds.), *From animals to animats 4*, pp. 13–22. Cambridge, MA: MIT Press.

Blythe, P., Todd, P. M., & Miller, G. F. (1999). How motion reveals intention: Categorizing social interactions. In G. Gigerenzer & P. M. Todd (eds.), *Simple heuristics that make us smart*, pp. 257–285. Oxford University Press.

Boas, Franz (1955). *Primitive art*. New York: Dover.

Bock, G. R., & Marsh, J. (eds.). (1993). *Experimental and theoretical studies of consciousness*. New York: John Wiley.

Boden, M. (1991). *The creative mind*. New York: Basic Books.

Boden, M. (ed.). (1994). *Explorations in creativity*. Cambridge, MA: MIT Press.

Boone, J. L. (1998). The evolution of magnanimity: When is it better to give than to receive? *Human Nature*, *9*(1), 1–21.

Borgia, G. (1986). Sexual selection in bowerbirds. *Scientific American*, *254*(6), 92–100.

Borgia, G. (1995). Complex male display and female choice in the spotted bowerbird: Specialized functions for different bower decorations. *Animal Behavior*, *49*, 1291–1301.

Botwin, M., Buss, D. M., & Schackelford, T. K. (1997). Personality and mate preferences: Five factors in mate selection and marital satisfaction. *Journal of Personality*, *65*, 107–136.

Bouchard, T. J. (1994). Genes, environment, and personality. *Science*, *264*, 1700–1701.

Bowler, P. J. (1983). *The eclipse of Darwinism: Anti-Darwinian evolution theories in the decades around 1900*. Baltimore, MD: Johns Hopkins University Press.

Box, J. F. (1978). *R. A. Fisher: The life of a scientist*. New York: John Wiley.

Boyce, M. S. (1990). The Red Queen visits sage grouse leks. *American Zoologist*, *30*, 263–270.

Boyd, R., & Richerson, P. J. (1985). *Culture and the evolutionary process*. University of Chicago Press.

Boyd, R., & Richerson, P. J. (1990). Group selection among alternative evolutionarily stable strategies. *Journal of Theoretical Biology*, *145*, 331–342.

Boyd, R., & Richerson, P. J. (1992). Punishment allows the evolution of cooperation (or anything else) in sizeable groups. *Ethology and Sociobiology*, *13*, 171–195.

Boyd, R., & Silk, J. B. (1997). *How humans evolved*. New York: Norton.

Boyer, P. (1994). *The naturalness of religious ideas: A cognitive theory of religion*. Berkeley: University of California Press.

Bradbury, J. W., & Andersson, M. B. (eds.). (1987). *Sexual selection: Testing the alternatives*. New York: John Wiley.

Bradbury, J. W., & Vehrencamp, S. L. (1998). *Principles of animal communication*. Sunderland, MA: Sinauer.

Brody, N. (1992). *Intelligence*. New York: Academic Press.

Brown, D. (1991). *Human universals*. New York: McGraw-Hill.

Browne, J. (1995). *Charles Darwin: Voyaging*. London: Random House.

Browne, K. (1998). *Divided labours: An evolutionary view of women at work*. London: Weidenfeld & Nicholson.

Bullock, S. G., & Cliff, D. (1997). The role of "hidden preferences" in the artificial co-evolution of symmetrical signals. *Proceedings of the Royal Society (London) B*, *264*, 505–511.

Burley, N. (1988). Wild zebra finches have band-color preferences. *Animal Behavior*, *36*, 1235–1237.

Burling, R. (1986). The selective advantage of complex language. *Ethology and Sociobiology*, 7, 1–16.

Buss, D. M. (1985). Human mate selection. *American Scientist*, *73*, 47–51.

Buss, D. M. (1989). Sex differences in human mate selection: Evolutionary hypotheses tested in 37 cultures. *Behavioral and Brain Sciences*, *12*, 1–49.

Buss, D. M. (1991). Evolutionary personality psychology. *Annual Review of Psychology*, *42*, 459–491.

Buss, D. M. (1994). *The evolution of desire: Strategies of human mating*. New York: Basic Books.

Buss, D. M. (1999). *Evolutionary psychology: The new science of mind*. New York: Allyn & Bacon.

Buss, D. M., *et al.* (1990). International preferences in selecting mates: A study of 37 cultures. *J. Cross-Cultural Psychology*, *21*, 5–47.

Buss, D. M., Malamuth, N. (eds.). (1996). *Sex, power, conflict: Evolutionary and feminist perspectives*. Oxford University Press.

Buss, D. M., & Schmitt, P. (1993). Sexual strategies theory: An evolutionary perspective on human mating. *Psychological Review*, *100*, 204–232.

Byrne, R. (1995). *The thinking ape: Evolutionary origins of intelligence.* Oxford University Press.

Byrne, R., & Whiten, A. (eds.). (1988). *Machiavellian intelligence: Social expertise and the evolution of intellect in monkeys, apes, and humans.* Oxford University Press.

Campbell, D. T. (1960). Blind variation and selective retention in creative thought as in other knowledge processes. *Psychological Review, 67,* 380–400.

Campbell, D. T. (1988). *Methodology and epistemology for social science: Selected papers.* University of Chicago Press.

Cant, J. (1981). Hypothesis for the evolution of human breasts and buttocks. *American Naturalist, 117,* 199–204.

Caro, T. M., & Sellen, D. W. (1990). The reproductive advantages of fat in women. *Ethology and Sociobiology, 11*(1), 51–66.

Cashdan, E. (1997). Women's mating strategies. *Evolutionary Anthropology, 5,* 134–142.

Catchpole, C. K., & Slater, P. J. B. (1995). *Bird song: Biological themes and variations.* Cambridge University Press.

Chadwick, W. (1990). *Women, art, and society.* London: Thames & Hudson.

Chagnon, N. (1983). *Yanomamo: The fierce people* (3rd ed.). New York: Holt, Rinehart, & Winston.

Chalmers, D. (1996). *The conscious mind: In search of a fundamental theory.* Oxford University Press.

Chance, M. R. A. (1957). The role of convulsions in behavior. *Behavioral Science, 2,* 30–45.

Chance, M. R. A., & Russell, W. M. S. (1959). Protean displays: A form of allaesthetic behavior. *Proceedings of the Zoological Society of London, 132,* 65–70.

Chernow, R. (1998). *Titan: The life of John D. Rockefeller, Sr.* London: Little, Brown.

Chomsky, N. (1980). Rules and representations. *Behavioral and Brain Sciences, 3,* 1–61.

Chorney, M. J., *et al.* (1997). A quantitative trait locus associated with cognitive ability in children. *Psychological Science, 9,* 1–8.

Clark, R. D., & Hatfield, E. (1989). Gender differences in receptivity to sexual offers. *Journal of Psychology and Human Sexuality, 2,* 39–55.

Clay, F. (1908). *The origin of the sense of beauty.* London: John Murray.

Cliff, D., & Miller, G. F. (1995). Tracking the Red Queen: Methods for measuring co-evolutionary progress in open-minded simulations. In F. Moran *et al.* (eds.), *Advances in artificial life,* pp. 200–218. Berlin: Springer-Verlag.

Clutton-Brock, T. H. (1991). *The evolution of parental care.* Princeton University Press.

Clutton-Brock, T. H., & Parker, G. A. (1995a). Sexual coercion in animal societies. *Animal Behavior, 49,* 1345–1365.

Clutton-Brock, T. H., & Parker, G. A. (1995b). Punishment in animal societies. *Nature, 373,* 209–216.

Coen, E. (1999). *The art of genes: How organisms make themselves.* Oxford University Press.

Conroy, G. C. (1997). *Reconstructing human origins: A modern synthesis.* New York: W. W. Norton.

Constable, J. (1997). Verse form: A pilot study in the epidemiology of representations. *Human Nature, 8,* 171–203.

Cooke, B., & Turner, F. (eds.). (1999). *Biopoetics: Evolutionary explorations in the arts.* St. Paul, MN: Icus Books.

Corballis, M. C. (1991). *The lopsided ape: Evolution of the generative mind.* Oxford University Press.

Cosmides, L. (1989). The logic of social exchange. Has natural selection shaped how humans reason? Studies with the Wason selection task. *Cognition, 31*, 187–276.

Cosmides, L., & Tooby, J. (1992). Cognitive adaptations for social exchange. In J. H. Barkow *et al.* (eds.), *The adapted mind*, pp. 163–228. Oxford University Press.

Cosmides, L., & Tooby, J. (1994). Origins of domain specificity: The evolution of functional organization. In L. A. Hirschfeld & S. A. Gelman (eds.), *Mapping the mind*, pp. 85–116. Cambridge University Press.

Cosmides, L., & Tooby, J. (1997). Dissecting the computational architecture of social inference mechanisms. In G. R. Bock & G. Cardew (eds.), *Characterizing human psychological adaptations*, pp. 132–161. New York: John Wiley.

Court, J. H. (1983). Sex differences in performance on Raven's Progressive Matrices: A review. *Alberta Journal of Educational Research, 29*, 54–74.

Crawford, C. B., & Krebs, D. (eds.). (1998). *Handbook of evolutionary psychology: Ideas, issues, and applications.* Mahwah, NJ: Erlbaum.

Crawford, V. (1998). A survey of experiments on communication via cheap talk. *Journal of Economic Theory, 78*, 286–298.

Cronin, H. (1991). *The ant and the peacock: Altruism and sexual selection from Darwin to today.* Cambridge University Press.

Crow, J. F. (1997). The high spontaneous mutation rate: Is it a health risk? *Proceedings of the National Academy of Sciences (USA), 94*, 8380–8386.

Crow, J. F. (1999). The odds of losing at genetic roulette. *Nature, 397*, 293–294.

Crow, T. J. (1998). Sexual selection, timing and the descent of man: A theory of the genetic origins of language. *Current Psychology of Cognition, 17*(6), 1079–1114.

Crystal, D. (1987). *The Cambridge encyclopedia of language.* New York: Cambridge University Press.

Cummings, D. D., & Allen, C. (eds.). (1998). *The evolution of mind.* Oxford University Press.

Cunningham, M. R., Druen, P. B., & Barbee, A. P. (1997). Angels, mentors, and friends: Trade-offs among evolutionary, social, and individual variables in physical appearance. In J. A. Simpson & D. T. Kenrick (eds.), *Evolutionary social psychology*, pp. 109–140. Mahwah, NJ: Erlbaum.

Dahlberg, F. (ed.). (1981). *Woman the gatherer.* New Haven, CT: Yale University Press.

Daly, M., & Wilson, M. (1983). *Sex, evolution, and behavior. Adaptations for reproduction* (2nd ed.). Boston: Willard Grant Press.

Daly, M., & Wilson, M. (1988). *Homicide.* New York: Aldine de Gruyter.

Daly, M., & Wilson, M. (1995). Discriminative parental solicitude and the relevance of evolutionary models to the analysis of motivational systems. In M. Gazzaniga (ed.), *The cognitive neurosciences*, pp. 1269–1286. Cambridge, MA: MIT Press.

Daly, M., & Wilson, M. (1998). *The truth about Cinderella: A Darwinian view of parental love.* London: Weidenfeld & Nicholson.

Daly, M., Salmon, C., & Wilson, M. (1997). Kinship: The conceptual hole in psychological studies of social cognition and close relationships. In J. A. Simpson & D. T. Kenrick (eds.), *Evolutionary social psychology*, pp. 265–296. Mahwah, NJ: Erlbaum.

Darwin, C. (1859). *On the origin of species by means of natural selection.* London: John Murray. (Reprinted in 1964 by Harvard University Press.)

Darwin, C. (1871). *The descent of man, and selection in relation to sex* (2 vols). London: John Murray. (Reprinted in 1981 by Princeton University Press.)

Dawkins, R. (1976). *The selfish gene.* Oxford University Press.
Dawkins, R. (1982). *The extended phenotype: The gene as the unit of selection.* Oxford: Freeman.
Dawkins, R. (1986). *The blind watchmaker.* New York: W. W. Norton.
Dawkins, R. (1993). Viruses of the mind. In B. Dahlbom (ed.), *Dennett and his critics,* pp. 13–27. Oxford: Blackwell.
Dawkins, R., & Krebs, J. R. (1978). Animal signals: Information or manipulation? In J. R. Krebs & N. B. Davies (eds.), *Behavioral ecology,* pp. 282–309. Oxford: Blackwell Scientific.
Dawkins, R., & Krebs, J. R. (1979). Arms races between and within species. *Proceedings of the Royal Society (London) B, 205,* 489–511.
Deacon, T. (1998). *The symbolic species: The co-evolution of language and the human brain.* London: Penguin.
Dennett, D. (1991). *Consciousness explained.* Boston: Little, Brown.
Dennett, D. (1995). *Darwin's dangerous idea: Evolution and the meaning of life.* New York: Simon & Schuster.
Desmond, A., & Moore, J. (1994). *Darwin.* New York: W. W. Norton.
Dessalles, J.-L. (1998). Altruism, status and the origin of relevance. In J. Hurford *et al.* (eds.), *Approaches to the evolution of language,* pp. 140–147. Cambridge University Press.
Diamond, J. (1988). Experimental study of bower decoration by the bowerbird *Amblyornis inornatus,* using colored poker chips. *American Naturalist, 131,* 631–653.
Diamond, J. (1992). *The third chimpanzee: The evolution and future of the human animal.* New York: HarperCollins.
Dijkstra, B. (1986). *Idols of perversity: Fantasies of feminine evil in fin-de-siècle culture.* Oxford University Press.
Dissanayake, E. (1988). *What is art for?* Seattle, WA: University of Washington Press.
Dissanayake, E. (1992). *Homo aestheticus: Where art comes from and why.* New York: Free Press.
Dixson, A. F. (1998). *Primate sexuality: Comparative studies of the prosimians, monkeys, apes, and human beings.* Oxford University Press.
Donald, M. (1991). *Origins of the modern mind: Three stages in the evolution of culture and cognition.* Cambridge, MA: Harvard University Press.
Driver, P. M., & Humphries, D. A. (1988). *Protean behavior: The biology of unpredictability.* Oxford: Clarendon Press.
Dugatkin, L. A. (1997). *Cooperation among animals: An evolutionary perspective.* Oxford University Press.
Dugatkin, L. A., & Reeve, H. K. (eds.). (1998). *Game theory and animal behavior.* Oxford University Press.
Dunbar, R. (1988). *Primate social systems.* London: Croom Helm.
Dunbar, R. (1996). *Grooming, gossip, and the evolution of language.* London: Faber & Faber.
Dunbar, R., Knight, C., & Power, C. (eds.). (1999). *The evolution of culture: An interdisciplinary view.* Edinburgh University Press.
Dunbar, R., Marriot, A., & Dundan, N. D. C. (1997). Human conversational behavior. *Human Nature, 8,* 231–246.
Eaton, S. B., Shostak, M., & Konner, M. (1988). *The paleolithic prescription: A program of diet and exercise and a design for living.* New York: Harper & Row.
Eberhard, W. G. (1985). *Sexual selection and animal genitalia.* Cambridge, MA: Harvard University Press.

Eberhard, W. G. (1996). *Female control: Sexual selection by cryptic female choice.* Princeton University Press.

Eibl-Eibesfelt, I. (1989). *Human ethology*. New York: Aldine de Gruyter.

Eigen, M. (1992). *Steps towards life: A perspective on evolution.* Oxford University Press.

Einon, D. (1998). How many children can one man have? *Evolution and Human Behavior, 19*, 413–426.

Ekman, P. (1992). *Telling lies: Clues to deceit in the marketplace, marriage, and politics* (2nd ed.). New York: Norton.

Eldredge, N. (1989). *Macroevolutionary dynamics: Species, niches, and adaptive peaks.* New York: McGraw-Hill.

Ellis, B., & Symons, D. (1990). Sex differences in sexual fantasy: An evolutionary psychological approach. *Journal of Sex Research, 27*, 527–556.

Ellis, H. (1905). *Sexual selection in man.* Philadelphia: F. A. Davis.

Ellis, H. (1934). *Man and woman: A study of secondary and tertiary sexual characters* (8th ed.). London: W. Heinemann.

Ellis, L. (1995). Dominance and reproductive success among nonhuman animals: A cross-species comparison. *Ethology and Sociobiology, 16*, 257–333.

Endler, J. A. (1992). Signals, signal conditions, and the direction of evolution. *American Naturalist, 139*, S125–S153.

Endler, J. A., & Basolo, A. L. (1998). Sensory ecology, receiver biases and sexual selection. *Trends in Ecology and Evolution, 13*(10), 415–420.

Enquist, M., & Arak, A. (1993). Selection of exaggerated male traits by female aesthetic senses. *Nature, 361*, 446–448.

Erdal, D., & Whiten, A. (1994). On human egalitarianism: An evolutionary product of Machiavellian status escalation? *Current Anthropology, 35*, 175–183.

Etcoff, N. (1999). *Survival of the prettiest: The science of beauty.* New York: Doubleday.

Eyre-Walker, A., & Keightley, P. D. (1999). High genomic deleterious mutation rates in hominids. *Nature, 397*, 344–346.

Eysenck, H. J. (1992). The definition and measurement of psychoticism. *Personality and Individual Differences, 13*, 667–673.

Falk, D. (1997). Brain evolution in females: An answer to Mr. Lovejoy. In L. Hager (ed.), *Women in human evolution*, pp. 114–136. London: Routledge.

Falk, D., Froese, N., Sade, D. S., & Dudek, B. C. (1999). Sex differences in brain/body relationships of Rhesus monkeys and humans. *Journal of Human Evolution, 36*, 233–238.

Farrell, J., & Rabin, M. (1996). Cheap talk. *Journal of Economic Perspectives, 10*(3), 103–118.

Feagin, S., & Maynard, P. (eds.). (1997). *Aesthetics.* Oxford University Press.

Feingold, A. (1992). Sex differences in variability in intellectual abilities: A new look at an old controversy. *Review of Educational Research, 61*(1), 61–84.

Feingold, A. (1994). Gender differences in variability in intellectual abilities: A cross-cultural perspective. *Sex Roles*, 30, 81–92.

Fisher, H. (1982). *The sex contract.* New York: Morrow.

Fisher, H. (1992). *Anatomy of love: The natural history of monogamy, adultery, and divorce.* New York: Simon & Schuster.

Fisher, H. (1999). *First sex: The natural talents of women and how they are changing the world.* New York: Random House.

Fisher, R. A. (1915). The evolution of sexual preference. *Eugenics Review, 7*, 184–192.

Fisher, R. A. (1930). *The genetical theory of natural selection.* Oxford: Clarendon Press.

Foley, R. (1996). The adaptive legacy of human evolution: A search for the EEA. *Evolutionary Anthropology, 4*(6), 194–203.

Foley, R. (1997). *Humans before humanity: An evolutionary perspective.* Oxford: Blackwell.

Foley, R. A., & Lee, P. C. (1991). Ecology and energetics of encephalization in hominid evolution. *Philosophical Transactions of the Royal Society (London) B, 334*, 223–232.

Frank, R. (1988). *Passions within reason: The strategic role of the emotions.* New York: Norton.

Frank, R. (1999). *Luxury fever: Why money fails to satisfy in an era of excess.* New York: Free Press.

Freud, S. (1910). *Leonardo da Vinci and a memory of his childhood.* In *Standard edition*, Vol. 10, 153–318.

Freud, S. (1930). *Civilization and its discontents.* In *Standard edition*, Vol. 21, 59–145.

Freud, S. (1953–1974). *The standard edition of the complete psychological works of Sigmund Freud* (Vols. 1–24; J. Strachey, ed. and trans.). London: Hogarth Press.

Frey, B. S., and Eichenberger, R. (1996) Marriage paradoxes. *Rationality and Society, 8*, 187–206.

Freyd, J. J. (1983). Shareability: The social psychology of epistemology. *Cognitive Science, 7*, 191–210.

Furlow, F. B., Armijo-Prewitt, T., Gangestad, S. W., & Thornhill, R. (1997). Fluctuating asymmetry and psychometric intelligence. *Proceedings of the Royal Society (London) B, 264*, 823–830.

Gallup, G. G. (1982). Permanent breast enlargement in human females: A sociobiological analysis. *Journal of Human Evolution, 11*, 597–601.

Gangestad, S. W. (1993). Sexual selection and physical attractiveness: Implications for mating dynamics. *Human Nature, 4*, 205–235.

Gangestad, S. W. (1997). Evolutionary psychology and genetic variation: Non-adaptive, fitness-related and adaptive. In G. R. Bock & G. Cardew (eds.), *Characterizing human psychological adaptations*, pp. 212–230. New York: John Wiley.

Gangestad, S. W., and Thornhill, R. (1997). Human sexual selection and developmental stability. In J. A. Simpson & D. T. Kenrick (eds.), *Evolutionary social psychology*, pp. 169–195. Mahwah, NJ: Erlbaum.

Gangestad, S. W., Thornhill, R., & Yeo, R. A. (1994). Facial attractiveness, developmental stability, and fluctuating symmetry. *Ethology and Sociobiology, 15*, 73–85.

Ghiselin, B. (ed.). (1952). *The creative process.* Berkeley, CA: University of California Press.

Gibbons, A. (1998). Which of our genes made us human? *Science, 281*, 1432–1434.

Gibson, K. R., & Ingold, T. (eds.). (1993). *Tools, language and cognition in human evolution.* Cambridge University Press.

Gigerenzer, G., & Goldstein, D. (1996). Reasoning the fast and frugal way: Models of bounded rationality. *Psychological Review, 103*, 650–669.

Gigerenzer, G., & Hug, K. (1992). Domain-specific reasoning: Social contracts, cheating, and perspective change. *Cognition, 43*, 127–171.

Gigerenzer, G., & Murray, J. L. (1987). *Cognition as intuitive statistics.* Hillsdale, NJ: Erlbaum.

Gigerenzer, G., & Todd, P. (eds.) (1999). *Simple heuristics that make us smart.* New York: Oxford University Press.

Gilligan, C. (1982). *In a different voice: Psychological theory and women's development.* Cambridge, MA: Harvard University Press.

Goldberg, S. (1993). *Why men rule: A theory of male dominance*. Chicago: Open Court.

Gombrich, E. H. (1984). *The sense of order: A study in the psychology of decorative art* (2nd ed.). London: Phaidon Press.

Goodall, J. (1986). *The chimpanzees of Gombe: Patterns of behavior*. Cambridge, MA: Harvard University Press.

Gould, J. L., & Gould, C. G. (1997). *Sexual selection*. New York: Scientific American Library.

Gould, S. J. (1977). *Ontogeny and phylogeny*. Cambridge, MA: Harvard University Press.

Gould, S. J. (1987). Freudian slip. *Natural History*, Feb., 14–19.

Gould, S. J., & Eldredge, N. (1977). Punctuated equilibria: The tempo and mode of evolution reconsidered. *Paleobiology*, *3*, 115–151.

Gowaty, P. A. (ed.). (1997). *Feminism and evolutionary biology*. New York: Chapman & Hall.

Grafen, A. (1990). Biological signals as handicaps. *Journal of Theoretical Biology*, *144*, 517–546.

Grammer, K. (1989). Human courtship behavior: Biological basis and cognitive processing. In A. E. Rasa *et al.* (eds.), *The sociobiology of sexual and reproductive strategies*, pp. 147–169. New York: Routledge.

Grammer, K. (1991). Strangers meet: Laughter and nonverbal signs of interest in opposite-sex encounters. *Journal of Nonverbal Behavior*, *14*(4), 209–236.

Grammer, K., & Thornhill, R. (1994). Human facial attractiveness and sexual selection: The role of symmetry and averageness. *Journal of Comparative Psychology*, *108*, 233–242.

Graziano, W. G., Jensen-Campbell, L. A., Todd, M., & Finch, J. F. (1997). Interpersonal attraction from an evolutionary psychology perspective: Women's reactions to dominant and prosocial men. In J. A. Simpson & D. T. Kenrick (eds.), *Evolutionary social psychology*, pp. 141–167. Mahwah, NJ: Erlbaum.

Gregor, T. (1985). *Anxious pleasures: The sexual lives of an Amazonian people*. University of Chicago Press.

Griffith, S. C., Owens, I. P. F., & Burke, T. (1999). Environmental determination of a sexually selected trait. *Nature*, *400*, 358–360.

Grosse, E. (1897). *The beginnings of art*. New York.

Guilford, T., & Dawkins, M. S. (1991). Receiver psychology and the evolution of animal signals. *Animal Behavior*, *42*, 1–14.

Haldane, J. B. S. (1932). *The causes of evolution*. London: Longman, Green.

Halpern, D. (1992). *Sex differences in cognitive abilities* (2nd ed.). Hillsdale, NJ: Erlbaum.

Hamer, D., & Copeland, P. (1998). *Living with our genes*. New York: Doubleday.

Hamilton, W. D. (1996). *Narrow roads of gene land*. Oxford: W. H. Freeman.

Hamilton, W. D., Axelrod, R., & Tanese, R. (1990). Sexual reproduction as an adaptation to resist parasites (A review). *Proceedings of the National Academy of Sciences (USA)*, *87*, 3566–3573.

Hamilton, W. D., & Zuk, M. (1982). Heritable true fitness and bright birds: A role for parasites? *Science*, *218*, 384–387.

Haraway, D. (1989). *Primate visions: Gender, race, and nature in the world of modern science*. New York: Routledge.

Hare, R. D. (1999). *Without conscience: The disturbing world of the psychopaths among us*. New York: Guilford Press.

Harris, J. R. (1998). *The nurture assumption*. London: Bloomsbury.

Harvey, P. H., & Bradbury, J. W. (1993). Sexual selection. In J. R. Krebs & N. B.

Davies (eds.), *Behavioral ecology: An evolutionary approach* (3rd ed.), pp. 203–233. London: Blackwell Scientific.

Harvey, P. H., & Harcourt, A. H. (1984). Sperm competition, testes size, and breeding systems in primates. In R. Smith (ed.), *Sperm competition and the evolution of animal mating systems*, pp. 589–659. New York: Academic Press.

Hasselquist, D., Bensch, S., & von Schantz, T. (1996). Correlation between male song repertoire, extra-pair paternity and offspring survival in the great reed warbler. *Nature*, *381*, 229–232.

Hasson, O. (1990). The role of amplifiers in sexual selection: An integration of the amplifying and Fisherian mechanisms. *Evolutionary Ecology*, *4*, 277–289.

Hatfield, E., & Rapson, R. L. (1996). *Love and sex: Cross-cultural perspectives*. Boston: Allyn & Bacon.

Hauser, M. (1996). *The evolution of communication*. Cambridge, MA: Harvard University Press.

Hausman, C. R. (1997). *Charles S. Peirce's evolutionary philosophy*. Cambridge University Press.

Haviland, W. A. (1996). *Cultural anthropology* (8th ed.). New York: Harcourt Brace Jovanovich.

Hawkes, K. (1991). Showing off: Tests of another hypothesis about men's foraging goals. *Ethology and Sociobiology*, *12*, 29–54.

Hawkes, K. (1993). Why hunter-gatherers work: An ancient version of the problem of public goods. *Current Anthropology*, *34*, 341–361.

Hersey, G. L. (1996). *The evolution of allure: Art and sexual selection from Aphrodite to the Incredible Hulk*. Cambridge, MA: MIT Press.

Hewlett, B. S. (ed.). (1992). *Father-child relations*. New York: Aldine de Gruyter.

Higgins, K. M. (ed.). (1996). *Aesthetics in perspective*. New York: Harcourt Brace.

Hilgard, E. R. (1987). *Psychology in America: A historical survey*. New York: Harcourt Brace Jovanovich.

Hill, K., & Hurtado, A. M. (1996). *Ache life history: The ecology and demography of a foraging people*. New York: Aldine de Gruyter.

Hinde, R. (1999). *Why gods persist: A scientific approach to religion*. London: Routledge.

Hirschfeld, L. A., & Gelman, S. (1994). *Mapping the mind: Domain specificity in cognition and culture*. Cambridge University Press.

Hoelzer, G. A. (1989). The good parent process of sexual selection. *Animal Behavior*, *38*(6), 1067–1078.

Hofstadter, D. (1995). *Fluid concepts and creative analogies*. New York: Basic Books.

Houle, D. (1998). How should we explain variation in the genetic variabilities of traits? *Genetica*, *102/103*, 241–253.

Houle, D. (on press). Is there a *g* factor for fitness? In J. Goode (ed.), *The nature of intelligence*. New York: John Wiley.

Houle, D., Hoffmaster, D., Assimacopolous, S., & Charlesworth, B. (1992). The genomic mutation rate for fitness in *Drosophila*. *Nature*, *359*, 58–60.

Houle, D., *et al.* (1994). The effects of spontaneous mutation on quantitative traits. I. Variances and covariances of life history traits. *Genetics*, *138*, 773–785.

Howard, D. J., & Berlocher, S. H. (eds.). (1998). *Endless forms: Species and speciation*. Oxford University Press.

Hrdy, S. B. (1981). *The woman that never evolved*. Cambridge, MA: Harvard University Press.

Hrdy, S. B. (1997). Raising Darwin's consciousness: Female sexuality and the prehominid origins of patriarchy. *Human Nature*, *8*, 1–50.

Hrdy, S. B. (1999). *Mother nature: A history of mothers, infants, and natural selection.* New York: Pantheon.

Humphrey, N. (1976). The social function of intellect. In P. P. G. Bateson & R. A. Hinde (eds.), *Growing points in ethology*, pp. 303–317. Cambridge University Press.

Humphrey, N. (1992). *A history of the mind.* New York: Simon & Schuster.

Hurford, J., Studdert-Kennedy, M., & Knight, C. (eds.). (1998). *Approaches to the evolution of language.* Cambridge University Press.

Huxley, J. S. (1938a). The present standing of the theory of sexual selection. In G. R. de Beer (ed.), *Evolution: Essays on aspects of evolutionary biology*, pp. 11–42. Oxford: Clarendon Press.

Huxley, J. S. (1938b). Darwin's theory of sexual selection and the data subsumed by it, in the light of recent research. *American Naturalist, 72*, 416–433.

Huxley, J. S. (1942). *Evolution: The Modern Synthesis.* New York: Harper.

Huxley, T. H. (1989). *Evolution and ethics.* Princeton University Press. (First published 1894).

Illouz, E. (1997). *Consuming the romantic utopia: Love and the cultural contradictions of capitalism.* Berkeley: University of California Press.

Iwasa, Y., & Pomiankowski, A. (1995). Continual change in mate preferences. *Nature, 377*, 420–422.

Iwasa, Y., Pomiankowski, A., & Nee, S. (1991). The evolution of costly mate preferences. II. The "handicap" principle. *Evolution, 45*, 1431–1442.

Jacobs, L. F. (1996). Sexual selection and the brain. *Trends in Ecology and Evolution, 11*(2), 82–86.

Jankowiak, W. (ed.). (1995). *Romantic passion: A universal experience?* New York: Columbia University Press.

Jensen, A. (1998). *The g factor: The science of mental ability.* London: Praeger.

Jensen-Campbell, L. A., Graziano, W. G., & West, S. (1995). Dominance, prosocial orientation, and female preferences: Do nice guys really finish last? *Journal of Personality and Social Psychology, 68*, 427–440.

Jerison, H. J. (1973). *Evolution of the brain and intelligence.* New York: Academic Press.

Jerison, H. J., & Jerison, I. (eds.). (1988). *Intelligence and evolutionary biology.* New York: Springer-Verlag.

Johanson, D. C., & Edgar, B. (1996). *From Lucy to language.* London: Weidenfeld & Nicholson.

Johanson, D. C., & O'Farrell, K. (1990). *Journey from the dawn: Life with the world's first family.* New York: Villard Books.

Johnstone, R. A. (1995). Sexual selection, honest advertisement and the handicap principle. *Biological Review, 70*, 1–65.

Johnstone, R. A. (1997). The tactics of mutual mate choice and competitive search. *Behavioral Ecology and Sociobiology, 40*(1), 51–59.

Johnstone, R. A., Reynolds, J. D., & Deutsch, J. C. (1996). Mutual mate choice and sex differences in choosiness. *Evolution, 50*, 1382–1391.

Jones, I. L., & Hunter, F. M. (1993). Mutual sexual selection in a monogamous seabird. *Nature, 36*, 238–239.

Jones, S., Martin, R., & Pilbeam, D. (eds.). (1992). *The Cambridge encyclopedia of human evolution.* New York: Cambridge University Press.

Kagel, J. H., & Roth, A. E. (1995). *The handbook of experimental economics.* Princeton University Press.

Kahane, H. (1995). *Contract ethics: Evolutionary biology and the moral sentiments.* London: Rowman & Littlefield.

Kalick, S. M., Johnson, R. B., Lebrowitz, L. A., & Langlois, J. H. (1998). Does human facial attractiveness honestly advertise health? Longitudinal data on an evolutionary question. *Psychological Science*, *9*(1), 8–13.

Keller, E. F., & Lloyd, E. A. (eds.). (1992). *Keywords in evolutionary biology*. Cambridge, MA: Harvard University Press.

Keller, M. C., Theissen, D., & Young, R. K. (1996). Mate assortment in dating and married couples. *Personality and Individual Differences*, *21*, 217–221.

Kenrick, D. T., & Keefe, R. C. (1992). Age preferences in mates reflects sex differences in reproductive strategies. *Behavioral and Brain Sciences*, *15*, 75–133.

Kenrick, D. T., Sadalla, E. K., Groth, G., & Trost, M. R. (1990). Evolution, traits, and the stages of human courtship: Qualifying the parental investment model. *Journal of Personality*, *58*, 97–116.

Kingdon, J. (1993). *Self-made man and his undoing*. New York: Simon & Schuster.

Kirkpatrick, M. (1982). Sexual selection and the evolution of female choice. *Evolution*, *36*, 1–12.

Kirkpatrick, M., Price, T., & Arnold, S. J. (1990). The Darwin-Fisher theory of sexual selection in monogamous birds. *Evolution*, *44*(1), 180–193.

Kirkpatrick, M., & Ryan, M. J. (1991). The evolution of mating preferences and the paradox of the lek. *Nature*, *350*, 33–38.

Knight, C. (1995). *Blood relations: Menstruation and the origins of culture*. New Haven, CT: Yale University Press.

Knight, C. (1998). Ritual/speech coevolution: A solution to the problem of deception. In J. Hurford *et al.* (eds.), *Approaches to the evolution of language*, pp. 68–91. Cambridge University Press.

Knight, C., Power, C., & Watts, I. (1995). The human symbolic revolution: A Darwinian account. *Cambridge Archaeological Journal*, *5*, 75–114.

Koestler, A. (1964). *The act of creation*. New York: Dell.

Köhler, W. (1925). *The mentality of apes*. New York: Harcourt, Brace, & World.

Kohn, M. (1999). *As we know it: Coming to terms with an evolved mind*. London: Granta.

Kohn, M., & Mithen, S. (1999). Handaxes: Products of sexual selection? *Antiquity*, *73*, 518–526.

Kondrashov, A. S. (1988a). Deleterious mutations and the evolution of sexual reproduction. *Nature*, *336*, 435–440.

Kondrashov, A. (1988b). Deleterious mutations as an evolutionary factor. III. Mating preference and some general remarks. *Journal of Theoretical Biology*, *131*, 487–496.

Kondrashov, A. (1995). Contamination of the genomes by very slightly deleterious mutations: Why have we not died 100 times over? *Journal of Theoretical Biology*, *175*, 583–594.

Krebs, J. R., & Davies, N. B. (eds.). (1997). *Behavioral ecology: An evolutionary approach* (4th ed.). Oxford: Blackwell Scientific.

Krebs, J. R., & Dawkins, R. (1984). Animal signals: Mind reading and manipulation. In J. R. Krebs & N. B. Davies (eds.), *Behavioral ecology: An evolutionary approach* (2nd ed.), pp. 380–402, Oxford: Blackwell Scientific.

Kreps, D. (1990). *Game theory and economic modelling*. Oxford University Press.

Krings, M., Stone, A., Schmitz, R. W., Krainitzki, H., Stoneking, M., & Pääbo, S. (1997). Neanderthal DNA sequences and the origin of modern humans. *Cell*, *90*, 19–30.

Kuklick, B. (1977). *The rise of American philosophy: Cambridge, Massachusetts, 1860–1930*. New Haven, CT: Yale University Press.

Lambert, D. M., & Spencer, H. G. (eds.). (1995). *Speciation and the recognition concept: Theory and application.* Baltimore, MD: Johns Hopkins University Press.

Lancaster, J. B. (1991). A feminist and evolutionary biologist looks at women. *Yearbook of Physical Anthropology, 34,* 1–11.

Landau, M. (1991). *Narratives of human evolution.* New Haven, CT: Yale University Press.

Lande, R. (1981). Models of specification by sexual selection on polygenic characters. *Proceedings of the National Academy of Sciences (USA), 78,* 3721–3725.

Lande, R. (1987). Genetic correlation between the sexes in the evolution of sexual dimorphism and mating preferences. In J. W. Bradbury & M. B. Andersson (eds.), *Sexual selection: Testing the alternatives,* pp. 83–95. New York: John Wiley.

Langlois, J. H., Roggmann, L. A., & Musselman, L. (1994). What is average and what is not average about attractive faces? *Psychological Science, 5,* 214–220.

Layton, R. (1991). *The anthropology of art* (2nd ed.). Cambridge University Press.

Lee, R. B. (1979). *The !Kung San: Men, women, and work in a foraging society.* Cambridge University Press.

Lee, R. B., & DeVore, I. (eds.). (1968). *Man the hunter.* Chicago: Aldine.

Lenain, T. (1997). *Monkey painting.* London: Reaktion Books.

Lieberman, P. (1991). *Uniquely human: The evolution of speech, thought, and selfless behavior.* Cambridge, MA: MIT Press.

Locke, J. (1998). *The devoicing of society: Why we don't talk to each other any more.* New York: Simon & Schuster.

Loehlin, J. C. (1992). *Genes and environment in personality development.* Newbury Park, CA: Sage.

Lovejoy, C. O. (1981). The origin of man. *Science, 211,* 341–350.

Low, B. S. (1979). Sexual selection and human ornamentation. In N. A. Chagnon & W. Irons (eds.), *Evolutionary biology and human social behavior,* pp. 462–487. Boston: Duxbury Press.

Low, B. S., Alexander, R. M., & Noonan, K. M. (1987). Human hips, breasts, and buttocks: Is fat deceptive? *Ethology and Sociobiology, 8,* 249–257.

Lubinski, D., & Benbow, C. P. (1992). Gender differences in abilities and preferences among the gifted: Implications for the math-science pipeline. *Current Directions in Psychological Science, 1*(2), 61–66.

Ludvico, L. R., & Kurland, J. A. (1995). Symbolic or not-so-symbolic wounds: The behavioral ecology of human scarification. *Ethology and Sociobiology, 16,* 155–172.

Lumsden, C. J., & Wilson, E. O. (1981). *Genes, mind, and culture.* Cambridge, MA: Harvard University Press.

Lykken, D. T. (1995). *The antisocial personalities.* Hillsdale, NJ: Erlbaum.

Lynch, M., & Walsh, B. (1998). *Genetics and analysis of quantitative traits.* Sunderland, MA: Sinauer.

MacDonald, K. (1995). The establishment and maintenance of socially imposed monogamy in western Europe. *Politics and the Life Sciences, 14,* 3–23.

Mackintosh, N. (1994). Intelligence in evolution. In J. Khalfa (ed.), *What is intelligence?,* pp. 27–48. Cambridge University Press.

Maes, P. (1991). A bottom-up mechanism for behavior selection in an artificial creature. In J. A. Meyer & S. W. Wilson (eds.), *From animals to animats,* pp. 238–246. Cambridge, MA: MIT Press.

Manning, J. T., Scutt, D., Whitehouse, G. H., and Leinster, S. J. (1997). Breast asymmetry and phenotypic quality in women. *Evolution and Human Behavior, 18*, 1–13.

Marcuse, H. (1956). *Eros and civilization: A philosophical inquiry into Freud.* London: Routledge & Kegan Paul.

Margulis, L., & Sagan, D. (1991). *Mystery dance: On the evolution of human sexuality.* New York: Summit Books.

Martin, R. D., Willner, L. A., & Dettling, A. (1994). The evolution of sexual size dimorphism in primates. In R. V. Short & E. Balaban (eds.), *The differences between the sexes*, pp. 159–200. Cambridge University Press.

Mascie-Taylor, C. G. N. (1988). Assortative mating from psychometric characters. In C. G. N. Mascie-Taylor & A. J. Boyce (eds.), *Human mating patterns*, pp. 62–82. Cambridge University Press.

Matthews, G., & Deary, I. J. (1998). *Personality traits.* Cambridge University Press.

May, R. M. (1992). How many species inhabit the earth? *Scientific American, 267*(4), 42–48.

Maynard Smith, J. (1956). Fertility, mating behavior and sexual selection in *Drosophila subobscura. Journal of Genetics, 54*, 261–279.

Maynard Smith, J. (1978). *The evolution of sex.* Cambridge University Press.

Maynard Smith, J. (1982). *Evolution and the theory of games.* Cambridge University Press.

Maynard Smith, J. (1985). Sexual selection, handicaps, and true fitness. *Journal of Theoretical Biology, 115*, 1–8.

Maynard Smith, J. (1998). *Evolutionary genetics* (2nd ed.). Oxford University Press.

Maynard Smith, J., & Szathmary, E. (1995). *The major transitions in evolution.* Oxford: W. H. Freeman.

Mayr, E. (1991). *One long argument: Charles Darwin and the genesis of modern evolutionary thought.* Cambridge, MA: Harvard University Press.

Mealey, L. (1995). The sociobiology of sociopathy: An integrated evolutionary model. *Behavioral and Brain Sciences, 18*, 523–599.

Michod, R. E. (1995). *Eros and evolution: A natural philosophy of sex.* New York: Addison-Wesley.

Michod, R. E., & Hasson, O. (1990). On the evolution of reliable indicators of fitness. *American Naturalist, 135*, 788–808.

Michod, R. E., & Levin, B. R. (eds.). (1988). *The evolution of sex: An examination of current ideas.* Sunderland, MA: Sinauer.

Miller, E. E. (1997). Could nonshared environmental variation have evolved to assure diversification through randomness? *Evolution and Human Behavior, 18*, 195–221.

Miller, G. A. (1996). *The science of words.* San Francisco: W. H. Freeman/Scientific American.

Miller, G. F. (1993). *Evolution of the human brain through runaway sexual selection: The mind as a protean courtship device* (2 vols). Ph.D. thesis, Psychology Department, Stanford University.

Miller, G. F. (1994a). Beyond shared fate: Group-selected mechanisms for cooperation and competition in fuzzy, fluid vehicles. *Behavioral and Brain Sciences, 17*(4), 630–631.

Miller, G. F. (1994b). Exploiting mate choice in evolutionary computation: Sexual selection as a process of search, optimization, and diversification. In T. C. Fogarty (ed.), *Evolutionary Computing*, pp. 65–79. Berlin: Springer-Verlag.

Miller, G. F. (1996). Political peacocks. *Demos Quarterly, 10*, 9–11.

Miller, G. F. (1997a). Protean primates: The evolution of adaptive unpredictability in competition and courtship. In A. Whiten & R. W. Byrne (eds.), *Machiavellian intelligence II*, pp. 312–340. Cambridge University Press.

Miller, G. F. (1997b). Mate choice: From sexual cues to cognitive adaptations. In G. R. Bock & G. Cardew (eds.), *Characterizing human psychological adaptations*, pp. 71–87. New York: John Wiley.

Miller, G. F. (1998a). How mate choice shaped human nature: A review of sexual selection and human evolution. In C. Crawford & D. Krebs (eds.), *Handbook of evolutionary psychology*, pp. 87–129. Mahwah, NJ: Erlbaum.

Miller, G. F. (1998b). Review of *The handicap principle* by Amotz Zahavi. *Evolution and Human Behavior, 19*(5), 343–347.

Miller, G. F. (1999a). Waste is good. *Prospect*, Feb., 18–23.

Miller, G. F. (1999b). Sexual selection for cultural displays. In R. Dunbar *et al.* (eds.), *The evolution of culture*, pp. 71–91. Edinburgh University Press.

Miller, G. F. (1999c). Evolution of human music through sexual selection. In N. L. Wallin *et al.* (eds.), *The origins of music.* Cambridge, MA: MIT Press.

Miller, G. F. (2000a). Mental traits as fitness indicators: Expanding evolutionary psychology's adaptationism. *Annals of the New York Academy of Sciences* (on press).

Miller, G. F. (2000b). Sexual selection for intelligence-indicators. In J. Goode (ed.), *The nature of intelligence.* New York: John Wiley (on press).

Miller, G. F., & Cliff, D. (1994). Protean behavior in dynamic games: Arguments for the co-evolution of pursuit-evasion tactics in simulated robots. In D. Cliff *et al.* (eds.), *From Animals to Animats 3*, pp. 411–420. Cambridge, MA: MIT Press.

Miller, G. F., & Todd, P. M. (1993). Evolutionary wanderlust: Sexual selection with directional mate preferences. In J.-A. Meyer *et al.* (eds.), *From Animals to Animats 2*, pp. 21–30. Cambridge, MA: MIT Press.

Miller, G. F., & Todd, P. M., (1995). The role of mate choice in biocomputation: Sexual selection as a process of search, optimization, and diversification. In W. Banzaf & F. Eeckman (eds.), *Evolution and biocomputation*, pp. 169–204. Berlin: Springer-Verlag.

Miller, G. F., & Todd, P. M. (1998). Mate choice turns cognitive. *Trends in Cognitive Sciences, 2*, 190–198.

Mitchell, M. (1998). *An introduction to genetic algorithms.* Cambridge, MA: MIT Press.

Mithen, S. (1996). *The prehistory of the mind: A search for the origins of art, religion, and science.* London: Thames & Hudson.

Mithen, S. (ed.). (1998). *Creativity in human evolution and prehistory.* London: Routledge.

Moller, A. P. (1994). *Sexual selection and the barn swallow.* Oxford University Press.

Moller, A. P., & Alatalo, R. V. (1999). Good-genes effects in sexual selection. *Proceedings of the Royal Society (London) B, 266*, 85–91.

Moller, A., Soler, M., & Thornhill, R. (1995). Breast asymmetry, sexual selection, and human reproductive success. *Ethology and Sociobiology, 16*, 207–216.

Moller, A. P., & Swaddle, J. P. (1998). *Asymmetry, developmental stability and evolution.* Oxford University Press.

Montagu, A. (1989). *Growing young.* Westport, CT: Bergin & Garvey.

Morris, D. (1962). *The biology of art.* New York: Knopf.

Morris, D. (1967). *The naked ape.* New York: Dell.

Morris, D. (1985). *Bodywatching: A field guide to the human species.* New York: Crown Books.

Nasar, S. (1998). *A beautiful mind.* New York: Simon & Schuster.

Neisser, U., *et al.* (1996). Intelligence: Knowns and Unknowns. *American Psychologist, 51*, 77–101.

Neuringer, A. (1986). Can people behave "randomly"? The role of feedback. *Journal of Experimental Psychology: General, 115*(1), 62–75.

Neuringer, A., & Voss, C. (1993). Approximating chaotic behavior. *Psychological Science, 4*(2), 113–119.

Nietzsche, F. (1968). *The will to power*. New York: Vintage. (Trans. W. Kaufmann & R. J. Hollingdale from Nietzsche's notebooks, 1883–1888.)

Nitecki, M. (ed.). (1990). *Evolutionary innovations*. Chicago: University of Chicago Press.

Noë, R., & Hammerstein, P. (1995). Biological markets. *Trends in Ecology and Evolution, 10*, 336–339.

O'Donald, P. (1980). *Genetic models of sexual selection*. Cambridge University Press.

Paglia, C. (1990). *Sexual personae: Art and decadence from Nefertiti to Emily Dickinson*. New Haven, CT: Yale University Press.

Pakkenberg, B., & Gundersen, J. G. (1997). Neocortical neuron number in humans: Effect of sex and age. *Journal of Comparative Neurology, 384*, 312–320.

Parish, A. R. (1993). Sex and food control in the "uncommon chimpanzee": How bonobo females overcome a phylogenetic legacy of male dominance. *Ethology and Sociobiology, 15*(3), 157–179.

Parker, S. T. (1987). A sexual selection model for hominid evolution. *Human Evolution, 2*, 235–253.

Paul, L., & Hirsch, L. R. (1996). Human male mating strategies: II. Moral codes of "quality" and "quantity" strategists. *Ethology and Sociobiology, 17*, 71–86.

Pawlowski, B. (1999). Loss of oestrus and concealed ovulation in human evolution: The case against the sexual selection hypothesis. *Current Anthropology, 40*, 257–275.

Perrett, D. I., *et al.* (1998). Sexual dimorphism and facial attractiveness. *Nature, 394*, 884–886.

Perusse, D. (1992). Cultural and reproductive success in industrial societies: Testing the relationship at the proximate and ultimate levels. *Behavioral and Brain Sciences, 16*, 267–322.

Peterson, D., & Goodall, J. (1993). *Vision of Caliban: On chimpanzees and people*. Boston: Houghton Mifflin.

Petrie, M. (1994). Improved growth and survival of offspring of peacocks with more elaborate trains. *Nature, 371*, 598–599.

Petrie, M., Halliday, T., & Sanders, C. (1991). Peahens prefer peacocks with elaborate trains. *Animal Behavior, 41*, 323–331.

Petrinovich, L. (1998). *Human evolution, reproduction, and morality*. Cambridge, MA: MIT Press.

Pfeiffer, J. (1982). *The creative explosion*. New York: Harper & Row.

Pinker, S. (1994). *The language instinct*. London: Allen Lane.

Pinker, S. (1997). *How the mind works*. New York: Norton.

Pitts, M., & Roberts, M. (1997). *Fairweather Eden: Life in Britain half a million years ago as revealed by the excavations at Boxgrove*. London: Century.

Plomin, R., DeFries, J., McClearn, G., & Rutter, M. (1997). *Behavioral genetics* (3rd ed.). San Francisco: W. H. Freeman.

Plotkin, H. (1998). *Evolution in mind: An introduction to evolutionary psychology*. Cambridge, MA: Harvard University Press.

Pollux, P. M. J., Wester, A., & de Haan, E. H. F. (1995). Random generation deficit in alcoholic Korsakoff patients. *Neuropsychologica, 33*(1), 125–129.

Pomiankowski, A., & Moller, A. (1995). A resolution of the lek paradox. *Proceedings of the Royal Society (London) B, 260*, 21–29.

Pomiankowski, A., Iwasa, Y., & Nee, S. (1991). The evolution of costly mate preferences. I. Fisher and biased mutation. *Evolution, 45*(6), 1422–1430.

Popper, K. (1972). *Objective knowledge: An evolutionary approach.* Oxford University Press.

Posner, R. (1992). *Sex and reason.* Cambridge, MA: Harvard University Press.

Poundstone, W. (1992). *Prisoner's dilemma: John von Neumann, game theory, and the puzzle of the bomb.* New York: Doubleday.

Power, C. (1999). "Beauty" magic: The origins of art. In R. Dunbar *et al.* (eds.), *The evolution of culture*, pp. 92–112. Edinburgh University Press.

Rapoport, A., & Budescu, D. (1992). Generation of random series in two-person strictly competitive games. *Journal of Experimental Psychology: General, 121*, 352–363.

Rapoport, A., Erev, I., Abraham, E. V., & Olson, D. E. (1997). Randomization and adaptive learning in a simplified poker game. *Organizational Behavior and Human Decision Processes, 69*, 31–49.

Rawls, J. (1971). *A theory of justice.* Cambridge, MA: Harvard University Press.

Rawson, P. (1978). *The art of tantra.* London: Thames & Hudson.

Real, L. (1991). Search theory and mate choice. II. Mutual interaction, assortative mating, and equilibrium variation in male and female fitness. *American Naturalist, 138*, 901–917.

Rentschler, I., Herzberger, B., & Epstein, D. (eds.). (1988). *Beauty and the brain: Biological aspects of aesthetics.* Berlin: Birkhauser.

Reynolds, J. D., & Gross, M. R. (1990). Costs and benefits of female choice: Is there a lek paradox? *American Naturalist, 136*, 230–243.

Rice, W. R. (1988). Heritable variation in fitness as a prerequisite for adaptive female choice: The effect of mutation-selection balance. *Evolution, 42*, 817–820.

Richards, I. A. (1943). *Basic English and its uses.* London: Kegan Paul.

Richards, R. J. (1987). *Darwin and the emergence of evolutionary theory of mind and behavior.* University of Chicago Press.

Ridley, Mark (ed.). (1997). *Evolution.* Oxford University Press.

Ridley, Matt (1993). *The red queen: Sex and the evolution of human nature.* New York: Viking.

Ridley, Matt (1996). *The origins of virtue.* London: Viking.

Roeder, K. D., & Treat, A. E. (1961). The detection and evasion of bats by moths. *American Scientist, 49*, 135–148.

Rogers, A. R., & Mukherjee, A. (1992). Quantitative genetics of sexual dimorphism in human body size. *Evolution, 46*, 226–234.

Rogers, C. R. (1995). *On becoming a person: A therapist's view of psychotherapy.* New York: Houghton Mifflin.

Rose, M. R., & Lauder, G. V. (eds.). (1996). *Adaptation.* London: Academic Press.

Rostand, E. (1993). *Cyrano de Bergerac* (Trans. Anthony Burgess). London: Nick Hern Books. (First published 1898 in French.)

Roth, A. E., & Sotomayor, M. (1990). *Two-sided matching.* Cambridge University Press.

Rowe, D. C. (1994). *The limits of family influence: Genes, experience, and behavior*. New York: Guilford Press.

Rowe, L., & Houle, D. (1996). The lek paradox and the capture of genetic variance by condition-dependent traits. *Proceedings of the Royal Society (London) B, 263*, 1415–1421.

Rowe, N. (1996). *The pictorial guide to the living primates*. East Hampton, NY: Pogonias Press.

Runciman, W. G. (1998). *The social animal*. London: HarperCollins.

Ryan, M. J. (1985). *The Tungara frog: A study in sexual selection*. Chicago University Press.

Ryan, M. J. (1990). Sexual selection, sensory systems, and sensory exploitation. *Oxford Surveys of Evolutionary Biology*, *7*, 156–195.

Ryan, M. J. (1997). Sexual selection and mate choice. In J. R. Krebs & N. B. Davies (eds.), *Behavioral ecology: An evolutionary approach* (4th ed.), pp. 179–202. Oxford: Blackwell Scientific.

Ryan, M. J., & Keddy-Hector, A. (1992). Directional patterns of female mate choice and the role of sensory biases. *American Naturalist*, *139*, S4–S35.

Samuelson, L. (1997). *Evolutionary games and equilibrium selection*. Cambridge, MA: MIT Press.

Sandars, N. K. (1985). *Prehistoric art in Europe*. London: Penguin.

Savage-Rumbaugh, S., & Lewin, R. (1994). *Kanzi: The ape at the brink of the human mind*. New York: Wiley.

Sawaguchi, T. (1997). Possible involvement of sexual selection in neocortical evolution of monkeys and apes. *Folia primatologica*, *68*, 95–99.

Scheib, J. (1994). Sperm donor selection and the psychology of female choice. *Ethology and Sociobiology*, *15*(3), 113–129.

Schmitt, B., & Simonson, A. (1997). *Marketing aesthetics*. New York: Free Press.

Searle, J. (1992). The problem of consciousness. In G. R. Bock & J. Marsh (eds.), *Experimental and theoretical studies of consciousness*, pp. 61–80. New York: John Wiley.

Searle, J. (1997). *The mystery of consciousness*. London: Granta Press.

Shackelford, T. K., & Larsen, R. J. (1999). Facial attractiveness and physical health. *Evolution and Human Behavior*, *20*, 71–76.

Sheets-Johnstone, M. (1990). Hominid bipedality and sexual selection theory. *Evolutionary Theory*, *9*(1), 57–70.

Shepard, R. N. (1987). Evolution of a mesh between principles of the mind and regularities of the world. In J. Dupré (ed.), *The latest on the best*, pp. 251–275. Cambridge, MA: MIT Press.

Shepard, R. N. (1997). The genetic basis of human scientific knowledge. In G. R. Bock & G. Cardew (eds.), *Characterizing human psychological adaptations*, pp. 23–38. New York: John Wiley.

Sherman, P. W. (1989). The clitoris debate and the levels of analysis. *Animal Behavior*, *37*(4), 697–698.

Shipman, P. (1998). *Taking wing: Archaeopteryx and the evolution of bird flight*. New York: Simon & Schuster.

Short, R. V., & Balaban, E. (eds.) (1994). *The differences between the sexes*. Cambridge University Press.

Shostak, M. (1981). *Nisa: The life and words of a !Kung woman*. Cambridge University Press.

Sigmund, K. (1993). *Games of life*. Oxford University Press.

Silver, L. M. (1998). *Remaking Eden: How genetic engineering and cloning will transform the American family*. New York: Avon.

Simon, H. (1957). *Models of men: Social and rational.* New York: Wiley.

Simonton, D. K. (1988). Age and outstanding achievement: What do we know after a century of research? *Psychological Bulletin, 104*, 251–267.

Simonton, D. K. (1993). Genius and chance: A Darwinian perspective. In J. Brockman (ed.), *Creativity*, pp. 176–201. New York: Simon & Schuster.

Simpson, J. A., & Kenrick, D. T. (eds.). (1997). *Evolutionary social psychology*. Mahwah, NJ: Erlbaum.

Singer, P. (ed.). (1994). *Ethics*. Oxford University Press.

Singh, D. (1993). Waist-to-hip ratio (WHR): A defining morphological feature of health and female attractiveness. *Journal of Personality and Social Psychology, 65*, 293–307.

Singh, D. (1995). Female health, attractiveness, and desirability for relationships: Role of breast asymmetry and waist-to-hip ratio. *Ethology and Sociobiology, 16*, 465–481.

Singh, D., & Bronstad, P. M. (1997). Sex differences in the anatomical locations of human body scarification and tattooing as a function of pathogen prevalence. *Evolution and Human Behavior, 18*, 403–416.

Skinner, B. F. (1953). *Science and human behavior*. New York: Macmillan.

Skinner, B. F. (1988). *Beyond freedom and dignity*. London: Penguin.

Skyrms, B. (1996). *Evolution of the social contract.* Cambridge University Press.

Sloman, S., & Sloman, L. (1988). Mate selection in the service of human evolution. *Journal of Social and Biological Structures, 11*, 457–468.

Small, M. (1993). *Female choices: Sexual behavior of female primates*. Ithaca, NY: Cornell University Press.

Smith, E. A., & Bird, R. L. B. (2000). Turtle hunting and tombstone opening: Public generosity as costly signalling. *Evolution and Human Behavior* (in press).

Smith, S. M., Ward, T. B., & Finke, R. A. (eds.). (1995). *The creative cognition approach.* Cambridge, MA: MIT Press.

Smuts, B. B. (1985). *Sex and friendship in baboons*. New York: Aldine de Gruyter.

Smuts, B. B. (1985). The evolutionary origins of patriarchy. *Human Nature, 6*, 1–32.

Sober, E., & Wilson, D. S. (1998). *Unto others: The evolution and psychology of unselfish behavior*. Cambridge, MA: Harvard University Press.

Sommers, C. H. (1994). *Who stole feminism? How women have betrayed women*. New York: Simon & Schuster.

Spatt, J., & Goldenberg, G. (1993). Components of random generation by normal subjects and patients with dysexecutive syndrome. *Brain and Cognition, 23*, 231–242.

Sperber, D. (1994). The modularity of thought and the epidemiology of representations. In L. A. Hirschfeld & S. A. Gelman (eds.), *Mapping the mind*, pp. 39–67. Cambridge University Press.

Stanford, C. (1999). *The hunting apes: Meat eating and the origins of human behavior.* Princeton University Press.

Stanyon, R., Consigliere, S., & Morescalchi, M. A. (1993). Cranial capacity in hominid evolution. *Human Evolution, 8*, 205–216.

Sterelny, K., & Griffiths, P. E. (1999). *Sex and death: An introduction to philosophy of biology*. University of Chicago Press.

Sternberg, R. J. (ed.). (1999). *Handbook of creativity*. Cambridge University Press.

Sternberg, R. J., & Barnes, M. L. (eds.). (1988). *The psychology of love*. New Haven, CT: Yale University Press.

Stringer, C., & McKie, R. (1996). *African exodus: The origins of modern humanity*. London: Jonathan Cape.

Studd, M. V. (1966). Sexual harassment. In D. M. Buss & N. M. Malamuth (eds.), *Sex, power, conflict: Evolutionary and feminist perspectives*, pp. 54–89. Oxford University Press.

Sulloway, F. J. (1979). *Freud, biologist of the mind: Beyond the psychoanalytic legend*. New York: Basic Books.

Symons, D. (1979). *The evolution of human sexuality*. Oxford University Press.

Symons, D. (1995). Beauty is in the adaptations of the beholder: The evolutionary psychology of human female sexual attractiveness. In P. R. Abrahamson & S. D. Pinker (eds.), *Sexual Nature/Sexual Culture*, pp. 80–118. University of Chicago Press.

Talbot, M. M. (1998). *Language and gender*. Cambridge: Polity Press.

Tannen, D. (1991). *You just don't understand: Women and men in conversation*. London: Virago.

Tassinary, L. G., & Hansen, K. A. (1998). A critical test of the waist-to-hip ratio hypothesis of female attractiveness. *Psychological Science*, *9*, 150–155.

Taylor, T. (1996). *The prehistory of sex*. London: Fourth Estate.

Tessman, I. (1995). Human altruism as a courtship display. *Oikos*, *74*(1), 157–158.

Thieme, H. (1997). Lower Paleolithic hunting spears from Germany. *Nature*, *385*, 807–810.

Thornhill, R. (1997). The concept of an evolved adaptation. In G. R. Bock & G. Cardew (eds.), *Characterizing human psychological adaptations*, pp. 4–22. New York: John Wiley.

Thornhill, R. (1998). Darwinian aesthetics. In C. Crawford & D. Krebs (eds.), *Handbook of evolutionary psychology*, pp. 543–572. Mahwah, NJ: Erlbaum.

Thornhill, R., and Gangestad, S. W. (1994). Fluctuating asymmetry and human sexual behavior. *Psychological Science*, *5*, 297–302.

Thornhill, R. and Gangestad, S. W. (1996). The evolution of human sexuality. *Trends in Ecology and Evolution*, *11*, 98–102.

Thornhill, R., & Grammer, K. (1999). The body and face of woman: One ornament that signals quality? *Evolution and Human Behavior*, *20*, 105–120.

Thornhill, R., & Thornhill, N. W. (1992). The evolutionary psychology of men's coercive sexuality. *Behavioral and Brain Sciences*, *15*, 363–421.

Tiger, L. (1992). *The pursuit of pleasure*. Boston: Little, Brown.

Todd, P. M., & Miller, G. F. (1991). On the sympatric origin of species: Mercurial mating in the Quicksilver Model. In R. K. Belew & L. B. Booker (eds.), *Proceedings of the Fourth International Conference on Genetic Algorithms*, pp. 547–554. San Mateo, CA: Morgan Kaufmann.

Todd, P. M. & Miller, G. F. (1993). Parental guidance suggested: How parental imprinting evolves through sexual selection as an adaptive learning mechanism. *Adaptive Behavior*, *2*(1), 5–47.

Todd, P. M., & Miller. G. F. (1997a). How cognition shapes cognitive evolution. *IEEE Expert: Intelligent systems and their applications*, *12*(4), 7–9.

Todd, P. M., & Miller, G. F. (1997b). Biodiversity through sexual selection. In C. G.

Langton & K. Shimohara (eds.), *Artificial Life V*, pp. 289–299. Cambridge, MA: MIT Press.

Todd, P. M., & Miller, G. F. (1999). From Pride and Prejudice to Persuasion: Satisficing in mate search. In G. Gigerenzer & P. Todd (eds.), *Simple heuristics that make us smart*, pp. 287–308. Oxford University Press.

Tooby, J. (1982). Pathogens, polymorphism, and the evolution of sex. *Journal of Theoretical Biology*, *97*, 557–576.

Tooby, J., & Cosmides, L. (1990a). On the universality of human nature and the uniqueness of the individual: The role of genetics and adaptation. *Journal of Personality*, *58*, 17–67.

Tooby, J., & Cosmides, L. (1990b). The past explains the present: Emotional adaptations and the structure of ancestral environments. *Ethology and Sociobiology*, *11*(4/5), 375–424.

Tooby, J., & Cosmides, L. (1992). The psychological foundations of culture. In J. H. Barkow *et al.* (eds.), *The adapted mind*, pp. 19–136. Oxford University Press.

Tooby J., & Cosmides, L. (1996). Friendship and the banker's paradox: Other pathways to the evolution of adaptations for altruism. *Proceedings of the British Academy*, *88*, 119–143.

Trail, P. W. (1990). Why should lek-breeders be monomorphic? *Evolution*, *44*(7), 1837–1852.

Trivers, R. (1971). The evolution of reciprocal altruism. *Quarterly Review of Biology*, *46*, 35–57.

Trivers, R. (1972). Parental investment and sexual selection. In B. Campbell (ed.), *Sexual selection and the descent of man 1871–1971*, pp. 136–179. Chicago: Aldine.

Trivers, R. (1974). Parent–offspring conflict. *American Zoologist*, *14*, 249–264.

Turing, A. M. (1950). Computing machinery and intelligence. *Mind*, *59*, 433–460.

Turner, F. (1991). *Beauty: The value of values*. Charlottesville, VA: University of Virginia Press.

Veblen, T. (1899). *The theory of the leisure class*. New York: Macmillan. (Reprinted by Dover.)

Vernon, P. E. (ed.). (1970). *Creativity: Selected readings*. Harmondsworth: Penguin.

Vining, D. R. (1986). Social versus reproductive success: The central theoretical problem of human sociobiology. *Behavioral and Brain Sciences*, *9*, 167–216.

Von Neumann, J., & Morgenstern, O. (1944). *Theory of games and economic behavior*. Princeton University Press.

de Waal, F. (1989). *Peacemaking among primates*. Cambridge, MA: Harvard University Press.

de Waal, F. (1996). *Good natured: The origins of right and wrong in humans and other animals*. New York: Harper Perennial.

de Waal, F., & Lanting, F. (1997). *Bonobo: The forgotten ape*. Berkeley, CA: University of California Press.

Wagenaar, W. A. (1972). Generation of random sequences by human subjects: A critical survey of literature. *Psychological Bulletin*, *77*, 65–72.

Wallace, R. A. (1980). *How they do it*. New York: William Morrow.

Wallin, N. L., Merker, B., & Brown, S. (eds.). (1999). *The origins of music*. Cambridge, MA: MIT Press.

Whiten, A. (ed.). (1991). *Natural theories of mind*. Oxford: Basil Blackwell.

Whiten, A., & Byrne, R. (1997). *Machiavellian intelligence II: Extensions and evaluations*. Cambridge University Press.

Wickett, J. C., Vernon, P. A., & Lee, D. H. (1994). *In vivo* brain size, head perimeter, and intelligence in a sample of healthy adult females. *Personality and Individual Differences*, *16*, 831–838.

Wiegmann, D. Real, L. A., Capone, T. A., & Ellner, S. (1996). Some distinguishing features of models of search behavior and mate choice. *American Naturalist*, *147*, 188–204.

Wilkinson, G. S. (1984). Reciprocal food sharing among vampire bats. *Nature*, *308*, 181–184.

Williams, G. C. (1966). *Adaptation and natural selection*. Princeton University Press.

Williams, G. C. (1975). *Sex and evolution*. Princeton University Press.

Williams, G. C. (1996). *Plan and purpose in nature*. London: Weidenfeld & Nicholson.

Williams, J. E., Satterwhite, R. C., & Saiz, J. L. (1998). *The importance of human psychological traits: A cross-cultural study*. New York: Plenum Press.

Wilson, D. S. (1994). Adaptive genetic variation and human evolutionary psychology. *Ethology and Sociobiology*, *15*, 219–235.

Wilson, D. S., Near, D., & Miller, R. R. (1996). Machiavellianism: A synthesis of the evolutionary and psychological literatures. *Psychological Bulletin*, *119*, 285–299.

Wilson, E. O. (1975). *Sociobiology: The new synthesis*. Cambridge, MA: Harvard University Press.

Wilson, E. O. (1998). *Consilience: The unity of knowledge*. New York: Knopf.

Wilson, M., & Daly, M. (1985). Competitiveness, risk taking, and violence: The young male syndrome. *Ethology and Sociobiology*, *6*, 59–73.

Wittelson, S. F., Glezer, I. I., & Kigar, D. L. (1995). Women have greater density of neurons in posterior temporal cortex. *Journal of Neuroscience*, *15*, 3418–3428.

Wolfe, J. B., Moore, A. J., & Brodie, E. D. (1997). The evolution of indicator traits for parental quality: The role of maternal and paternal effects. *American Naturalist*, *150*, 639–649.

Wolfe, T. (1982). *From Bauhaus to our house*. London: Jonathan Cape.

Wolpert, L. (1992). *The unnatural nature of science*. London: Faber & Faber.

Wrangham, R., & Peterson, D. (1996). *Demonic males: Apes and the origins of human violence*. New York: Houghton Mifflin.

Wright, R. (1994). *The moral animal: Evolutionary psychology and everyday life*. New York: Pantheon Books.

Wright, W. (1998). *Born that way: Genes, behavior, personality*. New York: Knopf.

Zahavi, A. (1975). Mate selection: A selection for a handicap. *Journal of Theoretical Biology*, *53*, 205–214.

Zahavi, A. (1978). Decorative patterns and the evolution of art. *New Scientist*, *19*, 182–184.

Zahavi, A. (1991). On the definition of sexual selection, Fisher's model, and the evolution of waste and of signals in general. *Animal Behavior*, *42*(3), 501–503.

Zahavi, A., & Zahavi, A. (1997). *The handicap principle: A missing piece of Darwin's puzzle*. Oxford University Press.

Ziman, J. (1978). *Reliable knowledge*. Cambridge University Press.

Ziman, J. (ed.). (2000). *Technological innovation as an evolutionary process*. Cambridge University Press.

Zuckerman, M. (1994). *Behavioral expressions and biosocial bases of sensation seeking*. Cambridge University Press.

Index